Development and Manufacture of Protein Pharmaceuticals

Pharmaceutical Biotechnology

Series Editor: Ronald T. Borchardt
The University of Kansas
Lawrence, Kansas

Recent volumes in this series:

Volume 7 PHYSICAL METHODS TO CHARACTERIZE PHARMACEUTICAL PROTEINS
Edited by James N. Herron, Wim Jiskoot, and Daan J. A. Crommelin

Volume 8 MODELS FOR ASSESSING DRUG ABSORPTION AND METABOLISM
Edited by Ronald T. Borchardt, Philip L. Smith, and Glynn Wilson

Volume 9 FORMULATION, CHARACTERIZATION, AND STABILITY OF PROTEIN DRUGS: Case Histories
Edited by Rodney Pearlman and Y. John Wang

Volume 10 PROTEIN DELIVERY: Physical Systems
Edited by Lynda M. Sanders and R. Wayne Hendren

Volume 11 INTEGRATION OF PHARMACEUTICAL DISCOVERY AND DEVELOPMENT: Case Histories
Edited by Ronald T. Borchardt, Roger M. Freidinger, Tomi K. Sawyer, and Philip L. Smith

Volume 12 MEMBRANE TRANSPORTERS AS DRUG TARGETS
Edited by Gordon L. Amidon and Wolfgang Sadée

Volume 13 RATIONAL DESIGN OF STABLE PROTEIN FORMULATIONS: Theory and Practice
Edited by John F. Carpenter and Mark C. Manning

Volume 14 DEVELOPMENT AND MANUFACTURE OF PROTEIN PHARMACEUTICALS
Edited by Steven L. Nail and Michael J. Akers

A Chronological Listing of Volumes in this series appears at the back of this volume

---

A Continuation Order Plan is available for this series. A continuation order will bring delivery of each new volume immediately upon publication. Volumes are billed only upon actual shipment. For further information please contact the publisher.

Development and Manufacture of Protein Pharmaceuticals

Edited by

Steven L. Nail

Purdue University
West Lafayette, Indiana

and

Michael J. Akers

Baxter Pharmaceutical Solutions LLC
Bloomington, Indiana

Kluwer Academic / Plenum Publishers
New York, Boston, Dordrecht, London, Moscow

Library of Congress Cataloging-in-Publication Data

Development and manufacture of protein pharmaceuticals/edited by Steven L. Nail and Michael J. Akers.
p. cm. — (Pharmaceutical biotechnology; v. 14)
Includes bibliographical references and index.
ISBN 0-306-46745-3
1. Protein drugs.
[DNLM: 1. Chemistry, Pharmaceutical—methods. 2. Technology, Pharmaceutical—methods. 3. Peptides—pharmacology. 4. Proteins—pharmacology. QV 744 D489 2002] I. Nail, Steven L. II. Akers, Michael J. III. Series.
RS431.P75 D485 2002
615′.19—dc21

2002016154

ISBN: 0-306-46745-3

233 Spring Street, New York, N.Y. 10013

http://www.wkap.nl/

10 9 8 7 6 5 4 3 2 1

A C.I.P. record for this book is available from the Library of Congress

Printed in the United States of America

Contributors

Michael J. Akers • Baxter Pharmaceutical Solutions LLC, Bloomington, Indiana 47402

Michael G. Beatrice • Corporate Regulatory and Quality Science, Abbott Laboratories, Abbott Park, Illinois 60064-6091

Daniel J. Brose • Chemica Technologies, Inc., Bend, Oregon 97701

Carl J. Burke • Department of Vaccine Pharmaceutical Research, Merck Research Laboratories, West Point, Pennsylvania 19486

Suchart Chongprasert • Drug Control Division, Food and Drug Administration, Muang, Nontaburi 11000, Thailand

Michael Dosmar • Separations Division, Sartorius Corp., Edgewood, New York 11717

John Geigert • Biopharmaceutical Quality Solutions, Carlsbad, California 92009

Shan Jiang • Bayer Corporation, Clayton, North Carolina 27606

Maik W. Jornitz • Product Management, Sartorius AG, D-37075 Goettingen, Germany

Shawn A. Knopp • Genetics Institute, Andover, Massachusetts 01810

C. Russell Middaugh • Department of Pharmaceutical Chemistry, University of Kansas, Lawrence, Kansas 66047

Steven L. Nail • School of Pharmacy, Purdue University, West Lafayette, Indiana 47907

Gautam Sanyal • AstraZeneca Research and Development Boston, Waltham, Massachusetts 02451

Mary Stickelmeyer • Lilly Research Laboratories, Indianapolis, Indiana 46285

Michael Townsend • Isis Pharmaceuticals, Inc., Carlsbad, California 92008

Vasu Vasudevan • Lilly Research Laboratories, Indianapolis, Indiana 46285

David B. Volkin • Department of Vaccine Pharmaceutical Research, Merck Research Laboratories, West Point, Pennsylvania 19486

Richard T. Wood • Pfizer, Inc., Terre Haute, Indiana 47808

Preface to the Series

A major challenge confronting pharmaceutical scientists in the future will be to design successful dosage forms for the next generation of drugs. Many of these drugs will be complex polymers of amino acids (e.g., peptides, proteins), nucleosides (e.g., antisense molecules), carbohydrates (e.g., polysaccharides), or complex lipids.

Through rational drug design, synthetic medicinal chemists are preparing very potent and very specific peptides and antisense drug candidates. These molecules are being developed with molecular characteristics that permit optimal interaction with the specific macromolecules (e.g., receptors, enzymes, RNA, DNA) that mediate their therapeutic effects. Rational drug design does not necessarily mean rational drug delivery, however, which strives to incorporate into a molecule the molecular properties necessary for optimal transfer between the point of administration and the pharmacological target site in the body.

Like rational drug design, molecular biology is having a significant impact on the pharmaceutical industry. For the first time, it is possible to produce large quantities of highly pure proteins, polysaccharides, and lipids for possible pharmaceutical applications. Like peptides and antisense molecules, the design of successful dosage forms for these complex biotechnology products represents a major challenge to pharmaceutical scientists.

Development of an acceptable drug dosage form is a complex process requiring strong interactions between scientists from many different divisions in a pharmaceutical company, including discovery, development, and manufacturing. The series editor, the editors of the individual volumes, and the publisher hope that this new series will be particularly helpful to scientists in the development areas of a pharmaceutical company (e.g., drug metabolism, toxicology, pharmacokinetics and pharmacodynamics, drug delivery, preformulation, formulation, and physical and analytical chemistry). In addition, we hope this series will help to build bridges between the development scientists and scientists in discovery (e.g., medicinal chemistry, pharmacology, immunology, cell biology, molecular biology) and in manufacturing (e.g., process chemistry, engineering). The design of successful dosage forms for the next generation of drugs will require not only a high level of expertise by individual scientists, but also a high degree of interaction between scientists in these different divisions of a pharmaceutical company.

Finally, everyone involved with this series hopes that these volumes will also be useful to the educators who are training the next generation of pharmaceutical scientists. In addition to having a high level of expertise in their respective disciplines, these young scientists will need to have the scientific skills necessary to communicate with their peers in other scientific disciplines.

RONALD T. BORCHARDT
Series Editor

Preface

In this era of biotechnology there have been many books covering the fundamentals of recombinant DNA technology and protein chemistry. However, not many sources are available for the pharmaceutical development scientist and other personnel responsible for the commercialization of the finished dosage forms of these new biopharmaceuticals and other products from biotechnology. This text will help to fill this gap.

Once active biopharmaceutical molecules are candidates for clinical trial investigation and subsequent commercialization, a number of other activities must take place while research and development on these molecules continues. The active ingredient itself must be formulated into a finished dosage form that can be conveniently used by health care professionals and patients. Properties of the biopharmaceutical molecule must be clearly understood so that the appropriate finished product formulation can be developed. Finished product formulation development includes not only the chemical formulation, but also the packaging system, the manufacturing process, and appropriate control strategies to assure such good manufacturing practice attributes as safety, identity, strength, purity, and quality.

This book contains eight chapters focused on the eventual commercialization of peptide and protein finished dosage forms. Authors for each chapter were carefully selected based on their knowledge, experience, and writing skill in their area of expertise. Each chapter topic was selected based on its importance in the successful development of commercially viable finished sterile dosage forms containing peptides and proteins. As those in the industry know, each piece of scientific and technical expertise must fit together with the rest for the product to be approved by the Food and Drug Administration and regulatory bodies worldwide.

To connect our book to other books in this series, we start with a chapter on preformulation studies. This chapter provides an excellent overview of the types of early studies on protein molecules essential to subsequent formulation development and manufacture of the finished dosage form. To those not well versed in the standard analytical techniques for early protein characterization relevant to eventual finished dosage form development, this chapter hits the mark for easy and clear understanding of these techniques and their applicability to understanding protein behavior and stability.

The formulation development chapter (Chapter 2) contains information similar to that appearing in other chapters and review articles written on this popular subject. What sets this chapter apart, however, is greater focus and breadth of coverage of the interrelationships among formulation development, packaging, manufacturing, and quality control. The chapter emphasizes a holistic approach to formulation development rather than being concerned only with stability and other pharmaceutics-related problems.

Four of the chapters deal with processing operations critical to manufacture of pharmaceutical products consisting of peptides and proteins: aseptic processing, thermal sterilization processes, membrane filtration, and freeze-drying. Principles of aseptic processing are very capably covered in Chapter 3. This chapter applies to both small-scale (clinical trial supplies) and commercial-scale aseptic processing and includes practical information on the processing of solution, suspension, and freeze-dried dosage forms.

Although thermal sterilization is not, of course, directly used for sterilization of protein formulations, it is an essential part of aseptic processing for sterilization of containers, closures, and processing equipment. Chapter 4 deals with basic principles of thermal sterilization and is intended primarily as introductory material for formulation and process development scientists not familiar with the fundamentals of thermal sterilization.

Chapter 5 deals with the use of membrane filtration, both crossflow filtration for downstream processing of proteins and dead-ended filtration for sterilization of finished pharmaceutical products. Pharmaceutical scientists should have a general understanding of both the theory and the practice of membrane filtration in order to select filtration membranes and size filtration systems properly and to implement reliable integrity testing procedures to provide adequate sterility assurance for protein pharmaceuticals. Filter fouling can be a troublesome aspect of filtration, whether during downstream protein processing or final sterile filtration. This chapter takes a quantitative, mechanistic approach to filter fouling and presents several case studies in the development of successful filtration schemes.

Because most pharmaceutically relevant proteins are not sufficiently stable in solution—either chemically or physically—to allow manufacture of ready-to-use sterile solutions, freeze-drying is an essential unit operation in the manufacture of protein formulations. Chapter 6 has been written largely in response to numerous requests from formulation and processing scientists for a sound basic introduction to the science and technology of freeze-drying where no previous background is assumed. The chapter is not specific to freeze-drying of proteins, but special considerations in protein freeze-drying are discussed, and areas in need of further research are identified.

Good manufacturing compliance takes on even more significance when very complex, highly potent, relatively unstable molecules derived from biotechnology processes are involved. Chapter 7 reviews the quality assurance and quality control concerns associated with biopharmaceutical manufacturing from bulk drug substance to finished product. The most current thinking and practice with regard to quality assurance and control of biopharmaceutical products is presented in this chapter.

Chapter 8 could be a book on its own. The author gives a historical look at the evolutionary process of the regulatory environment that resulted in the advent of biotechnology. This chapter is very thorough in detailing all of the changes in and current requirements for the registration of biopharmaceutical products in the United States.

The contents of this book meet a significant need of pharmaceutical scientists as well as others working in the development, manufacturing, control, and regulatory divisions of a biopharmaceutical company. We believe it to be unique in addressing many different aspects of the commercialization of peptide and protein sterile dosage forms.

Acknowledgments

We thank Prof. Ronald T. Borchardt for inviting us to edit this book. We also thank our coauthors for their excellent contributions and for their patience. We are grateful for the scientific and technological contributions of many scientists, engineers, and other experts around the world whose work is cited throughout the book. Most of all, we thank our wives, Lisa Nail and Mary Akers, for their love and support.

Steven L. Nail
Michael J. Akers

Contents

Chapter 1

Preformulation Studies as an Essential Guide to Formulation Development and Manufacture of Protein Pharmaceuticals
David B. Volkin, Gautam Sanyal, Carl J. Burke, and C. Russell Middaugh

Chapter 2

Formulation Development of Protein Dosage Forms
Michael J. Akers, Vasu Vasudevan, and Mary Stickelmeyer

Chapter 3

Aseptic Processing of Protein Pharmaceuticals
Michael Townsend

Chapter 5

Membrane Filtration
Daniel J. Brose, Michael Dosmar, and Maik W. Jornitz

Chapter 6

Fundamentals of Freeze-Drying

Steven L. Nail, Shan Jiang, Suchart Chongprasert, and Shawn A. Knopp

Chapter 7

Quality Assurance and Quality Control for Biopharmaceutical Products
John Geigert

Chapter 8

Regulatory Considerations in the Development of Protein Pharmaceuticals
Michael G. Beatrice

1

Preformulation Studies as an Essential Guide to Formulation Development and Manufacture of Protein Pharmaceuticals

David B. Volkin, Gautam Sanyal, Carl J. Burke, and C. Russell Middaugh

1. INTRODUCTION

Preformulation generally refers to a process in which a bulk drug material is characterized to a sufficient extent that it can be converted to a pharmaceutically acceptable drug substance. In the case of conventional, small-molecule (nonmacromolecular) pharmaceuticals, this procedure can usually be accomplished to a high degree of scientific rigor. For example, the atomic level structures of these compounds are usually known from a combination of X-ray crystallography, nuclear magnetic resonance (NMR), and mass spectrometry. Furthermore, the sensitivity of modern high-performance chromatography systems coupled to mass spectrometric detectors often permits altered forms of these molecules to be simply detected and identified. Unfortunately, the increased complexity of proteins, arising from a combination of their large size and higher order forms (secondary, tertiary, quaternary) of structure does not yet permit

David B. Volkin and Carl J. Burke • Department of Vaccine Pharmaceutical Research, Merck Research Laboratories, West Point, Pennsylvania 19486. *Gautam Sanyal* • AstraZeneca Research and Development Boston, Waltham, Massachusetts 02451. *C. Russell Middaugh* • Department of Pharmaceutical Chemistry, University of Kansas, Lawrence, Kansas 66047.

Development and Manufacture of Protein Pharmaceuticals, edited by Nail and Akers. Kluwer Academic/Plenum Publishers, New York, 2002.

the more straightforward type of analysis utilized for small molecules to be simply applied. Nevertheless, it is still possible to treat proteins primarily as chemical (rather than biological) entities through an understanding of their structure and stability and use of a more complex array of lower resolution methodologies.

Under ideal circumstances, one would like to have available an atomic-level-resolution structure of an early protein drug candidate determined by X-ray crystallography and/or NMR. This is often not the case, however, due to the labor-intensive nature of both approaches as well as the inherent difficulties in obtaining suitable crystals for the former and size limitations by the latter. As has previously been pointed out, however, once a structure has been obtained by one of these two methods, simple comparisons of either diffraction patterns or two-dimensional NMR proton spectra can be used to establish structural identity to standard proteins (Middaugh, 1994). More importantly, modern methods of mass spectrometry, especially the electrospray and laser desorption time-of-flight varieties, now permit the molecular weight of even large proteins to be determined to a high degree of accuracy (i.e., 0.01%). Very little information about protein conformation is available, however, from such measurements. For this reason, preformulation structural analysis of proteins usually consists of use of a wide variety of methods that examine various aspects of a protein's structure. For example, the secondary structure can be estimated by both far-ultraviolet (UV) circular dichroism (CD) and Fourier-transform infrared (FTIR) spectroscopies, and tertiary structure can be probed by near-UV CD and intrinsic fluorescence studies. The oligomeric state of a protein as well as potential aggregation phenomena can be explored by a combination of molecular sieve chromatography, analytical ultracentrifugation, static and dynamic light scattering, and chemical crosslinking. Less direct, but very sensitive, techniques used primarily in a comparative mode that sense various aspects of structure include capillary electrophoresis, ion-exchange and reversed-phase high-pressure liquid chromatography (HPLC), and isoelectric focusing. When a combination of the above techniques is employed, a fairly comprehensive picture of a protein can be constructed. Most importantly, these methods can be tweaked to detect alterations in protein structure. The use of these methods is considered below in more detail.

Probably the most critical portion of protein preformulation activities is what can best be described as finding the "weak points" in the molecule. Through the use of a combination of methods, a protein can be "stressed" and its susceptibility to various forms of physical and chemical degradation evaluated. Most often, temperature, pH, and/or redox conditions are used to induce structural alterations, because these variables have often been found to significantly elevate the rate of structural changes commonly seen in proteins. Although such studies cannot absolutely predict corresponding changes in real-time stability studies, they usually provide important clues in this regard.

The most common chemical changes encountered in proteins include deamidation of asparagine (and, to a lesser extent, glutamine) residues, cleavage of peptide bonds, cystine destruction, thiol–disulfide interchange, oxidation of cysteine and methionine, photodegradation of Trp, Tyr, and Cys, and glycation and carbamylation of amino groups (Volkin *et al.*, 1995b). Such changes are usually straightforward to detect and characterize using chromatographic methods combined with mass spectrometry, especially if they are present to significant extents. Conformational alterations, especially if they are of a subtle nature, can be much more difficult to detect, because one cannot be sure that a particular method such as CD or fluorescence is sensitive to the region of the structure in which the alteration occurs. This is perhaps the biggest uncertainty in the preformulation of proteins, but can be minimized by the use of multiple techniques and different environmental stresses, which may reveal small changes undetectable in the native state. A complete thermodynamic analysis of the folding and unfolding of a protein can be particularly useful in this regard if conditions of reversibility can be found. Similarly, the effect of stress on the aggregation state of the protein as well as on its bulk solubility is often useful.

This information should be immediately helpful in devising methods to increase the stability of a protein if this is necessary. Clues to potential problems with novel delivery methods and the necessity of lyophilization or even possibly the need to reengineer the protein should be immediately addressable after these types of analyses. In particular, we note three areas of special concern. First, if a protein contains a critical residue which is unusually labile, the creation of appropriate mutant forms needs to be immediately considered, although potential problems of immunogenicity must temper any such considerations. Second, if a protein contains specific binding sites for pharmaceutically acceptable ligands, such sites need to be quickly identified and extensively characterized, because this provides an immediate opportunity to dramatically stabilize a protein through inclusion of a simple excipient. Several examples of such a situation have been described (see below). Finally, if covalent modification of the native structure is being considered for stability or bioavailability reasons, similar preformulation studies of this new entity need to be quickly initiated, because previous work on the native form may be at least partially inapplicable. We now consider various aspects of these approaches in more detail.

2. PROTEIN STRUCTURAL CHARACTERIZATION AND CONFORMATIONAL STABILITY

In designing a stable protein formulation, the integrity and stability of the folded tertiary structure of the protein must be thoroughly examined, because proteins owe their specific biological functions to proper folding of the primary structure into a higher order structure with the minimum accessible free energy. Any successful formulation must ensure that this "native" structure is maintained for as long as possible under practical storage conditions. Ideally, to achieve this formulation goal, we should have at our disposal (a) sensitive biophysical assays for determining perturbations of native structure, (b) biological assays for assessing the significance of structural perturbations on function, and (c) methods for monitoring thermodynamics and kinetics of stabilization or destabilization of native structure. Most of the therapeutic proteins that are commercially available or in development are made using the recombinant DNA technology, and the accompanying purification process may involve unfolding and refolding of the protein. Biophysical characterization of such protein products must be performed to confirm that the correctly folded and biologically active structure is present. This may be considered either the final step in process development or the first step in preformulation research of protein therapeutics.

Several spectroscopic techniques can be used to examine the tertiary structure of proteins in solution and to assess the stability and effects of different formulation conditions on protein structure. These methods also provide the formulation scientist with tools for screening for suitable stabilizing agents if the inherent thermodynamic stability of a protein solution is inadequate for long-term storage. In the following discussion, we will focus primarily on optical techniques, i.e., UV-visible absorption, fluorescence (emission), circular dichroism (CD), and Fourier-transform infrared (FTIR) absorption spectroscopies. We will discuss the applications of these techniques to protein structure analysis and folding/unfolding studies, and discuss their relevance to pharmaceutical formulation development of proteins. In addition, the use of differential scanning calorimetry to monitor the thermal stability of protein formulations will be considered. A combination of these techniques will usually provide sufficiently detailed information on the inherent structural stability of a folded protein molecule to direct the formulation scientist to the methods which can be used to enhance a protein's stability sufficient for therapeutic use.

2.1. UV-Visible Absorption Spectroscopy

The spectral properties of the near-UV (250–320 nm) bands that originate from the three amino acids with aromatic side chains (i.e., phenylalanine, tryptophan, and tyrosine) depend upon their microenvironments, which are altered when a protein's tertiary structure is perturbed. In general, near-UV spectra of proteins manifest a small blue shift upon unfolding as the polarity of the surrounding environment is increased. As a result, the absorbance at a single wavelength in the region between 280 and 295 nm (the region of maximal slope in a spectrum containing contributions from tryptophan and tyrosine residues) usually changes and this can be simply monitored as a function of the perturbing condition. The magnitude of tryptophan or tyrosine absorbance change between folded and unfolded states is usually much smaller than the change observed by fluorescence, but the high precision and accuracy of UV spectroscopy makes folding/unfolding measurements feasible. The effects of pH, temperature, or chaotropic reagents on protein structure at equilibrium can thus be studied, as was nicely demonstrated for the model enzyme ribonuclease A by Fink and Painter (1987).

In employing changes in absorption intensity as a signal for monitoring unfolding or refolding, one needs to be aware of potential light scattering contributions, which may introduce a significant inaccuracy if the protein molecules self-associate to a size comparable to the wavelength of the incident light. Aggregation is a common occurrence when proteins unfold or form a partially unfolded "molten globule" structure, and is frequently encountered in accelerated thermal stability studies. Light scattering is typically indicated by gradually decreasing optical density values in the 300- to 350-nm range, where most proteins should have no intrinsic absorbance. The Rayleigh theory of light scattering suggests that the scattering intensity is proportional to the inverse fourth power of the wavelength of light. The presence of considerable light scattering can even produce a spurious shift in the near-UV absorbance maximum, which, if not corrected for, may be misinterpreted as a change in the microenvironments of aromatic residues and consequently conformational alterations. A simple log(optical density) versus log(wavelength) relationship is often used to obtain corrected absorbance values at a particular wavelength (Mach *et al.*, 1995).

In addition to this straightforward use of UV spectroscopy, second-derivative transformations of zero-order spectra have been shown to be sensitive to changes in protein tertiary structure that alter the polarities of the microenvironments of aromatic amino acids (Ragone *et al.*, 1984, 1987). Absorbance of tryptophan residues in the electronic ground state appears to be less sensitive to solvent polarity compared to the absorbance of tyrosine residues, whereas the reverse is true for fluorescence emission, which is more affected by the environment of the electronically excited state. Thus, the two techniques are nicely complementary approaches for monitoring changes in protein structure. Deconvolution of tyrosine and tryptophan contributions to second-derivative UV absorption spectra has been achieved (Mach and

Middaugh, 1994; Mach *et al.*, 1995). Furthermore, a major second-derivative phenylalanine negative peak at 258 nm is readily resolved from tyrosine and tryptophan contributions (Mach *et al.*, 1991). Changes in the average environment of these three residues in a protein can, therefore, be simultaneously measured in both equilibrium and kinetic experiments of protein unfolding and refolding. A major advantage of using second-derivative over conventional UV zero-order spectroscopy is that baseline inconsistencies originating from light scattering are minimized.

Although peptide bonds in α-helix, β-sheet, and disordered structures give rise to distinguishable absorption spectra in the far-UV, UV absorption is not usually the method of choice for monitoring changes in the secondary structures of proteins, for the following reasons: (a) The absorption bands occur between 185 and 205 nm, a wavelength range where interferences from buffer salts and amino acid side chains make accurate measurements difficult, (b) the extinction coefficients of β-sheet and disordered structures are very similar in this wavelength range, and (3) far-UV CD and FTIR spectroscopy provide higher resolution and more accurate methods for estimation of secondary structure content and associated changes.

2.2. Fluorescence Spectroscopy

The intrinsic fluorescence of proteins arises from the same three amino acids that manifest near-UV absorbance. Emission from tryptophan and to a lesser extent tyrosine residues usually dominates protein fluorescence spectra, whereas the phenylalanine contribution is usually weak and therefore is not widely used in protein structural and folding studies. Consequently, like near-UV absorbance, intrinsic fluorescence also yields information on the environment of these amino acids in the tertiary structure of a protein and changes in these environments when the structure is perturbed. Unlike absorbance spectroscopy, however, in which absorption of light by the ground electronic state of molecules is observed, fluorescence entails emission of light from the excited singlet electronic state that follows excitation of molecules from their ground state upon light absorption.

The excited state of the indole side chain of tryptophan residues is particularly sensitive to solvent polarity and, when contained in a protein, to the polarity of its microenvironment. The wavelength maxima λ_{max} of tryptophan side chains in proteins usually vary from approximately 320 to 350 nm, the lower and the upper ends of the range reflecting residues buried in the hydrophobic protein interior and those completely exposed to solvent, respectively. This offers a much wider window than the absorbance method for monitoring protein unfolding and refolding in both equilibrium and kinetic measurements as well as the ability to detect more subtle conformational changes. Furthermore, although not all of the light absorbed by a fluorophore is emitted as fluorescence, the sensitivity of steady-state fluorescence intensity measurements usually far exceeds that of absorbance measurements, thereby requiring only a fraction of the amount of the protein typically needed for near-UV absorbance experiments. For model tryptophan and tyrosine compounds, fluorescence quantum yield values are close to 0.2, suggesting that only about 20% of the photons absorbed by these fluorophores are emitted as fluorescence, the rest being lost to deexcitation of molecules by nonradiative processes. However, the measured fluorescence intensity is directly proportional to the intensity of the exciting light and to the quantum efficiency of the detector. This principle has been widely used to produce highly sensitive fluorometers that employ xenon-arc lamps or even laser sources and techniques to minimize optical loss. In many proteins, tyrosine fluorescence is quenched, in part, due to energy transfer to tryptophan, so that the effective quantum yield of tyrosine is lower. Furthermore, tyrosine fluorescence is relatively insensitive to solvent polarity. Tryptophan fluorescence spectra, in general, are much more sensitive to environment (polarity) than tyrosine spectra. Although unfolding of tertiary structure can also be studied by near-UV CD spectroscopy (250–320 nm), which also monitors changes in microenvironments of aromatic residues, the sensitivity of tryptophan and tryrosine fluorescence is much greater, allowing measurements to be made with relatively small amounts of proteins (typically a few micrograms per milliliter).

An example of using tryptophan fluorescence changes to monitor the kinetics of refolding of a potentially chemotherapeutic multidomain protein (TP40) upon dilution from a Gdn·HCl solution is shown in Fig. 1 (Gress *et al.*, 1994). TP40 is a fusion protein containing transforming growth factor-α (TGF-α) fused to a 40-kDa fragment of *Pseudomonas* exotoxin (PE). The refolding of the unfolded protein to its native state occurs via biphasic kinetics as measured by fluorescence spectroscopy (Fig. 1A). The faster phase of the apparently biphasic kinetics was quantitatively measured by stopped-flow spectrofluorometry (Fig. 1B). The tryptophan fluorescence of this protein was also utilized to monitor equilibrium unfolding as a function of Gdn·HCl concentration. Both steady-state fluorescence and fluorescence lifetime measurements were employed as complementary techniques to monitor unfolding, and very similar dependences of spectral shift and lifetime on Gdn·HCl concentration were observed (Fig. 1C). The decreases in tryptophan fluorescence intensity and lifetime as a function of increasing Gdn·HCl concentration are consistent with the tryptophan residues being exposed to a more polar environment as the protein unfolds.

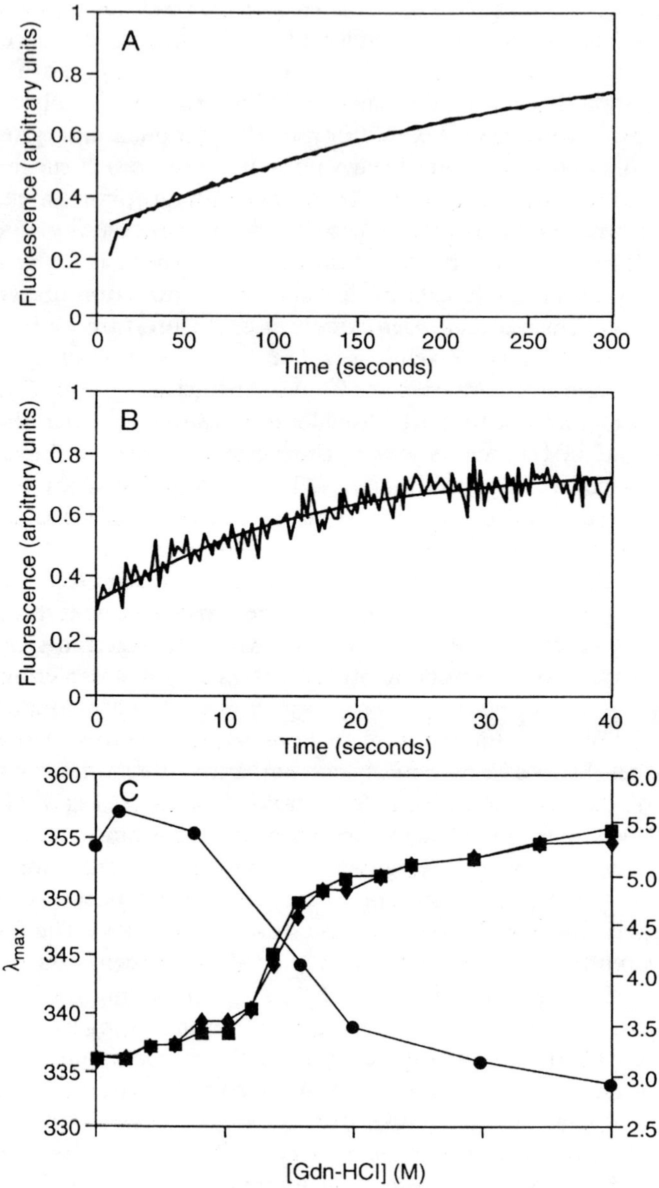
A
Fluorescence (arbitrary units)
Time (seconds)
B
Fluorescence (arbitrary units)
Time (seconds)
C
λ_{max}
τ_{av} (ns)
[Gdn-HCl] (M)

The use of steady-state fluorescence to study protein folding has been greatly aided by the use of extrinsic fluorescent probes with high quantum yields. Two kinds of probes are commonly used: (a) hydrophobic probes such as 2-*p*-toludinylnaphthalene-6-sulfonate (TNS) or 8-anilino-1-naphthalene-sulfonate (ANS), which bind to hydrophobic regions of proteins, and (b) probes that can be covalently attached to specific amino acid residues. Noncovalently bound hydrophobic probes have been extensively employed for detecting either partially folded or molten globule structures in protein folding pathways. These intermediate- and often high-aggregative structures are often brought about by acidic pH, high salt concentrations, and sub-denaturing concentrations of chaotropic agents. Because in such structures the hydrophobic regions become accessible to solvent, molten globules usually bind these probes to a much greater extent than native and unfolded proteins. As an example, the effect of urea on the intrinsic tyrosine fluorescence of GroEL has been compared with urea-induced changes in the fluorescence intensity of extrinsically bound bis-ANS (Gorovits *et al.*, 1995). This study revealed that changes in the intrinsic tyrosine fluorescence reflect a two-state unfolding transition as a function of urea concentration, whereas bis-ANS fluorescence shows the presence of residual tertiary structure in one or more regions of the protein even in the presence of 3–4 *M* urea (suggesting the presence of nucleation sites for folding). Common examples of extrinsic probes used for covalent labeling of proteins are fluorescein isothiocyanate (FITC) and dansyl chloride (which react primarily with amino groups) and iodoacetamide-based probes (which preferentially label thiol groups). The approach of using extrinsic probes, especially the hydrophobic-surface-seeking probes discussed above, may be useful for detecting subtle conformational changes in protein conformations in the process of formulation development of pharmaceutical proteins, provided that the excipients or experimental conditions used do not interfere with probe fluorescence.

Steady-state fluorescence quantum yield or intensity of both intrinsic tryptophan or tyrosine residues as well as extrinsic probes is attenuated by increased temperatures. However, a transition in the intensity versus temperature plot (e.g., due to a structural change) can usually be distinguished

Figure 1. Use of intrinsic tryptophan fluorescence of proteins to monitor unfolding and refolding. (A) Kinetics of refolding of TP40, upon dilution from a largely unfolded state in 2.5 *M* Gdn·HCl to 1.0 *M* Gdn·HCl, as measured by changes in tryptophan intensity (excitation, 280 nm, emission, 335 nm). (B) The kinetics of the initial fast phase (0–40 sec) of refolding of TP40 as quantitatively measured by stopped-flow spectrofluorometry. (C) Equilibrium unfolding of TP40 and PE40δcys (a 40–kDa fragment of PE containing Ala–Cys mutations) as monitored by changes in the fluorescence emission maximum (filled diamonds, TP40; filled squares, PE40δcys), and that of TP40 monitored by tryptophan fluorescence lifetime (filled triangles). Reprinted with permission from Gress *et al.* (1994). Copyright 1994 American Chemical Society.

from the monotonic decrease in fluorescence intensity that is produced by increasing temperatures. Furthermore, the shift in the wavelength maximum of a fluorescence spectrum is usually a good indicator of the exposure of the fluorophore to altered solvent polarity and can be monitored as a function of temperature to obtain thermal stability information. For example, acidic fibroblast growth factor (aFGF) contains a single tryptophan residue, the fluorescence of which is highly quenched in the native state by neighboring histidine residues. Thus, upon excitation at 280 nm, tyrosine fluorescence is observed (Copeland *et al.*, 1991). Thermal unfolding of this protein, however, results in the generation of a tryptophan-like spectrum with concomitant quenching of tyrosine fluorescence due to relaxation of the original Trp quenching in the native state. This phenomenon has been used as a sensitive indicator of unfolding of aFGF and to study stabilization of aFGF by various polyanions (Tsai *et al.*, 1993a; Volkin *et al.*, 1993).

Changes in exposure of tryptophan or tyrosine side chains due to conformational changes can also be studied by taking advantage of the collisional quenching of their fluorescence by reagents such as iodide or acrylamide (Eftink, 1994). Surface-exposed fluorophores are quenched by iodide, which cannot usually penetrate to fluorophores buried in protein interiors. In contrast, accessibility of tryptophan or tyrosine to acrylamide, a polar, neutral molecule, varies widely depending upon the extent of burial of the residue. For example, the accessibility of the single tryptophan residue of aFGF to acrylamide was unaltered upon heparin binding to the protein, suggesting that the stabilizing interaction of this polyanion with the positive charge cluster on the surface of aFGF did not alter the local conformation around the partially buried tryptophan residue (Copeland *et al.*, 1991). This observation is consistent with the lack of an effect of heparin on either the tryptophan fluorescence lifetime or emission spectrum of aFGF.

Another powerful application of fluorescence spectroscopy lies in the determination of distances or distance distribution (for a dynamic protein conformation) between specific sites on a protein. This technique usually involves covalently labeling two sites with two different probes that can act as a donor–acceptor energy transfer pair. A decrease in emission quantum yield or lifetime of the donor (or concomitant changes in the acceptor fluorescence) depends upon the distance between the labeled sites within the protein. Conformational changes in the protein would alter this distance and be reflected in altered efficiency of energy transfer between the donor and the acceptor. In any such study, one must be certain that the labeled protein retains native structure and biological function. For example, Scheraga and co-workers site-specifically labeled RNAse A with a donor–acceptor pair (Buckler *et al.*, 1995). Time-resolved nonradiative energy transfer measurements were used to elucidate the conformation of partially folded intermediates at sub-denaturing concentrations of Gdn·HCl. The doubly labeled RNAse was found to retain 87% of the enzymatic activity of the native RNAse. However, the midpoint of thermal unfolding of the labeled RNAse was 6°C lower than that of the wild-type protein, as detected by near-UV absorbance measurements, suggesting some destabilization of the tertiary structure upon chemical modification.

2.3. Circular Dichroism

Circular dichroism (CD) spectroscopy in the far-UV region (180–250 nm) is a powerful and convenient way to estimate the overall secondary structure content of proteins and related changes therein. Many excellent reviews have been written on this subject (e.g., Yang *et al.*, 1976; Johnson, 1990; Manning, 1989). The integrity of the native secondary structure in formulated proteins must be established, and CD spectroscopy is ideally suited for this purpose as long as the formulation excipients result in a clear solution and are not optically active in the far-UV region. Furthermore, the contribution of excipients to the optical density of the solution must be small in the wavelength range of measurement so that the dynode voltage applied to the detector stays within a reasonable range, thereby maintaining an acceptable signal-to-noise ratio.

The CD spectra results from a combination of optical absorption (electric dipole moment) and asymmetry (magnetic dipole moment). Both of these dipole moments must be nonzero for a measurable CD signal, a condition which is satisfied by peptide bonds. Furthermore, elements of protein secondary structure such as α helices, β sheets, turns, and disordered structures each produces a different magnetic dipole moment. This effect, coupled with small differences in the electronic absorption spectra of these structures, results in distinctive CD spectra from each. Different proteins contain varying contributions from these structures and therefore manifest differences in their spectra, which are made up of overlapping combinations of these individual contributions. For example, α helices manifest two distinct minima at wavelengths around 208 and 222 nm and a maximum at 192 nm. On the other hand, disordered structures produce a relatively featureless CD spectrum with a minimum around 200 nm, and β sheets yield a broad minimum at 217 nm and a maximum at 193 nm (Brahms and Brahms, 1980). Furthermore, CD intensities in the 180- to 230-nm range are different for different kinds of secondary structure. α Helices have the strongest magnetic dipole moment and therefore manifest the highest intensity bands.

When the ordered secondary structure of the peptide backbone is perturbed, such as by temperature-induced or denaturant-induced unfolding, changes in the far-UV CD are observed. A transition from an α-helical or a β-sheet structure to a disordered structure can be easily monitored by observing spectral changes or by simply following intensities at predetermined wavelengths as a function of increasing perturbation (e.g., temperature or denaturant concentration). Simple two- and multistate models exist which allow the data to be used to determine quantitative thermodynamic parameters for the transitions detected. These measurements are extremely useful for monitoring stabilities of protein pharmaceuticals. Although such measurements made at two or three wavelengths may not permit analysis of the exact nature of a structural transition, especially for proteins that contain significant contributions from all of the different types of secondary structure, they can certainly be used to reflect order–disorder transitions and provide initial estimates of overall conformational stability versus temperature or other perturbations. Accelerated thermal stability measurements using this technique have been applied in our laboratory to screen for stabilizers of aFGF, which, in the absence of a polyanionic stabilizer, unfolds and aggregates at temperatures slightly above ambient temperature (Volkin *et al.*, 1993). aFGF is a primarily β sheet-containing protein that binds heparin (Copeland *et al.*, 1991). Accelerated stability measurements employing CD, among other techniques, have enabled us to discover a number of polyanions which can be used as stabilizers in the pharmaceutically acceptable formulations of this protein (Tsai *et al.*, 1993a; Volkin *et al.*, 1993).

Determination of overall secondary structure content from CD spectra ideally requires that data with high signal-to-noise ratio be obtained in the short wavelength range of 180–240 nm, although for proteins containing relatively high concentrations (40% or higher) of α helix, a relatively good estimate of the helix content is often obtained by simply dividing the ratio of observed mean residue ellipticity at 222 nm by 30,000 deg cm^2/dmol (Chen and Yang, 1971). Reliable estimates of the four main types of structure require analyzing the entire CD spectrum (preferably containing data from as low as 180 nm) in terms of a basis set of proteins with independently determined (by X-ray crystallography) secondary structure content, and using one or more of the following algorithms: (a) the single value decomposition method (Manavalan and Johnson, 1987); (b) the self-consistent field method (Sreerama and Woody, 1993), and (c) a convex constraint analysis (Perczel *et al.*, 1991, 1992). These analyses do not, however, necessarily ensure that accurate values for secondary structure content will be obtained. A frequently encountered problem is that spectral contributions from aromatic side chains in the far-UV region are not accounted for in the structure analysis methods (Bolotina and Lugauskas, 1985; Manning and Woody, 1989). Furthermore, there are frequently practical difficulties in obtaining CD spectra with reasonable signal-to-noise ratio below 200 nm in the presence of salt and other excipients that must often be present in a protein formulation. This problem becomes particularly severe below 190 nm, because of high absorption from most commonly used inorganic and organic components. It should be emphasized that far-UV CD spectroscopy in the 200- to 240-nm range still provides valuable information regarding conformational changes, and this is an important tool for analyzing protein stability and developing protein formulations, even though quantitation in terms of actual secondary structure content may not be reliable.

CD spectra of proteins in the near-UV absorbance range (250–320 nm) provide information on the local environment of phenylalanine, tyrosine, and tryptophan residues in the tertiary structure of a protein. The CD intensities (ellipticity) of the bands in the ranges 250–260, 270–280, and 290–300 nm are usually indicative of the extent of induced asymmetry in the environments of phenylalanine, tyrosine, and tryptophan residues, respectively. When a protein is unfolded, these aromatic bands disappear or are greatly reduced in intensity because their immediate asymmetric environments are disrupted. Molten globule or partially folded structures of proteins usually manifest very weak or no aromatic CD bands. The lack of a near-UV CD band while the far-UV CD spectrum remains unchanged relative to the fully folded protein is an almost certain indication of the formation of a molten globule or partially folded structure. In this regard, near-UV CD and tryptophan fluorescence spectroscopy usually provide complementary information. However, CD offers an advantage, in that changes in the phenylalanine microenvironment can be readily observed. Furthermore, in most proteins containing tryptophan and tyrosine residues, tyrosine fluorescence is not clearly resolved. Near-UV CD offers a convenient means for probing the local environment of tyrosine residues for such proteins. The disadvantage of the near-UV CD technique is its low sensitivity with regard to the protein concentration requirement compared to fluorescence (one to three orders of magnitude higher concentration than the few micrograms per milliliter needed for fluorescence measurements). A good example of the complementary use of tryptophan fluorescence and near-UV CD for analyzing functionally significant subtle structural changes around tyrosine and tryptophan chromophores is described in detail for the chimeric protein TP40 (Gress *et al.*, 1994).

2.4. Fourier-Transform Infrared Absorption Spectroscopy

The amide I absorption region (1600–1700 cm^{-1}) of the IR spectra of proteins has been extensively used during the last decade to estimate the secondary structure content. The recent explosion in the use of this technique was brought about primarily by the following events: (a) the commercial availability of FTIR instrumentation, (b) the use of band narrowing techniques to enhance the resolution of the relatively broad amide I band (Kaupinnen *et al.*, 1981), and (c) the assignment of bands, following band fitting procedures, to ordered secondary structures based on the IR spectra of proteins for which high-resolution crystal structures are available (Byler and Susi, 1986). In general, β sheets and turns can be estimated with greater accuracy from amide I IR spectroscopy than from far-UV CD spectroscopy. In contrast, bands produced by the peptide backbone in disordered and α-helical conformations overlap extensively between 1645 and 1660 cm^{-1}, often prohibiting quantitative resolution of their individual contributions to the overall structure content. Therefore, far-UV CD spectroscopy frequently provides a more accurate estimate of the α-helix content of a protein than IR studies. In most IR solution studies, D_2O has been used as solvent in order to minimize the broad water absorption bands centered near 1640 cm^{-1}. The sensitivity of this technique is rather low in conventional transmission geometries requiring high protein concentrations, typically in the range of 1–10 mg/ml when D_2O is used as solvent and 10–100 mg/ml when H_2O is employed. FTIR spectroscopy has been recently extended to the study of protein folding and stability (reviewed in Middaugh *et al.*, 1995).

The potential use of this spectroscopic technique in the pharmaceutical development of protein formulations, however, goes far beyond its simple use as another "low-resolution" solution spectroscopic technique for monitoring secondary structure content and protein folding. This is because of the applicability of IR to different physical states of a sample, including dried and lyophilized materials. The effect of lyophilization on protein structure is a critical issue in the development of lyophilized formulations. Some proteins, for example, basic fibroblast growth factor and gamma interferon, have been found to irreversibly unfold upon lyophilization. Sugars have been observed, however, to be protective against conformational damage in some cases (Prestrelski *et al.*, 1993). These studies have employed use of KBr pellets and transmission geometry. It should be noted that KBr, at the very high concentrations used, could act as a chaotropic agent and care must be taken before assigning IR changes unambiguously to drying-induced structural perturbations of the proteins (Chan *et al.*, 1996). Two alternative approaches can be employed in this regard: (a) diffuse reflectance and (b) FTIR microscopy in either the reflectance or the transmission mode. An extensive study of the real-time kinetics of conformational changes of lysozyme during dehydration has been reported that employed attenuated total reflectance spectroscopy (Carpenter *et al.*, 1994). In this method, lyophilization was performed from a D_2O solution and the amide I′ and amide II′ bands were monitored as a function of decreasing temperature and dehydration.

In testing the effects of formulation conditions on a protein's secondary structure, it is necessary to be able to rapidly and quantitatively compare spectra. A recently introduced technique for quantifying the area of overlap of second-derivative spectra is very well-suited for this purpose (Kendrick *et al.*, 1996). This method is based on comparing area-normalized second-derivative spectra and assigning a correlation coefficient (originally introduced by Prestrelski *et al.*, 1993) in addition to a visual inspection (Fig. 2A). The Gdn·HCl-induced equilibrium unfolding profile of iso-1-cytochrome monitored by this technique has been found to be strikingly similar to that determined by CD (220 nm), as shown in Fig. 2B (Bowler *et al.*, 1993). In another pharmaceutically relevant application of FTIR spectroscopy, temperature-induced aggregation of chymotrypsinogen in D_2O solution has been systematically studied by Ismail *et al.* (1991), by monitoring the appearance of a relatively strong band at 1618 cm^{-1} and a weaker one at 1685 cm^{-1}. These bands, which have consistently been seen in aggregating protein systems, have been assigned to intermolecular β sheets and their presence permits a rapid identification of this phenomenon. As another example, the temperature-induced unfolding and refolding of ribonuclease A has been thoroughly studied in both H_2O and D_2O solutions (Fabian *et al.*, 1993).

It should be noted that the general validity of the resolution enhancement (Fourier self-deconvolution) and curve-fitting techniques most commonly employed for assignment of bands and their relative contributions to particular secondary structure types has been questioned (Surewicz *et al.*, 1993). In a series of studies, Venyaminov and Kalnin (1990a, b) identified the IR signals originating from the different amino acid side chains in the amide I and II absorption regions. In some cases, spectral contributions from certain side-chain residues need to be considered in secondary structure analyses. Interference from side-chain bands with peptide bond absorption, however, remains difficult to resolve quantitatively. In view of these difficulties in determining secondary structure content in terms of absolute percentages, the direct comparison of second-derivative spectra as described above is probably the most rapid and accurate way to test the effects of different formulation conditions on a protein's secondary structure (Kendrick *et al.*, 1996).

2.5. Differential Scanning Calorimetry

Differential scanning calorimetry (DSC) is the most direct technique available for measuring thermal unfolding and stability of proteins. Although

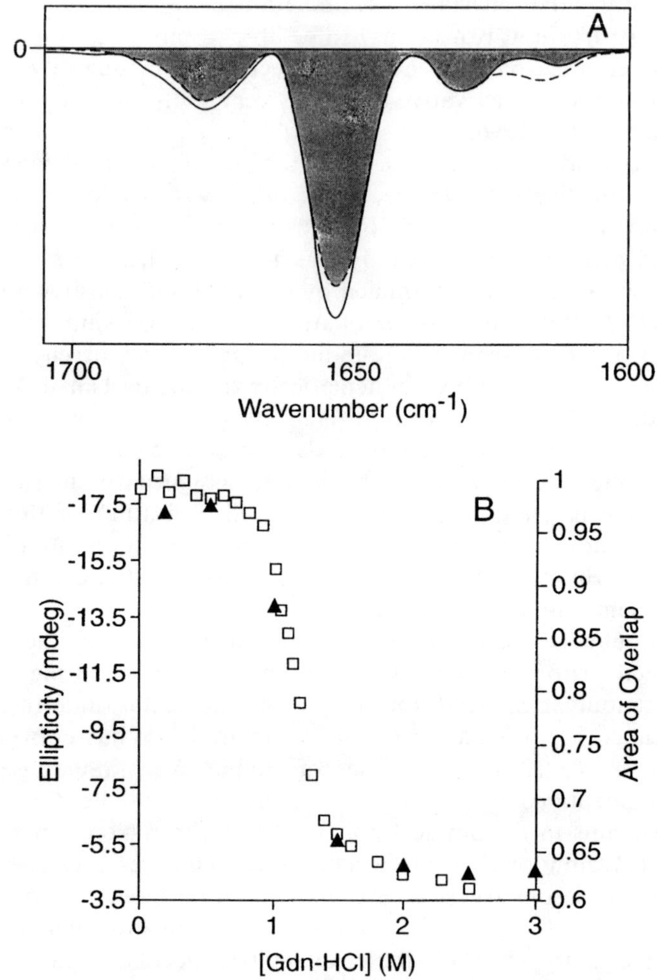

Figure 2. (A) Second-derivative area normalized FTIR spectra (amide I region) of γ-interferon in the "native" aqueous state (solid line) and rehydrated aqueous state following lyophilization (dashed line), the area of overlap being shown as the gray fill. (B) Guanidine hydrochloride-induced equilibrium unfolding transition of iso-1-cytochrome c as monitored by circular dichroism at 220 nm (open squares) and FTIR spectral area of overlap (filled triangles). Reprinted with permission from Kendrick *et al.* (1996). Copyright 1996 American Chemical Society.

thermodynamic calculations of the free energy of unfolding cannot be made when unfolding is irreversible, for example, due to aggregation of the unfolded state, a direct measure of the temperature range over which unfolding occurs is provided by heat capacity changes as a function of changing temperature. Potential use of various excipients as protein stabilizers can be evaluated by simply monitoring their effects on the heat flow profile of a protein sample scanned from low to high temperatures. For example, the effects of different sugars, amino acids, and salts on the thermal transition temperature T_m of aFGF unfolding was studied by DSC (Tsai *et al.*, 1993a). The moderately stabilizing effect of these reagents on the thermal unfolding of this protein, as detected by DSC, is in agreement with CD results. As an example, the effect of trehalose on the DSC profile of aFGF is shown in Fig. 3. The effects of nonionic detergents on the thermal stability of recombinant human growth hormone (hGH) was recently studied by DSC (Katakam *et al.*, 1995). Surfactants did not protect against temperature-induced unfolding of hGH, as judged by the protein denaturation endotherms, although they protected the protein against interfacial aggregation. Furthermore, high concentrations of these agents actually destabilized hGH, as reflected in the decrease in the onset temperature for the peak in the endotherm.

The effect of moisture on the thermal stability of lyophilized proteins can also be examined by DSC. For example, the introduction of moisture to freeze-dried bovine somatotropin has been proposed to alter the mechanism of temperature-induced structural changes (Bell *et al.*, 1994). In this case, a kinetic DSC approach was used in which the scan rate dependence of the thermal transition was monitored. The use of equilibrium thermodynamics for interpreting calorimetric results should in general be avoided when the thermal change is irreversible, as is often the case for proteins (Lepock *et al.*, 1989; Sanchez-Ruiz *et al.*, 1988).

3. PROTEIN SIZE, QUATERNARY STRUCTURE, AGGREGATION STATE, AND SOLUBILITY

The large-scale preparation of protein pharmaceuticals requires fermentation and isolation procedures in which the protein of interest must be properly folded either during *in vivo* expression or as a separate *in vitro* refolding step. Unfortunately, aggregation reactions often compete with the proper folding of a protein, especially upon *in vitro* refolding from the unfolded or partially folded state. Protein aggregation may be mediated by covalent or noncovalent interactions, such as intermolecular disulfide bond formation and apolar associations. Aggregation is a concern not only during the expression and isolation of a protein, but also in subsequent formulation processing steps. For example, the use of lyophilization to stabilize labile

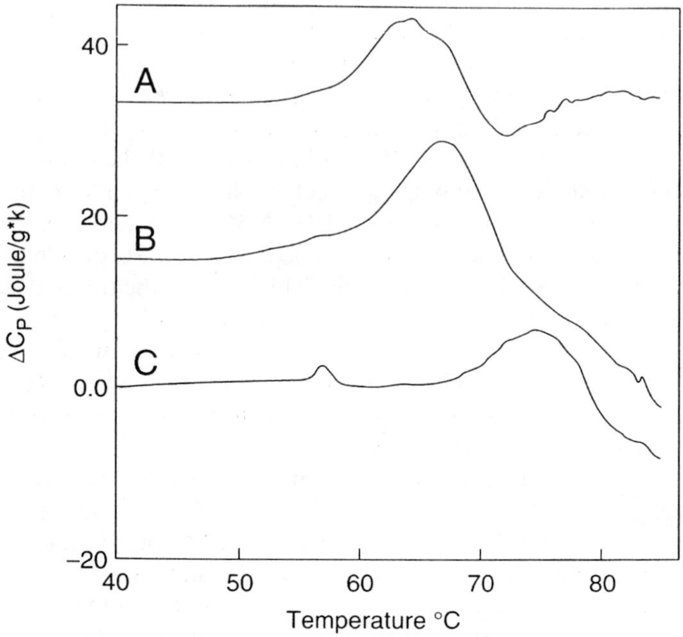

Figure 3. The effect of trehalose on the DSC endotherm of aFGF in the presence of $\frac{1}{3}\times$ heparin (by weight). All samples contained 1 mg/ml aFGF in a phosphate-buffered saline solution at pH 7.2. (A) No trehalose, (B) 0.5 *M* trehalose, (C) 1.5 *M* trehalose. Reprinted with permission from Tsai *et al.* (1993a). Copyright 1993 Plenum Press.

proteins during storage may also result in aggregation under certain conditions, because freezing and dehydration expose proteins to variations in pH and salt concentration. In addition, the aggregation of proteins storage in the dried state or during reconstitution has been observed (Constantino *et al.*, 1995). Aggregation of protein pharmaceuticals may lessen biological activity or increase the probability of an adverse antigenic response that will not occur with the native folded protein in its active monomeric or proper associative state (Cleland *et al.*, 1993). Thus, maintaining the biologically active form of a protein requires preserving its overall size, both by preventing aggregation and, in the case of oligomeric proteins, maintaining quaternary structure (Schein, 1994). A rational approach to the formulation of a stable pharmaceutical protein therefore involves a prior detailed understanding of a protein's overall size and quaternary structure as well as its association/aggregation behavior, in addition to its conformational stability and chemical integrity. Many methods are available for measuring the overall size of a protein and the most commonly used techniques will be discussed below. Complete biophysical characterization of a native, multimeric protein or an inactive, aggregated protein will depend on obtaining data acquired from a combination of methods, because various techniques possess different detection and discrimination capabilities as well as complementarity in this regard.

3.1. Electrophoresis

Polyacrylamide gel electrophoresis in the presence of sodium dodecyl sulfate (SDS–PAGE) is commonly used to assess the molecular weights of proteins. The excess of negatively charged sulfonate groups results in rodlike SDS/protein complexes of constant charge-to-mass ratio, which migrate proportionally to their molecular weight. Although this method is simple and rapid (especially with precast gels), the molecular weight estimate that is obtained is only accurate to approximately ±15% (Steele and Neilsen, 1978). If aggregation is mediated by disulfide bonds, SDS–PAGE can easily detect multimers if nonreducing conditions are employed. Noncovalent associations are difficult to detect, however, because exposure to SDS will dissociate and denature most proteins. Noncovalent aggregates can potentially be detected by covalently crosslinking the physically aggregated material with a suitable chemical reagent (e.g., glutaraldehyde) and then performing SDS–PAGE. Molecular weight estimates can be obtained with suitable standards, but discrimination of highly aggregated complexes is difficult. Large aggregates may not enter the gel matrix and may remain either undetected or merely seen as material at the gel top, whereas smaller compounds may migrate too rapidly and exit the gel undetected. Furthermore, heterogeneously aggregated proteins usually appear as a very broad, diffuse band, which also may complicate quantitation and comparisons. Because resolution and staining techniques (especially with silver) are not optimal, quantitation may also be troublesome.

Electrophoresis of chemically crosslinked oligomeric proteins has been successfully used, however, to produce a series of bands corresponding to species up to the largest oligomer present. This approach has been used to define subunit stoichiometries (Craig, 1988). For example, Herman *et al.* (1979) demonstrated varying ratios of monomer, dimer, and tetramer forms of lactate dehydrogenase by crosslinking the enzyme under different environmental conditions known to lead to inactivation or reactivation of the enzyme.

The use of capillary electrophoresis (CE) for the characterization of associative states of proteins offers many potential advantages compared to SDS–PAGE, such as attaining higher resolution in heterogeneous mixtures as well as more rapid analysis (minutes versus hours), yet in practice it also has some of the same limitations. Sample preparation is often similar, in which SDS and reducing agents are used in a manner analogous to SDS–PAGE. CE chromatography profiles can be monitored by a variety of methods, including absorbance, fluorescence, light scattering, and mass spectrometry (Kuhr, 1990). Fluorescence detection offers high sensitivity especially with the use of lasers and fluorescent derivatizing agents. Simultaneously, some limited information about protein structure can also be acquired using these types of detection (as described in previous sections) in addition to other techniques, such as Raman spectroscopy (Deyl and Struzinsky, 1991). Some currently available CE instrumentation allows collection of fractions, thus permitting identification and more extensive characterization of aggregated material. One disadvantage of CE is that the fine capillaries have the potential to become obstructed, thus preventing uniform mobility, especially if extensive protein aggregates are present. In summary, electrophoretic methods are readily performed, but yield qualitative or semiquantitative information. Therefore, further quantification and definition of protein aggregates is better suited to other analytical methods, as discussed below.

3.2. Chromatography

High-pressure size-exclusion chromatography (HP-SEC) and reversed–phase high–pressure liquid chromatography (RP-HPLC) are convenient methods for monitoring the size and quaternary structure of proteins by utilizing commonly available laboratory instrumentation. Chromatographic experiments are often well-suited to analysis of proteins in more complex formulations that contain polymers or other proteinaceous compounds that may interfere with spectroscopic methods of analysis. Under the appropriate chromatographic conditions, multimeric forms can be resolved, readily quantified, sized with the use of suitable standards, and isolated. These methods will generally yield different chromatographic profiles of the same protein preparation, because separation is based on different physical properties (i.e., size/shape and polarity).

Size-exclusion chromatography has the immediate advantage that elution position does suggest the presence of associated material. For example, as shown in Fig. 4, the HP-SEC elution of a partially aggregated preparation of basic fibroblast growth factor (bFGF) suggests a single aggregated species, whereas the RP-HPLC elution profile of the same preparation suggests that more than one nonnative form of the protein is present (Shahrokh *et al.*, 1994). With either chromatography method, protein-containing fractions can be easily collected and then further analyzed to better define the nature of the aggregated species. In addition, amino acid analysis, N-terminal sequencing, or peptide mapping may be performed at this point to ensure identity of the individual species.

Along with CE, the most common methods of detection for chromatography are absorbance and fluorescence, with refractive index or light scattering less frequently employed. Although absorbance detection generally uses a single wavelength in either the near-UV (280 nm for aromatic side chains) or the far-UV region (215 nm for peptide bonds), sufficient sample concentrations and longer spectroscopic path lengths allow multiple-wavelength and even complete spectral data to be collected with currently available chromatography detectors. This capability enables single experiments to yield information about not only the size, but also the nature and conformational state of the molecules in the individual fractions. For complex mixtures even within individual fractions, multiple-wavelength absorbance detection may permit discrimination of proteins from other components, such as nucleic acids, or even resolution of individual proteins within single fractions if their spectroscopic properties are sufficiently different. Fluorescence monitoring employing multiple wavelengths can also yield information about the extent of protein unfolding by consideration of the position of a protein's intrinsic fluorescence emission maximum relative to that of the protein in its native state.

In addition to spectroscopic monitoring, the availability of light scattering detectors enables estimation of size parameters. Combining gel permeation chromatography with intensity-based, low-angle laser light scattering detection provides a powerful technique for quantifying and defining the size of aggregates. For example, Endo *et al.* (1992) effectively used this method to examine the effect of temperature and pH on the aggregation state of erythropoietin. In the absence of a specialized light scattering detection, a fluorescence detector can be employed by monitoring the light scattered at 90° at the incident wavelength (Dollinger *et al.*, 1992). Standards of varying molecular weight can be used for size estimation, but specific proteins behave differently than common standard proteins due to shape differences and as a result yield inaccurate size values (Middaugh *et al.*, 1992). In general, it is difficult to identify suitable standards for larger protein aggregates, polymers, or viral particles. In the case of spherical protein particles, such as some types of viral capsids, spherical latex beads of known size can be utilized. It is

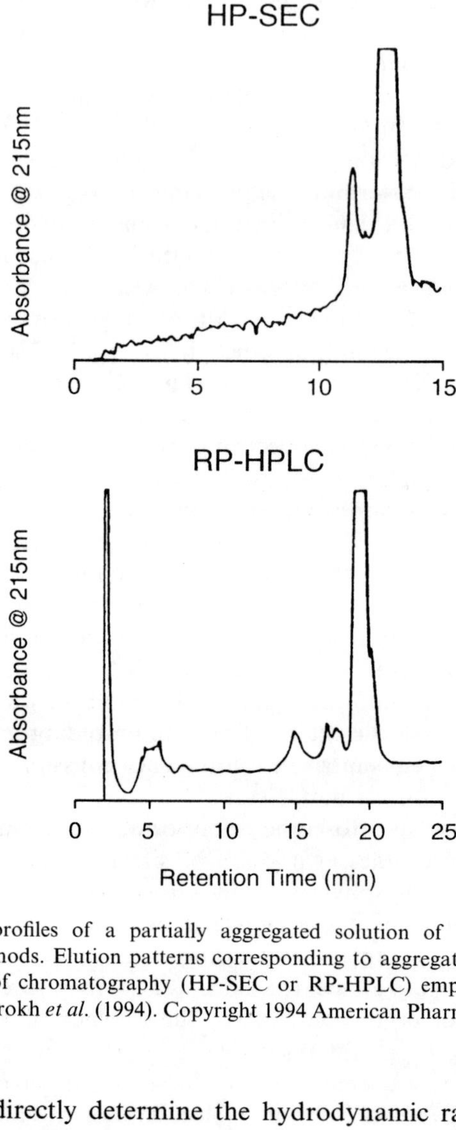

Figure 4. Elution profiles of a partially aggregated solution of bFGF using different chromatographic methods. Elution patterns corresponding to aggregate forms of the protein differ with the type of chromatography (HP-SEC or RP-HPLC) employed. Reprinted with permission from Shahrokh *et al.* (1994). Copyright 1994 American Pharmaceutical Association.

also possible to directly determine the hydrodynamic radius of a protein during chromatography by measuring fluctuations in the intensity of scattered light (dynamic light scattering) as well as radius of gyration and absolute molecular weights if the angular dependence of the scattered light is analyzed (see below). Specialized instrumentation has recently become available for both of these approaches in a chromatographic format.

3.3. UV Spectrophotometry

Although UV absorbance spectra are primarily used for protein mass quantitation, aggregation can be detected rather easily by changes in the absorbance baseline of the UV spectrum in nonabsorbing regions (e.g., >300 nm). Although it is only a qualitative technique (although in some cases the quantity "turbidity" can be quantitatively applied), the ready availability of UV spectrophotometers in many laboratories often enables one to rather easily obtain some information on the aggregation state of a protein preparation. Because light scattering is inversely proportional to the incident wavelength, protein aggregation results in an exponentially increasing baseline as the wavelength decreases (see Section 2.1). Consequently, baseline deviations due to light scattering are most evident in the UV region. One drawback is that the aggregates being monitored must be relatively large (>20 nm in diameter) before baseline changes are observed (Mach *et al.*, 1995). Kinetic data are easily collected on most common spectrophotometers, making this one of the most convenient methods for obtaining indirect information concerning the extent of protein aggregation over time (Eberlein *et al.*, 1994; Tsai *et al.*, 1993a; Volkin *et al.*, 1993). In such experiments, the appearance of optical density at an appropriate wavelength (e.g., 300–400 nm) is simply monitored as a function of time in the presence of aggregation-inducing stress, such as temperature or pH. Comparative experiments can then be performed in the presence and absence of pharmaceutical excipients to evaluate their stabilizing effects (Volkin *et al.*, 1993).

3.4. Analytical Ultracentrifugation

Although ultracentrifugation experiments are more time-consuming than the methods described above, they usually provide the most unambiguous determination of protein size and molecular weight under native solution conditions. Both sedimentation velocity and equilibrium sedimentation experiments can provide accurate measures of a protein's size and association state if the appropriate specialized instrumentation is available. Sedimentation velocity experiments can be performed in only a few hours, although absolute molecular weights cannot be obtained by this method. This technique also requires measurements (or estimates) of partial specific volume as well as diffusional or fractional coefficients. The sedimentation coefficient that is determined from such experiments also depends on the size and shape of the molecule or complex. In general, molecular weights calculated from velocity of sedimentation involve approximations that yield data comparable in accuracy to those obtained from PAGE estimates (Van Holde, 1975). Nevertheless, velocity experiments do provide at least semiquantitative information on the aggregation state of proteins in solution. Equilibrium experiments offer a more accurate way to determine the molecular weight of a protein and its complexes, although a measure of the protein's partial specific volume is again needed.

Detection systems are available that use either optical absorbance or interferometry, such that a wide range of concentrations and sample conditions can be examined. Detection of sedimentation profiles by interferometry has the advantage of being able to monitor preparations that have little optical absorbance, but multicomponent formulations severely complicate this type of data and may make quantitative analysis impossible. In comparison, optical absorbance has the advantage of monitoring only the protein itself in a complex mixture, if none of the other components have significant absorbance in the region of interest. For example, the association of acidic fibroblast growth factor (aFGF) in the presence of the stabilizing ligand heparin was successfully monitored in an equilibrium sedimentation experiment using a wavelength of 280 nm (Mach *et al.*, 1993). Because protein association is mediated in this example by a polysaccharide, the association state of aFGF can be easily followed, because heparin does not absorb light at this wavelength. High concentrations of salts or buffers in a protein solution may prohibit using UV detection to monitor the sedimentation profile of a protein at lower, more sensitive wavelengths. Formulation excipients may also absorb strongly in other regions of interest, which may be either a hindrance or help, depending on the information desired.

In complexes which are confined to a small number of molecules, such as subunit or multimeric proteins, the extent of association and the corresponding association constants can be well defined. As mentioned above, this requires knowledge of the partial specific volume of the protein of interest, which can be estimated by calculation or directly obtained from solution density measurements. Software for such analysis is available from the instrument manufacturer or from many academic sites (contact http://bioc02.uthscsa.edu/biochem/xla.html on the World Wide Web to obtain listings for available analysis software). When the stoichiometry of the associated state is small (ca. 3), analysis is relatively straightforward and can reveal the relative amounts of monomeric and associated populations (M. L. Johnson *et al.*, 1981). For example, Green and *et al.* (1991) employed equilibrium sedimentation analysis to extensively characterize the quaternary structure of the enzyme arginase, demonstrating a reversible monomer–

trimer equilibrium with a defined dissociation constant. Interestingly, the addition of specific metal ions dramatically stabilizes the association of the monomers to the oligomeric state. Protein quaternary structure can also be modulated by less specific interactions, such as ionic strength. Sedimentation velocity experiments have also been used to show that the relative amounts of the four major oligomeric forms of casein kinase 2 can be altered by varying the sodium chloride concentration (Valero *et al.*, 1995). Analytical ultracentrifugation has been used to define the association constants, molecular weights, and shapes of immunoglobulin complexes (Liu *et al.*, 1995). When examining more heterogeneous populations of aggregates, data analysis of equilibrium sedimentation data becomes complicated, but can be simplified, assuming uniform aggregation constants for all multimers (Cole *et al.*, 1993). Finally, using equilibrium sedimentation experiments, the extent of covalent modification of proteins can also be monitored. For example, the degree of glycosylation in a preparation of human stem cell factor was determined by direct comparison to the unglycoslyated form of the protein (Arakawa *et al.*, 1991).

3.5. Light Scattering

Light scattering was described in previous sections in terms of its use as a chromatographic detection technique, but it can also offer a rapid and accurate method for characterizing the size and shape of proteins when appropriate instrumentation is employed. The most straightforward experiment (in theory and practice) employs static light scattering. The intensity of scattered light by a particle is a complex function of its size and shape, but certain limiting assumptions allow size information to be experimentally obtained. Maximum information can be deduced by measuring the scattered intensity of light as a function of angle, but it is most common to perform such studies at a fixed angle, usually either 90° or at some low angle where scattering is more intense (e.g., 135°). Although small particles show uniform scattering intensity at all angles, larger particles, such as extensively aggregated proteins, tend to produce lower scattering intensity as the detection angle is increased. Analysis of such data permits both molecular weights and radii of gyration to be obtained as well as second virial coefficients to be calculated (a useful quantity beyond the scope of our discussion here). As particles become very large, light scattering behavior becomes a very complex function of particle shape, and data interpretation becomes strongly model dependent.

In addition to static light scattering, particle size can also be determined by dynamic light scattering (also sometimes referred to as photon correlation or quasielastic spectroscopy). This technique relies on the fluctuations in time of the intensity of scattered light. The degree of correlation of these fluctuations depends on the diffusion coefficient of the scattering particle, a quantity which is simply related to hydrodynamic radius through the Stokes–Einstein equation. This mathematical treatment (a "cumulant analysis") of these data also permits the degree of homogeneity (polydispersity index) to be accessed as well as size distributions to be constructed. In ideal circumstances (particles of very different diameters), the sizes of two or three components in a mixture can be deconvoluted. A major experimental limitation in both static and dynamic light scattering experiments is interference by dust. Because scattering is dominated by larger particles, the presence of even trace quantities of such material must be avoided if interpretable data are to be obtained. The simple filtration of protein samples prior to analysis is often effective in minimizing interferences from dust particles, however, this type of sample preparation may also effect the distribution and quantity of the protein particles themselves.

Although light scattering techniques offer the ability to obtain accurate molecular weights of relatively small macromolecules as well as large aggregates, perhaps its major use in preformulation work is as an empirical tool for investigating the effect of excipients upon the aggregation state of proteins. For example, characterization of the aggregation of insulin under varying conditions of concentration, pH, and ionic strength played a key role in formulation development and even permitted an estimate of critical association constants (Kadima *et al.*, 1993). Cleland and Wang (1990) used light scattering to characterize the aggregation of carbonic anhydrase during refolding as well as to optimize refolding conditions. Similar studies have been carried out with acidic fibroblast growth factor and were a key to the development of a stable formulation of this very unstable protein (Mach *et al.*, 1993; Middaugh *et al.*, 1992). Simple titration of pharmaceutical excipients into a protein preparation and simultaneous monitoring of light scattering allows formulation components to be conveniently evaluated in a relatively short period of time for their ability to inhibit protein aggregation. In addition, this approach can be used to monitor the interaction and association of protein–ligands (Mach *et al.*, 1993). As indicated above, because light scattering is exponentially proportional to the particle diameter, very small amounts of a larger species (e.g., aggregates) can easily be detected. Conversely, small amounts of large material will dominate the scattering intensity and mask the contributions of smaller sized populations. In such cases, other methods, such as SEC-HPLC or analytical ultracentrifugation, should be used in conjunction with light scattering measurements to ensure an accurate description of particle size distribution.

3.6. Mass Spectrometry

Compared to the methods discussed previously, mass spectrometry provides the most accurate molecular mass value. Electrospray (ES) and laser desorption time-of-flight (LD-TOF) mass spectrometry (MS) have emerged as powerful techniques well-suited for the analysis of pharmaceutical proteins (Hillenkamp and Karas, 1990; Smith *et al.*, 1991). ESMS offers higher resolution and accuracy than LD-TOF MS and is therefore the method of choice for detecting chemical modifications that result in relatively small mass changes (see Section 4.2). Covalent aggregates of large proteins, such as serum albumin, can be characterized as well (Smith *et al.*, 1991). However, a homogeneous sample is usually required to take maximum advantage of the technique. This will usually involve purification or separation before introduction into the instrument. Furthermore, instrumentation is more expensive to maintain and more difficult to operate compared to LD-TOF apparatus.

LD-TOF MS is typically more sensitive in the sense that it typically requires less sample (1–10 pmol is optimal). Moreover, sample homogeneity is not required, because only a few peaks are generated by an individual protein, in contrast to the complex charge spectrum seen in ESMS with similar samples. Furthermore, the simplicity of commercially available instrumentation and its operation has allowed this method to become widely accessible to process and formulation scientists. This method requires, however, the presence of a laser absorbing matrix to induce volatilization, which, along with the desorption process itself, could dissociate aggregated or oligomeric proteins, thereby precluding the ability to obtain the desired information on the aggregation state of the protein in solution (Hillenkamp and Karas, 1990). Nevertheless, in some cases, multimeric states can be observed with careful sample preparation and analysis. For example, as shown in Fig. 5, Suter *et al.* (1992) were able to observe tetrameric as well as smaller associated forms of interleukin-2 (IL-2). The use of crosslinking reagents should also enable noncovalently associated, multimeric states of proteins to be observed, as previously described with SDS–PAGE (Farmer and Caprioli, 1991). In general, protein aggregates having a mass of less than 200 kDa can reliably be identified with this technique, although relative peak signal intensities may not be proportional to the populations of each

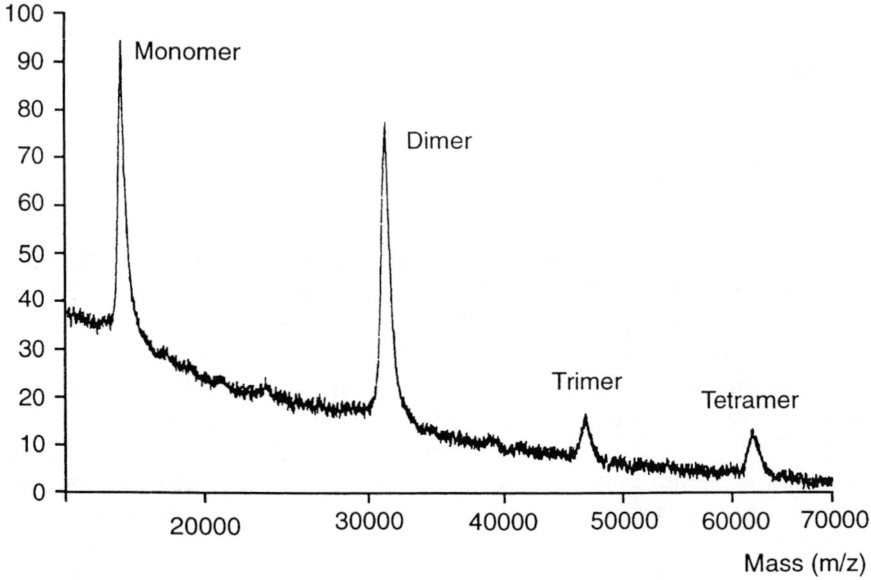

Figure 5. LD-TOF MS of aggregated forms of recombinant IL-2. The spectrum was acquired using 20 pmol of IL-2 in a sinapinic acid matrix. The oligomerized protein peaks can be eliminated by reduction and blocking of cysteine residues (not shown). Reprinted with permission from Suter *et al.* (1992). Copyright 1992 Academic Press.

particular species, because of potential differences in desorption or detection efficiency. Thus, quantitation is best evaluated by the methods previously described, such as chromatography or analytical ultracentrifugation.

3.7. Other Methods

The presence of aggregates can often be detected by a number of other optical techniques commonly used to study protein conformation and structure. The consequences of light scattering on UV absorption spectra have already been discussed, and a similar effect may be seen in circular dichroism spectra. Depending on the scattering characteristics of the high molecular weight material, spectral peaks may decrease in intensity due to absorption flattening (a particle screening effect). In the event of preferential scattering of one type of circularly polarized light, peaks may also shift position (Wallace and Mao, 1984). Fourier-transform infrared spectra of aggregated proteins usually manifest additional peaks that can be attributed to intermolecular β structure (Dong *et al.*, 1995; van Stokkum *et al.*, 1995). In addition, fluorescence spectra may also show decreased emission due to scattering and an inner-filter effect in which densely packed chromophores prevent uniform exposure to incident light (Lakowicz, 1983). Furthermore, time-resolved fluorescence experiments can be employed to characterize protein aggregation (Brochon *et al.*, 1993).

Nonspectral methods can also be used to determine protein size and molecular weight. For instance, osmotic pressure measurements can provide an accurate measurement of molecular weight if the concentration of the protein in solution is accurately known. Another technique gaining increased acceptance is field flow fractionation (Giddings, 1989). Litzén *et al.* (1993) showed that field flow fractionation of a monoclonal antibody preparation provided better resolution of various aggregated protein species than did gel permeation chromatography.

3.8. Solubility

A critical physical property of a protein which is often neglected in a quantitative sense, but is of manifest importance in a practical pharmaceutical one is aqueous solubility. Low solubility often presents a significant obstacle during formulation development, and must be directly addressed at an early stage of development. Measurement of solubility is often complicated by the tendency of proteins to form poorly defined suspensions and gels. These phenomena do not permit an equilibrium solubility to be directly determined by simple addition of dry (e.g., lyophilized) protein to solvent until saturation is obtained (via measurement of protein concentration or, ideally, thermodynamic activity, in the saturated liquid phase). An alternative method employs sequential precipitation of protein by addition of high molecular weight (e.g., 8 kDa) polyethylene glycol (PEG). Plots of the logarithms of protein solubility versus PEG concentration are often linear and an estimate of the activity of the protein in the absence of PEG can be obtained by extrapolation to zero PEG content. For reasons described elsewhere (Middaugh and Volkin, 1992), however, this approach is best used in a comparative manner. Further information about the protein solubility problem can be found in Middaugh and Volkin (1992).

An indirect method that has been increasingly employed to monitor the effect of solution conditions on protein solubility uses turbidity (light scattering) measurements. In this approach, it is most common to follow the rate of turbidity appearance as a function of variables such as pH, temperature, and excipient. Because only a simple absorption spectrophotometer (or fluorometer, if scattering at right angles is monitored) is necessary for such studies, this method has been extensively employed in this early screening phase of formulation development (Volkin *et al.*, 1993). Again, however, its primary use is as a comparative measure, because no absolute values of solubilities are derivable from such methods.

4. DEGRADATIVE COVALENT REACTIONS AND THEIR EFFECT ON PHYSICOCHEMICAL AND BIOLOGICAL PROPERTIES OF PROTEINS

Our discussion concerning preformulation studies of protein pharmaceuticals has focused on the effect of environmental conditions on these macromolecules' conformational integrity and stability. Exposure to elevated temperature, acidic/basic conditions, extreme redox potentials, or high ionic strength can all lead to conformational changes in protein structure (local alterations in secondary or tertiary structure), reversible unfolding (cooperative loss of most higher ordered structure), or irreversible alterations (permanent changes in the structural or chemical integrity of the protein molecule). The interplay between changes in protein structure able to facilitate chemical alterations of amino acid side chains (or peptide backbone) and the effect of degradative covalent reactions on the loss of native protein structure should be a primary focus of preformulation studies. For example, partial unfolding of a protein molecule leads to both increased solvent accessibility of previously unexposed amino acid residues as well as increased flexibility of the polypeptide chain. In many cases these types of effects facilitate the reactivity of both peptide bonds and amino acid side chains. In fact, the partial loss of conformational integrity is frequently observed to be the first step in the irreversible inactivation of proteins and enzymes at high temperatures (Volkin and Klibanov, 1989; Volkin and Middaugh, 1992). Conversely, covalent reactions altering either the amino acid side chains or the polypeptide backbone may lead to the loss of conformational stability. For example, the reduction of disulfide bonds or hydrolysis of polypeptide chains can induce protein unfolding leading to aggregation and inactivation. Obviously, a coupled interaction between these two processes can also occur and further complicate the goal of a complete temporal, physicochemical description of a protein pharmaceutical.

4.1. Covalent Alterations in Protein Structure Commonly Observed during Formulation and Storage

During formulation and *in vitro* storage, protein pharmaceuticals in liquid solution are typically exposed to pH values ranging from 4 to 8 and temperatures of 2–37°C for time periods up to 1–2 years. Under these conditions, the most commonly observed changes in amino acid side chains are deamidation and isoaspartate formation of Asn and Asp residues and the oxidation of Met and Cys residues (Powell *et al.*, 1996). The mechanistic details of these reactions have been studied extensively in both model proteins and peptides as well as protein therapeutics over the past decade. Many excellent reviews can be consulted for a detailed discussion of the reaction pathways of protein deamidation and isoaspartate formation (Aswad and Guzzetta, 1995; Bischoff and Kolbe, 1994; Clarke *et al.*, 1992; Wright, 1991) as well as protein oxidation reactions (Li *et al.*, 1995; Nguyen, 1994).

Briefly, the spontaneous deamidation of Asn residues under typical formulation conditions proceeds via an intramolecular mechanism as outlined in Fig. 6. The asparaginyl side-chain carbonyl group undergoes nucleophilic attack from the peptide bond nitrogen adjacent to the asparaginyl side chain, resulting in ring closure and release of ammonia. The resulting five-membered ring (succinimide or cyclic imide) is susceptible to hydrolysis, leading to a mixture of aspartyl and isoaspartyl residues. The isomerization of aspartyl residues to isoaspartyl can also occur via the same cyclic imide intermediate mechanism. The primary structure (presence of specific neighboring groups) as well as the conformational flexibility of the local higher ordered structure surroun-

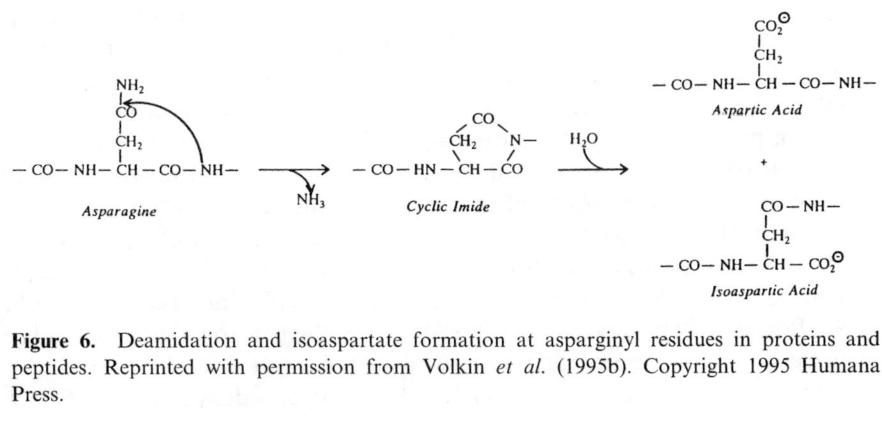

Figure 6. Deamidation and isoaspartate formation at asparginyl residues in proteins and peptides. Reprinted with permission from Volkin *et al.* (1995b). Copyright 1995 Humana Press.

ding the asparaginyl and aspartyl residues affect the rate at which deamidation and isoaspartate formation can occur. In general, Asn–Gly and Asn–Ser are the most labile sequences as a consequence of the minimum steric hindrance to cyclic imide formation provided by absent and small Gly and Ser side chains, respectively. Increased conformational flexibility also enhances deamidation by providing conformational mobility, which permits the peptide bond to more readily participate in the cyclic imide formation.

The oxidation of protein amino acid side chains under typical formulation conditions occurs primarily at the sulfur-containing amino acid residues cysteine and methionine. Protein-bound cysteine residues can oxidize to form inter- or intramolecular disulfide bonds or monomolecular byproducts, such as sulfenic and sulfinic acid (only harsher oxidation conditions can lead to irreversible formation of sulfonic acid):

$$\underset{\text{Cysteine}}{2\ \text{Protein–}CH_2SH} \rightarrow \underset{\text{Cystine}}{\text{Protein–}CH_2S\text{–}SCH_2\text{–Protein}}$$

$$\underset{\text{Cysteine}}{\text{Protein–}CH_2SH} \rightarrow \underset{\text{Sulfenic acid}}{\text{Protein–}CH_2SOH} \rightarrow \underset{\text{Sulfinic acid}}{\text{Protein–}CH_2SO_2H} \rightarrow \underset{\text{Sulfonic acid}}{\text{Protein–}CH_2SO_3H}$$

Similarly, the oxidation of protein–bound methionine residues results in the conversion of the thioether to its sulfoxide counterpart (whereas only more severe oxidation conditions lead to the formation of the sulfone):

$$\underset{\text{Methionine}}{\text{Protein–}CH_2CH_2SCH_3} \rightarrow \underset{\text{Methionine sulfoxide}}{\text{Protein–}CH_2CH_2S(O)CH_3} \rightarrow \underset{\text{Methionine sulfone}}{\text{Protein–}CH_2CH_2S(OO)CH_3}$$

Protein oxidation can occur through two general mechanisms, which involve (1) site-specific (metal-catalyzed) and (2) non-site-specific oxidation. Metal-ion-catalyzed oxidation occurs near protein metal-ion-binding sites, whereas non-site-specific oxidation is initiated by light or exogenous oxidants generated from solubilized metal ions or impurities from buffer salts or formulation excipients (Li *et al.*, 1995). The rate and extent of the metal-catalyzed oxidation of proteins are greatly influenced by the primary structure of a protein (metal-ion-binding sites). The location of non-site-specific oxidation of Met and Cys residues within proteins often correlates with surface and solvent exposure of the reactive amino acid residues.

Upon exposure to elevated temperatures at neutral pH or to acidic/basic conditions, other amino acid residues can also begin to degrade (Manning *et al.*, 1989; Volkin and Klibanov, 1989; Volkin and Middaugh, 1992; Powell, 1994; Volkin *et al.*, 1995b). At acidic pH, the hydrolysis of the peptide bond can occur nonenzymatically at Asp residues with particular lability noted at Asp–Pro sequences. In addition, deamidation of Asn and Gln can also occur through an acid-catalyzed hydrolytic mechanism (Wright, 1991). At neutral pH, racemization at Asp-X and pyroglutamic acid formation at N-terminal Gly residues can occur, especially at elevated temperatures (Powell, 1994). In addition, nonenzymatic cyclization of the first two amino acid residues at the N-terminus and subsequent cleavage can result in the formation of a diketopiperazine and a truncated protein. Recombinant human growth hormone undergoes this reaction when incubated in the dry state at 50°C for 4 days at pH 7.4 (Battersby *et al.*, 1994). Prolonged exposure to basic conditions renders disulfide bonds chemically unstable, leading to a variety of degradation products, including beta-elimination reactions and disulfide interchange. In general, high pH accelerates both the oxidation of cysteine residues and the deamidation of Asn residues as discussed above. Exogenous oxidants can lead to the degradation of aromatic amino acid residues such as histidine, tryptophan, and tyrosine. The free amino groups of lysine residues as well as the N-terminus can react with reducing sugars to form condensation products via a series of chemical reactions known as Malliard reactions. Again, we describe these degradative covalent reactions only in passing; numerous reviews have recently appeared describing them in more depth (Aswad and Guzzetta, 1995; Bischoff and Kolbe, 1994; Clarke *et al.*, 1992; Cleland *et al.*, 1993; Li *et al.*, 1995; Nguyen, 1994; Powell, 1994; Volkin *et al.*, 1995b).

4.2. Methods for Detecting Covalently Altered Proteins and Effect of Covalent Modification on Protein Conformational Integrity and Biological Activity

Analysis of the chemical degradation of proteins can be performed at the level of the intact protein or at the individual peptide level after enzymatic cleavage of the protein. In intact proteins, one can determine to what extent and at what rate these reactions have occurred. Furthermore, one can isolate the chemically altered protein to ascertain if the chemical alteration perturbs its physical or biological properties. These types of studies should be the focus of preformulation characterization of protein pharmaceuticals. Analytical tools such as isoelectric focusing, ion exchange chromatography, and matrix-assisted laser desorption time-of-flight (MALD-TOF) mass spectrometry are commonly utilized to examine intact proteins for chemical changes. At the peptide level, identification of the specific site of the covalently altered amino acid can be more rigorously quantified (Cottrell, 1994). Proteins are enzymatically digested and the peptides analyzed directly by mass spectrometry. Alternatively, the peptides can be separated by reversed-phase chromatography or capillary electrophoresis. The chromatographs of the individual peptides (peptide map) from native and chemically altered proteins are then compared to identify which peptide contains the altered amino acid and the peptide(s) identified by amino acid sequencing. Currently, identification of the peptide is most often made "on-line" during the RP-HPLC employing electrospray mass spectrometry detector (LC/MS) (Chait and Kent, 1992; Siuzdak, 1994). Alternatively, the intact protein can be either chemically or enzymatically modified at the site of specific types of chemical changes in the protein, permitting either direct the detection of covalent changes in the intact protein or the tagging of the corresponding peptide for identification during peptide mapping studies. For example, as discussed above, the deamidation of Asn results in the formation of both aspartic acid and isoaspartic acid residues. The carboxylic acid residues of either residue can be chemically esterified with methanolic hydrochloric acid, whereas the isoaspartate residues can be specifically labeled with S-adenosyl-L-(methyl-^{3}H)-methionine by the enzyme protein isoaspartyl methyltransferase. These approaches have been applied successfully to identify the sites of deamidation in human growth hormone and hirudin (Aswad and Guzzetta, 1995; Bischoff and Kolbe, 1994).

Preformulation studies typically emphasize the detection of chemical alterations of protein pharmaceuticals at the intact protein level, thereby allowing the effects of covalent changes on the activity and conformational integrity of proteins to be examined. In the case of deamidation of asparagine residues, a neutral asparagine residue is converted to charged aspartic acid with the concomitant release of ammonia (see above). Analytical techniques that separate protein molecules on the basis of charge are therefore commonly used to monitor deamidation. Isoelectric focusing electrophoresis (Gianazza, 1995), ion-exchange chromatography, and capillary electrophoresis are frequently used to monitor the accumulation of chemically modified charged species.

For example, a combination of these techniques has been used to characterize the deamidation-induced microheterogeneity of monoclonal immunoglobulins (Tsai *et al.*, 1993b). As can be seen in Fig. 7, the identification of varying degrees of chemically altered antibody was ascertained by a combination of anion exchange chromatography and agarose gel isoelectric focusing. The cause of microheterogeneity cannot be readily determined, because any number of posttranslational events or chemical modifications can lead to the observed charge differences. In this case, deamidation of the immunoglobulin was identified as the cause of the microheterogeneity by a variety of direct and indirect experimental observations. Differential glycosylation was ruled out by a combination of carbohydrate analysis and MALD-TOF mass determinations of the isolated

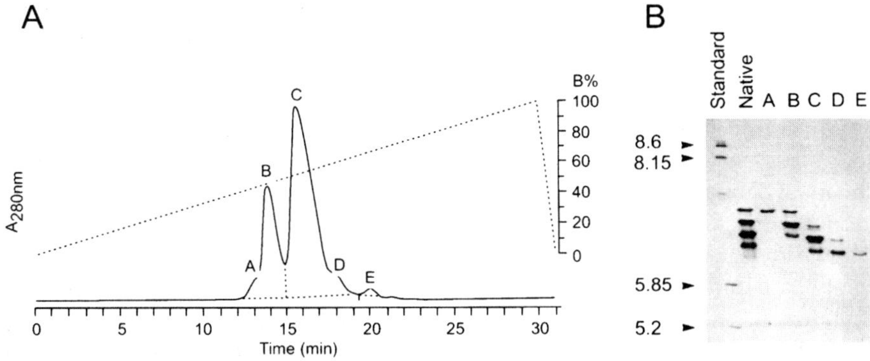

Figure 7. Deamidation-induced isoforms of h1B4 monoclonal antibody as measured by (A) chromatographic separation by anion exchange chromatography and (B) isoelectric focusing. Reprinted with permission from Tsai *et al.* (1993b). Copyright 1993 Plenum Press.

species. In addition, during incubation under conditions favoring deamidation, the isoelectric focusing bands were observed to accumulate more acidic species. This process also correlated with the kinetics of ammonia release as determined by an enzymatic assay (Tsai *et al.*, 1993b). Each charged species separated by anion exchange chromatography was collected and then analyzed by a variety of biochemical and biophysical techniques. This particular deamidation reaction was found to have no effect on either the overall conformation of the antibody as determined by circular dichroism and fluorescence spectroscopy or the binding activity of antibody as measured by a competitive binding assay.

Similar results have been observed with the deamidation of aFGF, in which the initial deamidation of Asn residues does not affect the overall conformation, conformational stability (as analyzed by thermal unfolding experiments), or mitogenic (biological) activity of the protein. One interesting challenge with this protein was that aFGF tightly binds highly negatively charged polyanions such as the sulfated polysaccharide heparin and these polyanions are required as stabilizers during formulation of the protein to prevent unfolding and aggregation during storage (Volkin and Middaugh, 1996). Due to the high charge density of the complex, conventional isoelectric focusing (IEF) cannot distinguish multiply deamidated, species, because a deamidation event does not dramatically alter the complex's charge density. An IEF-urea method was developed to unfold the protein and dissociate the aFGF-heparin complex, thus allowing quantitation of the deamidation events in aFGF during storage in the presence of polyanions (Volkin *et al.*, 1995a).

In contrast to the previous examples, deamidation of Asn residues has been shown to cause some degree of biological inactivation or loss of activity in a variety of other proteins and enzymes (Teshima *et al.*, 1995). Recombinant soluble CD4 was observed to have reduced binding capacity for the human immunodeficiency virus-1 (HIV-1) envelope glycoprotein gp120 due to a specific deamidation of Asn-52 residue in the V1 domain of the molecule. This chemical change was caused by storage of the protein in the liquid state for 6 months at 25°C at pH 7.2 (Teshima *et al.*, 1991). Interestingly, screening the deamidated protein with a panel of conformational antibodies indicated that the overall conformation of the protein was unchanged. This result suggests that Asn-52 of soluble CD4 might be directly involved in the binding of the protein to gp120. Deamidation has also been shown to contribute to the irreversible thermal inactivation of ribonuclease A, in which the enzymatic activity was shown to decrease with increasingly deamidated species (isolated from preparative IEF gels) (Zale and Klibanov, 1986). More recently, ribonuclease A specifically deamidated at Asn-67 was resolved into aspartic acid- and isoaspartic acid-containing protein fractions by hydrophobic interaction chromatography (Di Donato *et al.*, 1993). It was shown that the deamidation at Asn-67 to either Asp-67 or isoAsp-67 lowered the catalytic activity of the enzyme. The effect of deamidation on the renaturation of the unfolded and reduced enzyme was then examined. The Asp-67 deamidated species refolded at a rate equivalent to the native enzyme, whereas the Isoasp-67 molecule refolded 50% more slowly. These results directly demonstrate that deamidation of an asparagine residue can affect both the biological activity and the physical properties of a protein.

The detection of protein oxidation reactions such as the conversion of methionine to its sulfoxide or the formation of disulfide bonds from cysteine residues can also be identified at the intact protein level. Unlike the deamidation reaction described above, these oxidation reactions do not necessarily alter the charge state of the protein, but instead often lead to measurable molecular weight differences between native and chemically altered proteins. For example, intermolecular disulfide bond formation can be easily followed by accumulation of oligomer bands on unreduced SDS–PAGE. SDS–PAGE under reducing conditions is typically run at the same time to confirm the disulfide-linked nature of the higher molecular weight species. MALD-TOF mass spectrometry can also be utilized to detect the 2-Da mass change resulting from disulfide bond formation, although this analysis is typically used more for peptides than larger proteins because of its increased sensitivity for the former (Cotter, 1992). The oxidation of methionine residues results in a mass change of 16 Da, which can be more easily detected by MALD-TOF mass spectrometry (Geisow, 1992). For example, the H_2O_2-induced oxidation of the two methionine residues of relaxin was successfully quantitatively characterized by mass spectrometric analysis of the intact protein (Nguyen *et al.*, 1993).

Protein oxidation products are typically isolated by reversed-phase or hydrophobic interaction chromatography, so that the effect of oxidation on the conformation and activity of purified, altered protein can be examined. For example, the autooxidation of three free cysteine residues of aFGF leads to the formation of intra- and intermolecular disulfide species as identified by SDS–PAGE migration patterns. Using a site-directed Cys to Ser mutant (containing two of the three cysteine residues found in native aFGF), intramolecular disulfide species were generated upon incubation with copper salts and then purified by reversed-phase HPLC (Linemeyer *et al.*, 1990). The oxidized form of aFGF was inactive in a cell-based mitogenic assay, although mitogenic activity was fully recovered upon reduction of the oxidized species with dithiothreitol. As another example, human growth hormone contains three methionine residues whose rate of reactivity toward oxidation by H_2O_2 correlates with their solvent-accessible surface area (Nguyen, 1994). Partially oxidized protein retains most of its biological activity, whereas complete oxidation of the three methionine residues leads to the loss of most of the biological activity of the protein. Interestingly, conformational studies by circular dichroism show that partially oxidized human growth hormone retains its native secondary structure, whereas fully oxidized protein exhibits an altered structure (reviewed by Nguyen, 1994). No simple or general correlation, however, exists between methionine surface exposure and reactivity toward oxidation, probably due to the fact that metal-catalyzed oxidation reactions can occur in hydrophobic protein interiors if metal-ion-binding sites are present (Li *et al.*, 1995).

It is clear from these studies that no straightforward correlation exists between the chemical lability of amino acid side chains and the conformational integrity and biological activity of proteins. In the case of Asn deamidation and Met oxidation, numerous examples of specific covalent changes have been documented in proteins, some leading to complete inactivation and others having little or no effect (Powell *et al.*, 1996). The major emphasis during the preformulation stage of protein drug development should generally be to characterize the interplay of the conformational integrity of a protein and the degradation of the labile amino acid residues described above. The effect of these covalent reactions on the biological activity must also be considered at these very early stages of preformulation studies. Although space does permit us to consider it here, the role of specific bioassays in the analysis of protein-based pharmaceuticals is critical in this regard. Even if these reactions are found to have no effect on the physicochemical or biological properties of a protein, the quantitative characterization of the location and kinetics of deamidation and oxidation reactions occurring during storage must still be performed in order to ensure process and product consistency. Furthermore, the effect of these degradative reactions on the immunogenicity and pharmacokinetics of a protein drug as measured by animal and potentially human studies is also of utmost importance during the development of protein-based pharmaceuticals (Cleland *et al.*, 1993). These studies, however, are typically beyond the scope of the initial preformulation work.

5. CONCLUSIONS

Preformulation studies of protein pharmaceuticals entail a series of detailed biophysical and biochemical characterization experiments with the bulk drug substance in the presence and absence of various formulation excipients. In this chapter, we reviewed various experimental techniques commonly applied to quantitatively monitor the primary, secondary, tertiary, and higher order structure of protein molecules. The interplay between a protein's chemical integrity and higher order structural stability is a key parameter to be followed during these studies. By determining the effect of various environmental factors such as pH, temperature, and ionic strength on the conformational and covalent integrity of a protein molecule, an extremely useful set of protein stability "maps" can be generated, thereby providing profiles of the conformational and covalent integrity (and ideally biological activity) of a protein molecule as a function of environmental conditions. These data can be then be applied to the subsequent formulation development steps required both to determine the composition of the final drug product (dosage form, excipient composition, and container/closure) and ensure adequate storage stability (typically 2 years at 2–8°C).

A detailed determination of the biophysical and biochemical properties of a protein pharmaceutical enables the proper setup and performance of the formulation processing steps in the manufacturing environment by guiding and shaping the engineer's design criteria. Thus, carefully obtained preformulation data provide the type of information required to prevent stability problems from occurring during the large-scale production of final container-formulated protein pharmaceuticals. By investing the time and effort to obtain a detailed understanding of the causes and mechanisms of protein instability at the molecular level, the formulation scientist is, at a minimum, buying an insurance policy. This protection is quite valuable if or when something unexpected or atypical arises during formulation development scaleup work or during the actual manufacturing of a protein pharmaceutical, because it provides the tools and information necessary to solve problems in an expeditious manner.

REFERENCES

Arakawa, T., Yphantis, D. A., Lory, J. W., Narhi, L. O., Lu, H. S., Prestrelski, S. J., Clogston, C. L., Zsebo, K. M., Mendiaz, E. A., Wypych, J., and Langley, K. E., 1991, Glycosylated and unglycosylated recombinant derived human stem cell factors are dimeric and have extensive secondary structure, *J. Biol. Chem.* **266:** 18942–18948.

Aswad, D. W., and Guzzetta, A. W., 1995, Methods for analysis of deamidation and isoaspartate formation in peptides and proteins, in: *Deamidation and Isoaspartate Formation in Peptides and Proteins* (D. W. Aswad, ed.), CRC Press, Boca Raton, PL, pp. 7–29.

Battersby, J. E., Hancock, W. S., Canova-David, E., Oeswein, J., and O'Connor, B., 1994, Diketopiperazine formation and N-terminal degradation of recombinant human growth hormone, *Int. J. Peptide Protein Res.* **44:**215–222.

Bell, L. N., Hageman, M. J., and Bauer, J. M., 1995, Impact of moisture on thermally induced denaturation and decomposition of lyophilized bovine somatotropin, *Biopolymers* **35:**201–209.

Bischoff, R., and Kolbe, H. V. J., 1994, Deamidation of asparagine and glutamine residues in proteins and peptides: Structural determinants and analytical methodology, *J. Chromatogr. B* **662:**261–278.

Bolotina, I. A., and Lugauskas, V. Yu, 1985, Detemination of the secondary structure of proteins from the circular dichroism spectra. IV. Consideration of the contribution of aromatic amino acid residues to the circular dichroism spectra of proteins in the peptide region, *Molek. Biol.* (USSR) **19:**1409–1421.

Bowler, B. E., May, K., Zaragoza, T., York, P., Dong, A., and Caughey, W. S., 1993, Destabilizing effects of replacing a surface lysine of cytochrome *c* with aromatic amino acids: implications for the denatured state, *Biochemistry* **32:**183–190.

Brahms, S., and Brahms, J., 1980, Determination of protein secondary structure in solution by vacuum ultraviolet circular dichroism, *J. Mol. Biol.* **138:**149–178.

Brochon, J. C., Tauc, P., Mérola, F., and Schoot, B. M., 1993, Analysis of arecombinant protein preparation on physical homogeneity and state of aggregation, *Anal. Chem.* **65:**1026–1034.

Buckler, D. R., Haas, E., and Scheraga, H. A., 1995, Analysis of the structure of ribonuclease A in native and partially denatured states by time-resolved nonradiative dynamic excitation energy transfer between site-specific extrinsic probes, *Biochemistry* **34:**15965–15978.

Byler, D. M., and Susi, H., 1986, Examination of the secondary structure of proteins by deconvoluted FTIR spectra, *Biopolymers* **25:**469–487.

Carpenter, J. F., Prestrelski, S. J., Anchordoguy, T. J., and Arakawa, T., 1994, Interactions of stabilizers with proteins during freezing and drying, in: *Formulation and Delivery of Proteins and Peptides* (J. L. Cleland and R. Langer, eds.), American Chemical Society, Washington, DC, pp. 134–147.

Chait, B. T., and Kent, S. B. H., 1992, Weighing naked proteins: Practical, high accuracy mass measurement of peptides and proteins, *Science* **257:**1885–1894.

Chan, H. K., Ongpipattanakul, B., and Au-Yeung, J., 1996, Aggregation of rhDNase occurred during the compression of KBr pellets used for FTIR spectroscopy, *Pharm. Res.* **13:**238–242.

Chen, Y. H., Yang, J. T., 1971 A new approach to the calculation of secondary structures of globular proteins by optical rotatory dispersion and circular dichroism, *Biochem. Biophys. Res. Comm.* **44:**1285–1291.

Clarke, S., Stephenson, R. C., and Lowenson, J. D., 1992, Lability of asparagine and aspartic acid residues in proteins and peptides: Spontaneous deamidation and isomerization reactions, in: *Stability of Protein Pharmaceuticals; Chemical and Physical Pathways of Protein Degradation* (T. J. Ahern and M. Manning, eds.), Plenum Press, New York, pp. 1–29.

Cleland, J. L., and Wang, D. I. C., 1990, Refolding and aggregation of bovine carbonic anhydrase B: Quasi-elastic light scattering analysis, *Biochemistry* **29:** 11072–11078.

Cleland, J. L., Powell, M. F., and Shire, S. J., 1993, The development of stable protein formulations: A close look at protein aggregation, deamidation and oxidation, *Crit. Rev. Ther. Drug Carrier Sys.* **10:**307–377.

Cole, J. L., Gehman, J. D., Shafer, J. A., and Kuo, L. C., 1993, Solution oligomerization of the rev protein of HIV-1: Implications for function, *Biochemistry* **32:**11769–11775.

Constantino, H. R., Langer, R., and Klibanov, A. M., 1995, Aggregation of a lyophilized pharmaceutical protein, recombinant albumin: Effect of moisture and stabilization by excipients, *Bio/Technology* **13:**493–496.

Copeland, R. A., Ji, H., Halfpenny, A. J., Williams, R. W., Thompson, K. C., Herber, W. K., Thomas, K. A., Bruner, M. W., Ryan, J. A., Marquis-Omer, D., Sanyal, G., Sitrin, R. D., Yamazaki, S., and Middaugh, C. R., 1991, The structure of human acidic fibroblast growth factor and its interaction with heparin, *Arch. Biochem. Biophys.* **289:**53–61.

Cotter, R. J., 1992, Time-of-flight mass spectrometry for the structural analysis of biological molecules, *Anal. Chem.* **64:**1027–1039.

Cottrell, J. S., 1994, Protein identification by peptide mass fingerprinting, *Peptide Res.* **7:**115–124.

Craig, W. S., 1988, Determination of quaternary structure of an active enzyme using chemical cross-linking with glutaraldehyde, *Meth. Enzymol.* **156:**333–345.

Deyl, Z., and Struzinsky, R., 1991, Capillary zone electrophoresis: Its applicability and potential in biochemical analysis, *J. Chromatogr.* **569:**63–122.

Di Donato, A., Ciardiello, M. A., Nigris, M. D., Piccoli, R., Mazzarella, L., and D'Alessio, G., 1993, Selective deamidation of ribonuclease A: Isolation and characterization of the resulting isoaspartyl and aspartyl derivatives, *J. Biol. Chem.* **268:**4745–4751.

Dollinger, G., Cunico, B., Kunitani, M., Johnson, D., and Jones, R., 1992, Practical on-line determination of biopolymer molecular weights by high-performance liquid chromatography with classical light-scattering detection, *J. Chromatogr.* **592:**215–228.

Dong, A., Prestrelski, S. J., Allison, S. D., and Carpenter, J. F., 1995, Infrared spectroscopic studies of lyophilization- and temperature-induced protein aggregation, *J. Pharm. Sci.* **84:**415–424.

Eberlein, G. A., Stratton, P. R., and Wang, Y. J., 1994, Stability of rhbFGF as determined by UV spectroscopic measurements of turbidity, *PDA J. Pharm. Sci. Technol.* **48:**224–230.

Endo, Y., Nagai, H., Watanabe, Y., Ochi, K., and Takagi, T., 1992, Heat-induced aggregation of recombinant erythropoietin in the intact and deglycosylated states as monitored by gel permeation chromatography combined with a low-angle laser light scattering technique, *J. Biochem.* (Tokyo) **112:**700–706.

Eftink, M., 1994, The use of fluorescence methods to monitor unfolding transitions in proteins, *Biophys. J.* **66:**482–501.

Fabian, H., Schultz, C., Naumann, D., Landt, O., Hahn, U., and Saenger, W., 1993, Secondary structure and temperature-induced unfolding and refolding of ribonuclease T_1 in aqueous solution, *J. Mol. Biol.* **232:**967–981.

Farmer, T. B., and Caprioli, R. M., 1991, Assessing the multimeric states of proteins: Studies using laser desorption mass spectrometry, *Biol. Mass Spectrom.* **20:** 796–800.

Fink, A. L., and Painter, B., 1987, Characterization of the unfolding of ribonuclease A in aqueous methanol solvents, *Biochemistry* **26:**1665–1671.

Geisow, M. J., 1992, Mass measurement at high molecular weight—New tools for biotechnologists, *TIBTECH* **10:**432–441.

Gianazza, E., 1995, Isoelectric focusing as a tool for the investigation of post-translational processing and chemical modification of proteins, *J. Chromatogr. A* **705:**67–87.

Giddings, J. C., 1989, Field flow fractionation of macromolecules. *J. Chromtogr.* **470:**327–335.

Gorovits, B. M., Seale, J. W., and Horowitz, P. M., 1995, Residual structure in urea-denatured chaperonin GeoEL, *Biochemistry* **34:**13928–13933.

Green, S. M., Ginsburg, A., Lewis, M. S., and Hensley, P., 1991, Roles of metal ions in the maintenance of the tertiary and quaternary structure of arginase from *Saccharomyces cerevisiae, J. Biol. Chem.* **266:**21474–21481.

Gress, J. O., Marquis-Omer, D., Middaugh, C. R., and Sanyal, G., 1994, Evidence for an equilibrium intermediate in the folding–unfolding pathway of a transforming growth factor-α–*Pseudomonas* exotoxin hybrid protein, *Biochemistry* **33:**2620–2627.

Hermann, R., Rudolph, R., and Jaenicke, R., 1979, Kinetics of *in vitro* reconstitution of oligomeric enzymes by cross-linking, *Nature* **277:**243–245.

Hillenkamp, F., and Karas, M., 1990, Mass spectrometry of peptides and proteins by matrix-assisted ultraviolet laser desorption/ionization. *Meth. Enzymol.* **193:** 280–295.

Ismail, A. A., Mantsch, H., and Wong, P. T. T., 1992, Aggregation of chymotrypsinogen: Portrait by infrared spectroscopy. *Biochim. Biophys. Acta* **1121:** 183–188.

Johnson, M. L., Correia, J. J., Yphantis, D. A., and Halvorson, H. R., 1981, Analysis of data from the analytical ultracentrifuge by nonlinear least-squares techniques, *Biophys. J.* **36:**575–588.

Johnson, W. C., 1990, Protein secondary structure and circular dichroism: A practical guide. *Proteins Struc. Funct. Genet.* **7:**205–214.

Kadima, W., Øgendal, L., Bauer, R., Kaarsholm, N., Brodersen, K., Hansen, J. F., and Porting, P., 1993, The influence of ionic strength and pH on the aggregation properties of zinc-free insulin studied by static and dynamic light scattering, *Biopolymers* **33:**1643–1657.

Katakam, M., Bell, L. N., and Banga, A., 1995, Effect of surfactants on the physical stability of recombinant human growth hormone, *J. Pharm. Sci.* **84:**713–716.

Kauppinen, J. K., Moffatt, D. J., Mantsch, H. H., and Cameron, D. G., 1981, Fourier transforms in the computation of self-deconvoluted and first-order derivative spectra of overlapped band contours, *Anal. Chem.* **53:**1454–1457.

Kendrick, B. S., Dong, A., Allison, S. D., Manning, M. C., and Carpenter, J. F., 1996, Quantitation of the area of overlap between second-derivative amide I infrared spectra to determine the structural similarity of a protein in different states, *J. Pharm. Sci.* **85:**155–158.

Kuhr, W. G., 1990, Capillary electrophoresis, *Anal. Chem.* **62:**403R–414R.

Lakowicz, J. R., 1983, *Principles of Fluorescence Spectroscopy*, Plenum Press, New York, pp. 40–47.

Lepock, J. R., Frey, H. E., Bayne, H., and Markus, J., 1989, Relationship of hypothermia-induced hemolysis of human erythrocytes to the thermal denaturation of membrane proteins, *Biochim. Biophys. Acta* **980:**191–201.

Li, S., Schoneich, C., and Borchardt, R. T., 1995, Chemical instability of protein pharmaceuticals: Mechanisms of oxidation and strategies for stabilization, *Biotechnol. Bioeng.* **48:**90–500.

Linemeyer, D. L., Menke, J. G., Kelly, L. J., DiSalvo, J., Soderman, D., Schaeffer, T., Ortega, S., Gimenez-Gallego, G., and Thomas, K. A., 1990, Disulfide bonds are neither required, present, nor compatible with full activity of human recombinant acidic fibroblast growth factor, *Growth Factors* **3:**287–298.

Litzén, A., Walter, J. K., Krischollek, H., and Wahlund, K. G., 1993, Separation and quantitation of monoclonal antibody aggregates by asymmetrical flow field flow fractionation and comparison to gel permeation chromatography. *Anal. Biochem.* **212:**469–480.

Liu, J., Lester, P., Builder, S., and Shire, S. J., 1995, Characterization of complex formation of humanized anti-IgE monoclonal antibody and monoclonal human IgE, *Biochemistry* **34:**10474–10482.

Mach, H., and Middaugh, C. R., 1994, Simultaneous monitoring of the environment of tryptophan, tyrosine, and phenylalanine residues in proteins by near-ultraviolet second-derivative spectroscopy, *Anal. Biochem.* **222:**323–331.

Mach, H., Thomson, J. A., Middaugh, C. R., and Lewis, R. V., 1991, Examination of phenylalanine microenvironments in proteins by second-derivative absorption spectroscopy, *Arch. Biochem. Biophys.* **287:**33–40.

Mach, H., Volkin, D. B., Burke, C. J., Middaugh, C. R., Linhardt, R. J., Fromm, J. R., Loganathan, D., and Mattson, L., 1993, Nature of the interaction of heparin with acidic fibroblast growth factor, *Biochemistry* **32:**5480–5489.

Mach, H., Volkin, D. B., Burke, C. J., and Middaugh, C. R., 1995, Ultraviolet absorption spectroscopy, in: *Protein Stability and Folding: Theory and Practice* (B. A. Shirley, ed.), Humana Press, Totowa, NJ, pp. 91–114.

Manavalan, P., and Johnson, W. C., Jr., 1987, Variable selection method improves the prediction of protein secondary structure from circular dichroism spectra, *Anal. Biochem.* **167:**76–85.

Manning, M. C., 1989, Underlying assumptions in the estimation of secondary structure content in proteins by circular dichroism spectroscopy—A critical review, *J. Pharm. Biomed. Anal.* **7:**1103–1119.

Manning, M. C., and Woody, R. W., 1989, Theoretical study of the contribution of aromatic side chains to the circular dichroism of basic bovine pancreatic trypsin inhibitor, *Biochemistry* **28:**8609–8613.

Manning, M. C., Patel, K., and Borchardt, R. T., 1989, Stability of protein pharmaceuticals, *Pharm. Res.* **6:**903–918.

Middaugh, C. R., 1994, The analysis of protein pharmaceuticals: Near future advances, *Cytotechnology* **15:**187–194.

Middaugh, C. R., and Volkin, D. B., 1992, Protein solubility, in: *Stability of Protein Pharmaceuticals: Chemical and Physical Pathways of Protein Degradation* (T. J. Ahern and M. Manning, eds.), Plenum Press, New York, pp. 109–134.

Middaugh, C. R., Mach, H., Burke, C. J., Volkin, D. B., Dabora, J. M., Tsai, P. K., Bruner, M. W., Ryan, J. A., and Marfia, K., 1992, The molecular basis of the anti-growth factor activity of suramin, *Biochemistry* **31:**9016–9024.

Middaugh, C. R., Mach, H., Ryan, J., Sanyal, G., and Volkin, D., 1995, Infrared spectroscopy, in: *Protein Stability and Folding: Theory and Practice* (B. A. Shirley, ed.), Humana Press, Totowa, NJ, pp. 137–156.

Nguyen, T. H., 1994, Oxidation degradation of protein pharmaceuticals, in: *Formulation and Delivery of Proteins and Peptides* (J. L. Cleland and R. Langer, eds.), American Chemical Society, Washington, DC, pp. 59–71.

Nguyen, T. H., Burnier, J., and Meng, W., 1993, The kinetics of relaxin oxidation by hydrogen peroxide, *Pharm. Res.* **10:**1563–1571.

Perczel, A., Hollosi, M., Tusnady, G., and Fasman, G. D., 1991, Convex constraint analysis; a natural deconvolution of circular dichroism curves of proteins, *Protein Eng.* **4:**669–679.

Perczel, A., Park, K., and Fasman, G. D., 1992, Analysis of the circular dichroism spectrum of proteins using the convex constraint algorithm: A pratical guide, *Anal. Biochem.* **203:**89–93.

Powell, M. F., 1994, Peptide stability in aqueous parenteral formulations: Prediction of chemical stability based on primary sequence, in: *Formulation and Delivery of*

Proteins and Peptides (J. L. Cleland and R. Langer, eds.), American Chemical Society, Washington, DC, pp. 100–119.

Powell, M. F. *et al.*, 1996, A compendium and hydropathy/flexibility analysis of common reactive sites in proteins: Reactivity at Asn, Asp, Gln and Met motifs in neutral pH solution, in: *Formulation, Characterization and Stability of Protein Drugs:Case Histories* (Y. J. Wang and R. Pearlman, eds.), Plenum Press, New York, pp. 1–140.

Prestrelski, S. J., Arakawa, T., Wu, C-S. C., O'Neil, K. D., Westcott, K. R., and Narhi, L. O., 1992, Solution structure and dynamics of epidermal growth factor and transforming growth factor-α, *J. Biol. Chem.* **267:**319–322.

Prestrelski, S. J., Tedeschi, N., Arakawa, T., and Carpenter, J. F., 1993, Dehydration-induced conformational transitions in proteins and their inhibition by stabilizers, *Biophys. J.* **65:**661–671.

Ragone, R., Colonna, G., Balestrieri, C., Servillo, L., and Irace, G., 1984, Determination of tyrosine exposure in proteins by second-derivative spectroscopy, *Biochemistry* **23:**1871–1875.

Ragone, R., Colonna, G., Bismuto, E., and Irace, G., 1987, Unfolding pathway of myoglobin: Effect of denaturants on solvent accessibility to tyrosyl residues detected by second-derivative spectroscopy, *Biochemistry* **26:** 2130–2134.

Sanchez-Ruiz, J. M., Lopez-Lacomba, J. L. Cortijo, M., and Mateo, P. L., 1988, Differential scanning calorimetry of the irreversible thermal denaturation of thermolysin, *Biochemistry* **27:**1648–1652.

Sanyal, G., Marquis-Omer, D., and Middaugh, C. R., 1996, Biophysical characterization and formulation of TP40: A chimeric protein that requires a pH-dependent conformational change for its biological activity, in: *Formulation, Characterization and Stability of Protein Drugs: Case Histories* (Y. J. Wang and R. Pearlman, eds.), Plenum Press, New York, pp. 365–392.

Schein, C. H., 1994, Controlling oligomerization of pharmaceutical proteins. *Pharm. Acta Helv.* **69:**119–126.

Shahrokh, Z., Stratton, P. R., Eberlein, G. A., and Wang, Y. J., 1994, Approaches to analysis and demonstrating mass balance in pharmaceutical protein (basic fibroblast growth factor) formulations, *J. Pharm. Sci.* **83:**1645–1650.

Siuzdak, G., 1994, The emergence of mass spectrometry in biochemical research, *Proc. Natl. Acad. Sci. USA* **91:**11290–11297.

Smith, R. D., Loo, J. A., Barinaga, C. J., Edmonds, C. G., and Udseth, H. R., 1989, Capillary zone electrophoresis and isotachophoresis–mass spectrometry of polypeptides and proteins based upon an electrospray ionization interface, *J. Chromatogr.* **480:**211–232.

Smith, R. D., Loo, J. A., Loo, R. R. O., Busman, M., and Udseth, H. R., 1991, Principles and practice of electrospray ionization–mass spectrometry for large polypeptides and proteins, *Mass Spectrom. Rev.* **10:**359–451.

Sreerama, N., and Woody, R. W., 1993, Self-consistent algorithm for secondary structure analysis from CD data, *Anal. Biochem.* **209:**32–44.

Steele, J. C. H., and Neilsen, T. B., 1978, Evaluation of cross-linked polypeptides in SDS gel electrophoresis, *Anal. Biochem.* **84:**218–224.

Surewicz, W. K., Mantssch, H. H., and Chapman, D., 1993, Determination of protein secondary structure by Fourier transform infrared spectroscopy: A critical assessment, *Biochemistry* **32:**389–394.

Suter, M. J.-F., Moore, W. T., Farmer, T. B., Cottrell, J. S., and Caprioli, R. M., 1992, Analysis of protein–protein binding with laser desorption mass spectrometry: Can multimeric forms be studied? in: *Techniques in Protein Chemistry III* (J. Villafranca, ed.), Academic Press, San Diego, CA, pp. 447–455.

Teshima, G., Porter, J., Yim, K., Ling, V., and Guzzetta, A., 1991, Deamidation of soluble CD4 at asparagine-52 results in reduced binding capacity for the HIV-1 envelope glycoprotein gp120, *Biochemistry* **30:**3916–3922.

Teshima, G., Hancock, W. S., and Canova-David, E., 1995, Effect of deamidation and isoaspartate formation on the activity of proteins, in: *Deamidation and Isoaspartate Formation in Peptides and Proteins* (D. W. Aswad, ed.), CRC Press, Boca Raton, FL, pp. 168–191.

Tsai, P. K., Volkin, D. B., Dabora, J. M., Thompson, K. C., Bruner, M. W., Gress, J. O., Matuszewska, B., Keogan, M., Bondi, J. V., and Middaugh, C. R., 1993a, Formulation design of acidic fibroblast growth factor, *Pharm. Res.* **10:**649–659.

Tsai, P. K., Bruner, M. W., Irwin, J. I., Ip, C. C. Y., Oliver, C. N., Nelson, R. W., Volkin, D. B., and Middaugh, C. R., 1993b, Origin of the isoelectric heterogeneity of a monoclonal immunoglobulin, *Pharm. Res.* **10:**1580–1586.

Valero, E., DeBonis, S., Filhol,O., Wade, R. H., Langowski, J., Chambaz, E. M., and Cochet, C., 1995, Quaternary structure of casein kinase 2—Characterization of multiple oligomeric states and relation with its catalytic activity. *J. Biol. Chem.* **270:**8345–8352.

Van Holde, K. E., 1975, Sedimentation analysis of proteins, in: *The Proteins*, 3rd ed. (H. Neurath, R. L. Hill, and C.-L. Boeder, eds), Academic Press, New York, pp. 225–291.

Van Stokkum, I. H. M., Linsdell, H., Hadden, J. M., Haris, P. I., Chapman, D., and Bloemendal, M., 1995, Temperature-induced changes in protein structure studied by Fourier transform infrared spectroscopy and global analysis, *Biochemistry* **34:**10508–10518.

Venyaminov, S. Y., and Kalnin, N. N., 1990a, Quantitative IR spectrophotometry of peptide compounds in water (H_2O) solutions. I. Spectral parameters of amino and residue absorption bands, *Biopolymers* **30:**1243–1257.

Venyaminov, S. Y., and Kalnin, N. N., 1990b, Quantitative IR spectrophotometry of peptide compounds in water (H_2O) solutions. II. Amide absorption bands of polypeptides and fibrous proteins in α-, β- and random coil conformations, *Biopolymers* **30:**1259–1271.

Volkin, D. B., and Klibanov, A. M., 1989, Minimizing protein inactivation, in *Protein Function:A Practical Approach* (T. E. Creighton, ed.), Oxford University Press, Oxford, pp. 1–24.

Volkin, D. B., and Middaugh, C. R., 1992, The effect of temperature on protein structure, in: *Stability of Protein Pharmaceuticals: Chemical and Physical Pathways of Protein Degradation* (T. J. Ahern and M. Manning, eds.), Plenum Press, New York, pp. 215–247.

Volkin, D. B., and Middaugh, C. R., 1996, The characterization, stabilization and formulation of acidic fibroblast growth factor, in: *Formulation, Characterization and Stability of Protein Drugs: Case Histories* (Y. J. Wang and R. Pearlman, eds.), Plenum Press, New York, pp. 181–217.

Volkin, D. B., Tsai, P. K., Dabora, J. M., Gress, J. O., Burke, C. J., Linhardt, R. J., and Middaugh, C. R., 1993, The physical stabilization of acidic fibroblast growth factor by polyanions, *Arch. Biochem. Biophys.* **300:**30–41.

Volkin, D. B., Verticelli, A. M., Bruner, M. W., Tsai, P. K., Marfia, K. E., Sardana, M., and Middaugh, C. R., 1995a, Deamidation of polyanion-stabilized acidic fibroblast growth factor, *J. Pharm. Sci.* **84:**7–11.

Volkin, D. B., Mach, H., and Middaugh, C. R., 1995b, Degradative covalent reactions important to protein stability, in: *Protein Stability and Folding: Theory and Practice* (B. A. Shirley, ed.), Humana Press, Totowa, NJ, pp. 35–63.

Wallace, B. A., and Mao, D., 1984, Circular dichroism analyses of membrane proteins: An examination of differential light scattering and absorption flattening effects in large membrane vesicles and membrane sheets, *Anal. Biochem.* **142:**317–328.

Wright, H. T., 1991, Nonenzymatic deamidation of asparaginyl and glutaminyl residues in proteins, *CRC Crit. Rev. Biochem. Mol. Biol.* **26:**1–52.

Yang, J. T., Chen, G. C., and Jirgensons, B., 1976 Optical rotatory dispersion and circular dichroism of protein, in: *Handbook of Biochemistry and Molecular Biology*, 3rd ed., Vol. III (G. D. Fasman, ed.), CRC Press, Boca Raton, FL, pp. 3–140.

Zale, S.E., and Klibanov, A.M., 1986, Why does ribonuclease irreversibly inactivate at high temperatures? *Biochemistry* **25:**5432–5444.

2

Formulation Development of Protein Dosage Forms

Michael J. Akers, Vasu Vasudevan, and Mary Stickelmeyer

1. INTRODUCTION

A formulation scientist assigned the task of developing a stable, elegant, and manufacturable dosage form of a therapeutic protein drug has been given a significant challenge. Most proteins, as natural physiological molecules, are inherently unstable outside the human or animal body. Stability challenges in protein formulation development are typically enormous. The instability of these reactive and complex molecules must be considered not only in the *formulation* process, but also in development of the *packaging system* and the *manufacturing* process. These three areas are intimately and inseparably connected.

Protein dosage forms are also sterile dosage forms. Sterile dosage forms must be essentially free* from microbial contamination (sterile), free from pyrogenic (including endotoxin) contamination, and free from particulate

*The term "essentially free" is preferable over the more absolute term "free" when dealing with the subject of microbiological contamination. Except for products which can be terminally sterilized, which do not include proteins, there is no total and absolute assurance that each unit of product is, in fact, sterile.

Michael J. Akers • Baxter Pharmaceutical Solutions LLC, Bloomington, Indiana 47402. *Vasu Vasudevan and Mary Stickelmeyer* • Lilly Research Laboratories, Indianapolis, Indiana 46285.

Development and Manufacture of Protein Pharmaceuticals, edited by Nail and Akers. Kluwer Academic/Plenum Publishers, New York, 2002.

matter contamination (ready-to-use and reconstitutable solutions). Depending on the route of administration, sterile dosage forms must also be isotonic. For example, the intravenous route of administration can tolerate fairly wide ranges of "tonicity" (osmolality or osmolarity) or oncocity* whereas subcutaneous and intramuscular routes may require tighter control of product tonicity. Sterile products administered into spinal fluid or topically applied to the eye must be as close to isotonic as possible because of the potential of irreparable damage of spinal or corneal cells due to extremes in the osmolar concentrations of administered products. In addition, products administered by the injectable or ophthalmic topical routes should be as close to physiological pH (7.4) as possible to minimize pain and tissue irritation or damage.

This chapter was written to provide the basic approaches and techniques used to design and develop dosage forms of proteins. To develop dosage forms means not only to generate a viable formulation, but also to identify a final packaging system, to design and scale up a quality manufacturing method, and to employ valid measurements to assure product quality. In addition, in this era of globalization, formulations must be developed that are acceptable from a regulatory standpoint throughout the world.

Since protein stabilization has already been extensively discussed in many excellent references (Table I), we also intend to cover other issues essential in the complete formulation development of protein products yet not covered elsewhere, such as antimicrobial preservation, packaging components, container–closure integrity, clinical trial manufacturing, and development history reports.

We have reviewed the literature and have selected the articles which provide both intensive analysis and extensive information on solving protein formulation and other product development problems. Advanced injectable (e.g., controlled release, implantable devices, gene delivery) and noninjectable (e.g., pulmonary, oral, buccal) protein formulation research will not be covered in this chapter, but other references are available that deal with these advances (Baker, 1980; Davis *et al.*, 1986; Senior and Radomsky, 2000; Hillery *et al.*, 2001).

2. WHY PROTEINS PRESENT UNIQUE CHALLENGES TO THE DEVELOPMENT SCIENTIST

Many texts and articles already discuss the great difficulties scientists experience in protein dosage formulation because of the significant

* Oncotic pressure = osmotic pressure exerted by colloids (e.g., plasma proteins) in a solution.

Table I
Major Protein Formulation References

Stability of Protein Pharmaceuticals, Parts A and B, T. J. Ahern and M. C. Manning, (eds.), *Pharmaceutical Biotechnology*, vols 2 and 3, Plenum Press, New York 1992
Formulation concerns of protein drugs, T. Chen *Drug Dev Ind. Pharm.* **18:**1311–1354 (1992)
The formulation of proteins and peptides, M. J. Groves, M. H. Alkan, and A. J. Hickey, in: *Pharmaceutical Biotechnology* (M. E. Klegerman and M. J. Groves, eds.), Interpharm Press, Englewood, CO, 1992
Protein stability and degradation mechanisms, B. L. Currie and M. J. Groves, in: *Pharmaceutical Biotechnology* (M. E. Klegerman and M. J. Groves eds.), Interpharm Press, Englewood, CO, 1992
Stability and Characterization of Protein and Peptide Drugs: Case Histories, Y. J. Wang and R. Pearlman, Plenum Press, New York, 1993
Factors affecting short-term and long-term stabilities of proteins, T. Arakawa, S. J. Prestrelski, W. C. Kenney, and J. F. Carpenter, *Adv. Drug Deliv. Rev.* **10:**1–28, 1993
Formulation and Delivery of Proteins and Peptides, Design and Development Strategies, J. L. Cleland and R. Langer, (eds.), ACS Symposium Series 567, American Chemical Society, Washington, DC, 1994
Formulation, Characterization, and Stability of Protein Pharmaceuticals, R. Pearlman and Y. J. Wang, Plenum Press, New York, 1996

instabilities of these molecules. Depending on the amino acid types and sequence, proteins are subject to various types of degradation mechanisms, including hydrolysis, oxidation, racemization, and interaction with a variety of solutes and surfaces. These mechanisms are especially critical, because pharmaceutical proteins are very pure and removed from their natural environments where they are most stable (Hanson and Rouan, 1992). Dealing with physical instability (e.g., denaturation, aggregation, and adsorption) often is more a problem with proteins than dealing with their chemical stabilization. Physical instability actually involves solubility problems with large molecules. Although proteins generally contain many polar groups capable of ionization and hydrogen bonding with water, they also can contain many hydrophobic amino acids that under various conditions will preferentially self-associate, leading to aggregation and decreased solubility. Therefore, the development scientist needs to know the structure of the protein and its conformation in solution in order to anticipate potential chemical and physical stability difficulties and then, using principles outlined in this chapter, develop formulation strategies which will overcome these instabilities. Protein formulations may also have significant potential for supporting microbial growth as compared to smaller molecules. The problems associated with protein microbial growth promotion properties are covered in Chapter 3. Table II summarizes some of the primary differences one must recognize in developing protein dosage forms compared to nonprotein dosage forms.

Table II
Protein versus Small Molecule Comparison from a Stability Standpoint[a]

Protein	Small molecule
Many potential reactive sites	Few reactive sites
Many ionizable sites	Few ionizable sites
Buffer effect usually a unique acid/base catalysis	Buffer effect usually general acid/base catalysis
Secondary/tertiary/quarternary structure	Lacks "higher order" structure
Disperse (colloidal) aqueous systems	Single and continuous phase in solution
Temperature effects can be discontinuous (denaturation)	Temperature effects are continuous
Readily supports microbial growth	

[a] Courtesy, in part, of Dr. Lee Kirsch, University of Iowa, Iowa City, IA.

3. GENERAL FORMULATION PRINCIPLES FOR PROTEINS

Protein stability, both in the dry state and in solution, is the main reason why formulation science has such a presence in the commercial development of protein dosage forms. Proteins are complex in size and structure and, as macromolecules, contain a large number of functional groups. Generally, their biological activity in solution depends on a specific three-dimensional conformation. Almost every conceivable environmental factor (e.g., temperature, light, water, pH, presence of glass, rubber, or plastic, shear, presence of salts and other solutes, both macromolecules and low molecular weight compounds, detergents, or sanitizing agents, nature of the filling processes, freeze-thawing, freeze-drying) can effect conformational changes and lead to denaturation, aggregation, or adsorption to surfaces. The challenge to formulation scientists is to develop a stable formulation that can be consistently manufactured and is stable in a given packaging system over the shelf life of the product. A reasonable target expiration dating is 18 months to 2 years at ambient temperature or, failing this, at refrigerated conditions. Aqueous, ready-to-use solutions are preferable dosage forms for many reasons (convenience, cost, customer acceptance), but most proteins are not sufficiently stable in solution to allow practical expiration dating. Therefore, most protein dosage forms are solid forms in the commercial package with the solid form being produced by freeze-drying. Stability data should include not only the freeze-dried solid, but also solution stability after reconstitution with an appropriate vehicle.

Most additives in protein formulations are needed for stability purposes. These include buffers to enhance stability against specific acid/base-catalyzed hydrolysis, antioxidants, chelating agents, and inert gases to enhance stability against oxidative degradation, cryoprotectants/lyoprotectants to enhance stability during freeze-drying of the protein product, surface-active agents to minimize interfacial denaturation, and excipients (e.g., albumin) to minimize protein adsorption to inert surfaces such as glass. The best formulation strategy is to keep the formulation as simple as possible, to have a clear reason for including each additive, and, if possible, to use excipients that have previously been used in Food and Drug Administration (FDA)-approved formulations. Hundreds of articles have appeared in the literature just in the last 20 or so years that report the stabilizing effects of additives on various proteins. We will reference those we feel have the most relevance to the industrial formulation scientist.

It is also important in this age of globalization that the formulation scientist develops a formulation that is acceptable worldwide. This is not an easy assignment, because there are many commonly used additives acceptable in one country, but not another. For example, disodium ethylenediaminetetraacetic acid (DSEDTA) is acceptable for use in injectable products in the United States and Europe, but is not acceptable in Japan. Levels of antimicrobial preservative agent(s) needed to pass the United States Pharmacopeia (USP) preservative efficacy test are much lower than levels required to pass the European Pharmacopoeia (EP) test.

Other additives in protein products (not including controlled drug delivery systems) serve one or more of the following functions:

- Agents for antimicrobial preservation
- Agents for solubility enhancement
- Bulking agents for freeze-dried products
- Agents for achieving isotonicity

Most proteins alone and in final product formulations support the growth of microorganisms. The microbial growth properties of proteins alone and in the final product formulation should be well known and steps should be taken to assure that the antimicrobial properties of the final formulation meet the appropriate acceptance criteria. For multiple-dose products, the addition of an antimicrobial preservative system is required to provide antimicrobial properties to the final product. Although including an antimicrobial preservative in a single-dose product has the advantage of providing additional assurance against introduction of microorganisms during manufacturing, this practice is generally frowned upon by regulatory agencies. Therefore, strict microbial control during manufacturing and the integrity of the packaging system all must be optimized in order to minimize the risk of inadvertent microbial contamination of the final product.

Table III
Common Stability and Compatibility Problems with Proteins and Possible Solutions

Stability problem	Possible solutions
Hydrolysis, deamidation (e.g., asparagine deamidation)	pH control, buffers, low ionic strength
Oxidation (e.g., methionine oxidation)	Antioxidants, chelating agents, low pH, oxygen-free processing and packaging
β-Elimination	Low pH, chelating agents
Transpeptidation	pH control, lower concentration
Racemization	pH control, buffers
Disulfide exchange	Thiol scavengers (e.g., cysteine)
Denaturation during freeze-drying	Cryo-, lyoprotectants
Aggregation, precipitation	pH control, surface-active agents, minimize mechanical stress
Adsorption to surfaces	Surface-active agents, albumin, presaturation

The formulation scientist must be aware of the potential for adverse effects of low-level impurities in formulation components and packaging materials on physical and chemical stability of proteins. Impurities such as peroxides from surface-active agents and other polymeric agents, aldehydes from polymer synthesis and degradation, and extractables from rubber closures must be known and controlled to avoid both short-term and long-term adverse effects on product quality.

Table III summarizes common stability and/or compatibility issues with protein dosage forms and suggested approaches for solving these issues. These approaches will be covered in more detail later in this chapter.

4. WHY PACKAGING, PROCESSING, AND FORMULATION ARE INTERRELATED

Most publications that deal with protein formulation do not cover aspects of the manufacturing process or packaging. Yet the three are interrelated. A formulation is not stable unless the product can be manufactured consistently at a large scale and packaged in a container/closure system that can maintain sterility and stability for a relatively long period of time. Packaging of proteins is especially challenging because of the inherent interactive nature of proteins with inert surfaces such as glass, rubber, and plastic. For many proteins, adsorption at these surfaces sometimes results in the surface denaturation and subsequent aggregation of the protein (Cleland *et al.*, 1993). This includes interfacial denaturation at the air–water interface (e.g., the headspace in a vial containing a protein solution). Minimizing foaming caused by agitation during manufacture, as well as during use of the product, may be critical in order to avoid significant loss of protein activity or generation of visible particulate matter.

It is well known now that processing of protein formulations can affect protein stability. Examples include adverse effects of freezing and/or drying that occur during the lyophilization process, mixing/agitation processes, the filtration process, complicated manufacturing procedures requiring longer filling or hold times, and the movement of intermediate product from one location or site to another. In all these examples, the protein formulation must be designed to resist changes in potency, purity, and other physical–chemical characteristics of the protein itself and the finished formulation.

The bottom line message here is simple: A formulation scientist developing a protein (or, for that matter, any) dosage form must consider the formulation, process, and package together, not focus on one aspect exclusive of others. The smart formulation scientist, in fact, not only will consider all aspects of the formula, process, and package, but also will develop close interactions with packaging engineers, polymer scientists, manufacturing experts, and other experts in areas outside of the formulation scientist's direct expertise. ("None of us is as smart as all of us"—Satchel Paige.)

5. COMMERCIALLY AVAILABLE PROTEIN DOSAGE FORMS

Table IV summarizes U.S. marketed protein dosage forms approved by the FDA through 2000. The table contains information from the *Physicians' Desk Reference* (2001) on the dosage form, route of administration, and types and quantities of additives. Although preferential interaction experiments (e.g., Arakawa and Timasheff, 1982) can predict which solutes can serve as protein stabilizers, the majority of protein formulation research and development requires a great amount of trial and error to finalize the type and amount of formulation components. Prior "art," in the sense of knowing what has worked before and, particularly for injectable formulations, which additives have a history of safety and regulatory acceptance, greatly assists the protein formulation scientist in developing stable, elegant, and manufacturable dosage forms. Characterization of protein structure, as well as collecting preformulation data as described in Chapter 1, will provide supporting data for stabilizers and other additives that are most

Table IV
Current commercial protein forms

Generic name[a]	Trade name	Manufacturer	Physical form[b]	Route of administration[c]	Excipients[d]
Abciximab (*Antiplatelet prevention of blood clots*)	ReoPro	Centocor Lilly	Solution	IV	Sodium phosphate Sodium chloride Polysorbate 80 (0.001%)
Alglucerase (*Gaucher disease*)	Cerdase	Genzyme	Solution	IV infusion	Sodium citrate Human serum albumin (1%)
Alteplase (*Acute MI; pulmonary embolism*)	Activase	Genentech	Freeze-dried (*recons. with D5W or NS*)	IV	L-Arginine Polysorbate 80 (0.011%) Phosphoric acid Incompatible with preservatives
Antihemophilic factor	Koate-HP Recombinate	Miles Baxter Genetics Institute	Freeze-dried (*recons. with sterile WFI*)	IV	PEG (1500 ppm) Glycine (0.05 M) Polysorbate 80 (25 ppm) Tri-n-butyl phosphate (5 ppm) Calcium chloride (3 mM) Albumin (10 mg/ml)
Anti-inhibitor coagulant complex (*Vitamin k dependent clotting factors [Heparin]*)	Autoplex T	Nabi	Freeze-dried (*recons. with sterile WFI*)	IV	Polyethylene glycol (2 mg/ml) Sodium citrate
Domase alpha DNAse (*Cystic fibrosis*)	Pulmozyme	Genentech	Aerosol	Inhalation	Calcium chloride Sodium chloride

Eptifibatide (*Acute coronary syndrome inhibits platelet aggregation*)	Integrilin	Cor	Solution	IV	Citric acid Sodium hydroxide to pH 5.25
Erythropoietin (*Anemia*)	Epogen Procrit	Amgen Ortho Biotech	Solution	IV or SC	Sodium chloride Sodium citrate Human serum albumin (0.25%)
Filgrastim (Granulocyte CSF)	Neupogen	Amgen	Solution	IV or SC	Sodium acetate Mannitol Polysorbate 80 (0.004%)
Hepatitis B vaccine	Engerix-B	GlaxoSK	Solution	IM	20 mcg adsorbed on 0.5 mg aluminum Thimerosal Sodium chloride Sodium phosphates
	Recombinax	Merck	Suspension	IM	5.40 mcg adsorbed on 0.5 mg aluminum Thimerosal
Human chorionic gonadotropin	A.P.L.	Wyeth-Ayerst	Freeze-dried (*recons. with special diluent*)	IM	Benzyl alcohol (2%) (diluent) Phenol (0.2%) (diluent) Lactose
Human chorionic gonadotropin	Pregnyl	Organon	Freeze-dried (*recons. with speical diluent*)	IM	Sodium phosphate Sodium chloride (diluent) Benzyl alcohol (0.9%) (diluent)
	Profasi	Serono	Freeze-dried (*recons. with bact. WFI*)	IM	Sodium phosphate Mannitol Benzyl alcohol (0.9%) (diluent)

(*continued*)

Table IV
(continued)

Generic name[a]	Trade name	Manufacturer	Physical form[b]	Route of administration[c]	Excipients[d]
Human growth hormone	Humatrope	Lilly	Freeze-dried (*recons. with special diluent*)	SC	Mannitol Glycine Sodium phosphate Meta cresol (0.3%) (diluent) Glycerin (diluent)
	Nutropin	Genentech	Freeze-dried (*recons. with bact. WFI*)	SC	Mannitol Sodium phosphates Benzyl alcohol (0.9%) (diluent)
	Nutropin NQ	Genentech	Solution	SC	Sodium chloride Phenol (0.5%) Polysorbate (0.4%) Sodium citrate 10 mM
Human insulin	Humulin R	Lilly Squibb-Novo	Solution	SC	Glycerin *m*-Cresol (0.25%)
Human insulin	Humulin N	Lilly Squibb-Novo	Suspension	SC	Protamine *m*-Cresol (0.16%) Phenol (0.065%) Sodium phosphate
Human insulin	Humulin L	Lilly Squibb-Novo	Suspension	SC	Methylparaben (0.1%) Sodium chloride Sodium acetate Zinc oxide
Immune globulin (*Gamma globulin IgG*)	WinRho SDF	Nabi	Freeze-dried (*recons. with accompanying 0.9% sodium chloride*)	IV or IM	Glycine Sodium chloride Polysorbate 80 (0.01%)

Interferon α-2a (*Hairy cell leukemia*)	Roferon-A	Roche	Freeze-dried and RTU Solution	IM or SC	Sodium chloride Phenol (0.3%) Human serum albumin
Interferon α-n3 (*Genital warts*)	Alferon N	Interferon	Solution	Intralesional	Phenol (0.33%) Human albumin (0.1%) Sodium chloride Potassium chloride Sodium phosphates
Interferon α-2b (*Hairy cell leukemia*)	Intron A	Schering	Freeze-dried (*recons. with bact WFI)*	IM or SC Intralesion	Glycine Sodium phosphates Human serum albumin Benzyl alcohol (0.9%)
Interferon α-26 (*Hairy cell leukemia*)	Intron A	Schering	Solution	IM or SC Intralesion	Glycine Sodium phosphates Human serum albumin Parabens
Interferon β-1B (*Multiple Sclerosis*)	Betaseron	Berlex Chiron	Freeze-dried	SC	Human serum albumin Mannitol Sodium chloride
Interferon γ-1B	Actimmune	Genentech	Solution	SC	Mannitol Sodium succinate Polysorbate 20 (0.01%) Sodium citrate
Interleukin-2 (*Metastatic renal cell carcinoma*)	Proleukin	Chiron	Freeze-dried (*recons. with sterile WFI*)	IV Infusion	After reconstitution, dilute in 50 mL D5W Incompatible with preservatives

(*continued*)

Table IV
(*continued*)

Generic name[a]	Trade name	Manufacturer	Physical form[b]	Route of administration[c]	Excipients[d]
Interleukin-11 (*Prevention of severe thrombocytopenia; reducing need for platelet transfusions*)	Neumega	Genetics Institute	Freeze-dried (*recons. with sterile WFI*)	SC	Glycine Phosphate buffer to pH 7
Leuprolide Acetate	Lupron	TAP Pharma	Solution	SC	Sodium chloride Benzyl alcohol Acetic acid
Leuprolide Acetate	Lupron Depot	TAP Pharma	Freeze-dried	IM	Gelatin DL-Lactic/glycolic acids D-mannitol Reconstituted with diluent containing CMC sodium, D-mannitol, Polysorbate 80
Muromonab-CD3 (*Immuno suppressant*)	Orthoclone OKT3	Ortho Biotech	Solution	IV	Polysorbate 80 (0.1%) Sodium phosphates Soduim chloride

Rituximab (*Treatment of B-cell non-Hodgkins lymphoma*)	Rituxan	Genentech	Solution pH 6.5	IV	Sodium chloride Sodium citrate Polysorbate (0.07%)
Sargramostim (*Granulocyte macrophage CSF for bone marrow transplant*)	Leukine	Immunex	Freeze-dried (*recons. with sterile WFI*)	IV Infusion	Mannitol Sucrose TRIS Incompatible with preservatives
Sermorelin acetate (*Stimulates pituitary gland to release growth hormone*)	Geref Ampuls	Serono	Freeze-dried (*recons. with accompanying 0.9% sodium chloride*	SC	Mannitol pH adjusted with sodium phosphate buffer
Urokinase	Abbokinase	Abbott	Freeze-dried (*recons. with. sterile WFI*)	IV, Intra-coronary Infusion	Mannitol Sodium chloride Human serum albumin

[a]MI, myocardial infarction; CSF, colony stimulating factor.
[b]recons., reconstituted; D5W, dextrose 5% in water; NS, 0.9% sodium chloride; WFI, water for injection; bact., bacterial; RTU, ready to use.
[c]IV, intravenous; SC, subcutaneous; IM, intramuscular.
[d]CMC, carboxymethylcellulose.

likely to be effective. For example, pH–solubility/stability studies will provide direction on what type of buffers to use, if any. Having knowledge of the structural conformation of a protein in order to predict which amino acids in the protein sequence may be particularly vulnerable to degradation because of exposure to the environment may give the formulation scientist some direction on the stabilizers required. The final selection of excipients, unfortunately, must be a result of much empirical evaluation. However, information such as that given in Table IV summarizes what others have done with their protein products, and thus can give significant guidance to formulation scientists facing the development and stabilization of new peptide and protein dosage forms.

Note that protein dosage forms primarily are divided into three types

1. Ready-to-use solutions
2. Freeze-dried powders that are reconstituted into solutions immediately before administration
3. Ready-to-use suspensions

Proteins are commonly formulated at very low doses (very dilute solutions), although there are examples of relatively high dose protein products, such as formulations of immunoglobulin G (IgG) at 50 mg/ml. In general, dilute solutions are less physically stable than more concentrated solutions (Hanson and Rouan, 1992) and adsorption to surfaces will result in a higher fractional loss of protein. However, in the case of the Neutral Protamine Hagedorn (NPH) formulation of insulin, the rate of formation of higher molecular weight polymers increases as a function of concentration (Brange *et al.*, 1992b). Also, for interleukin 1β (IL-1β), aggregation/precipitation was shown to demonstrate biphasic kinetics (slower rate followed by a more rapid rate) at temperatures lower than 55°C and to be dependent on concentration. When the concentration was increased from 100 to 500 mg/ml, the slower rate was observed to be suppressed and a more rapid degradation was observed (Gu *et al.*, 1991). In general, however, there are surprisingly few literature reports of protein stability as a function of concentration.

6. CHEMICAL STABILIZATION

6.1. pH, Hydrolysis, and Buffers

The effect of solution pH on stability is probably the most important factor to study in early protein dosage form development. Figure 1 schematically depicts expected stability problems of proteins as a function of pH. pH–stability studies are conducted very early to understand relative protein stability over a pH range, typically from about pH 3 to about pH 10. The relationship of stability and solubility to pH usually follows a pattern of higher solubility resulting in lower chemical stability and lower solubility resulting in lower physical stability. Protein solubility is usually at a minimum at its isoelectric point. Insulin, for example, has an isoelectric point of 5.4, and at this pH it is quite insoluble in water (< 0.1 mg/ml). Adjusting the solution pH to less than 4 or greater than 7 greatly increases insulin solubility (> 30 mg/ml, depending on zinc concentration and species source of insulin), but also increases the rate of deamidation at these pH ranges (Brange, 1992). An example of the effect of pH on deamidation and polymerization of insulin is shown in Fig. 2 (Brange and Langkjaer, 1993). In dosage form development, the scientist must first determine what pH range provides acceptable solubility of the protein for proper dosage, then determine whether this pH range also provides acceptable stability. There is usually a trade-off between solubility and stability and it is up to the scientist to identify what pH is optimal for both. When an acceptable trade-off does not exist for a solution formulation, a freeze-dried formulation is usually indicated.

Hydrolysis or *deamidation* occurs with proteins containing susceptible Asn and Gln amino acids, the only two amino acids that are primary amides. The side-chain amide linkage in a Gln or an Asn residue has been shown to undergo deamidation to form free carboxylic acid. Deamidation

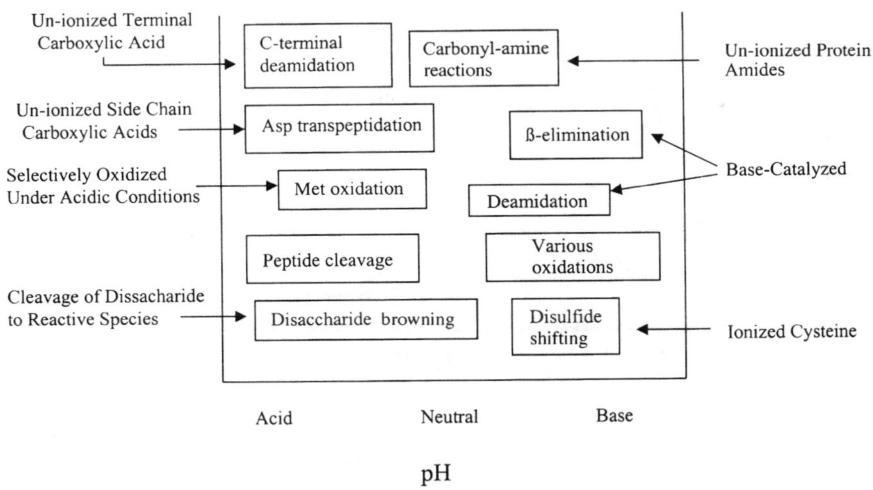

Figure 1. Protein reactions as a function of pH. Figure courtesy of Dr. Lee Kirsch.

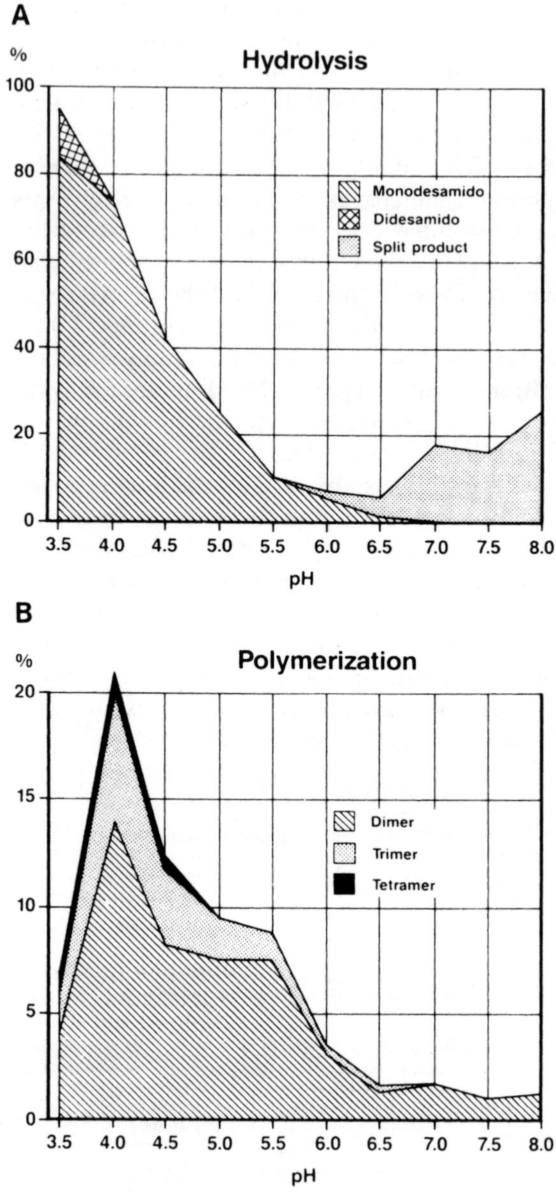

Figure 2. Chemical transformation of insulin during storage of rhombohedral insulin crystals (bovine insulin crystals: 0.7% NaCl, 0.2% phenol) as a function of pH during storage at 25°C for 12 months. (A) Formation of the hydrolysis products mono and didesamido insulins and the insulin split product (AB-A9). (B) Formation of covalent dimers and oligomers. Reprinted with permission from Brange and Langkjaer (1993). Copyright 1993 Plenum Press.

can be promoted by a variety of factors, including extremes in pH, temperature, and ionic strength. Many investigators have observed altered forms of proteins which have been attributed to this deamidation process. Robinson and Rudd (1974), Geiger and Clarke (1987), and Clarke *et al.* (1992) reviewed the chemical conditions necessary for hydrolysis of Asn and Gln residues for many proteins. It was shown that neutral or alkaline conditions enhanced the rate of deamidation of proteins mainly at the Asn–Gly sequence. This rate was also found to be higher than the hydrolysis of the amino acid Asn itself. This deamidation proceeds through a five-membered cyclic imide intermediate formed by intramolecular attack of the succeeding peptide nitrogen at the side-chain carbonyl carbon of the Asn residue. The cyclic imide then spontaneously hydrolyzes to give a mixture of peptides in which the backbone is attached via either an Asp (α-carboxyl) or an iso-Asp (β-carboxyl) linkage. Haley *et al.* (1966) showed that Gln also undergoes a similar deamidation reaction via the formation of a six-membered ring. Peptide backbone hydrolysis also has been shown to occur from this cyclic imide (Tyler-Cross and Schirch, 1991).

One approach that has been used to lower the rate of deamidation is to reduce pH, because in general, deamidation is slower at acidic pH than at neutral or alkaline pH. However, caution must be exercised here, because a reduction in pH may lead to cleavage or cyclization at Asp–X residues, where X is usually a residue with a small side chain, such as Gly or Ser. Proteins with Asp–X degradation must be formulated at a higher pH to avoid cyclization (Manning *et al.*, 1991). However, higher pH conditions (i.e., greater than pH 8) may catalyze oxidation, thiol–disulfide exchange, and β-elimination reactions. In the case of insulin, multiple deamidation sites are observed, where deamidation at the A21 position predominates at acidic pH and deamidation at B3 predominates at neutral pH (Brange, 1992).

Buffers. Buffers are used to prevent small changes in solution pH which can affect protein solubility and stability. Buffers are composed of salts of ionic compounds, the most common of which are acetate, citrate, and phosphate. Buffer systems acceptable for use in parenteral solutions are listed in Table V.

The proper selection of buffer type and concentration is done by performing solubility and stability studies as a function of pH and buffer species. In general, it is good practice to keep the buffer concentration as low as practical.

Potential problems associated with using buffers include the following.

1. It may be difficult to meet the pH target with a buffer system while preparing solutions during scaleup and full-scale manufacturing. Dilute solutions of strong acids (hydrochloric acid) or bases (sodium hydroxide) usually are required, which may alter the buffer capacity

Table V
Buffers Used in Protein Formulations

Buffer system	pK_a	pH range of use
Acetate	4.76	2.5–6.5
Citrate	3.14, 4.8, 5.2	2.5–6.0
Glutamate	9.67(pK_{a3})	8.2–10.2
Glycinate	2.4,9.8	6.5–7.5
Histidine	1.8, 6.0, 9.2	6.2–7.8
Lactate	3.8	3.0–6.0
Maleate	1.92, 6.23	2.5–5.0
Phosphate	7.2 (pK_{a2})	6.0–8.2
Succinate	4.2, 5.64	4.8–6.3
Tartrate	2.93, 4.23	3.0–5.0
Tris	6.2 (pK_b 7.8)	6.8–7.7

of the buffer system and the ionic strength of the solution depending on the buffer type, the concentration of the buffer, and the pH of the formulation (Niebergall, 1990).

2. Increasing buffer capacity to better control pH probably also will significantly increase ionic strength, which, in turn, may cause increase in frequency of pain upon injection due to the increased osmolality.
3. General acid and/or general base buffer catalysis may accelerate the hydrolytic degradation of the protein. An example is given in Fig. 3 (Yoshioka *et al.*, 1993), where the inactivation rate of β-galactosidase increases with increasing concentration of phosphate buffer up to 0.5 *M*, then decreases, presumably because of higher buffer concentration causing a reduction in water activity. Cleland *et al.* (1993) cite several examples where the rate of protein deamidation was markedly dependent on the buffer anion. One interesting example was reported by Capasso *et al.* (1991) in which the deamidation rate of a small peptide was measured using different buffers. The peptide was most unstable in a phosphate buffer and most stable in Tris buffer.
4. Buffer crystallization during freezing and freeze-drying may change the pH of the freeze-concentrated solution and could affect drug/protein stability (Murase and Franks, 1989; Szkudlarek *et al.*, 1996); (van den Berg and Rose, 1959). The dibasic phosphate salt tends to crystallize during freezing and the degree of pH change depends on a variety of factors, including the original pH, the buffer concentration, and whether the cation of the phosphate buffer is sodium or phosphate. A freezing-induced pH shift may occur with mono-

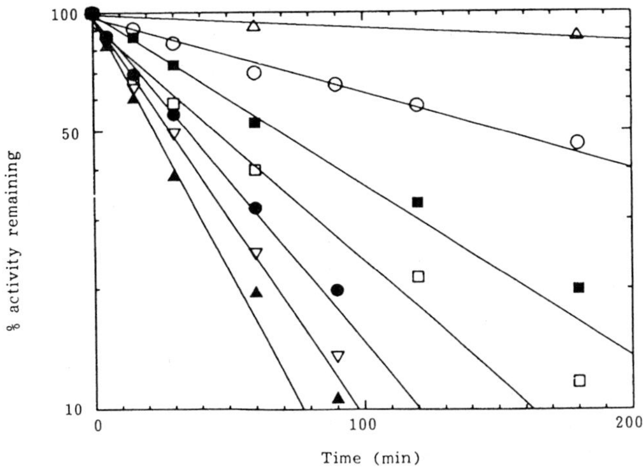

Figure 3. Inactivation of β-galactosidase in pH 7.4 phosphate buffer solution at 50°C as a function of phosphate buffer concentration: (△) 10, (○) 50, (□) 100, (▽) 200, (▲) 500, (●) 700, and (■) 900 m*M*. The concentration of β-galactosidase was 0.1 mg/ml. Reprinted with permission from Yoshioka *et al.* (1993). Copyright 1993 Plenum Press.

sodium succinate buffer, causing a loss of activity of interferon gamma during freeze-drying (Lam *et al.*, 1996). This was overcome by using glycolate buffer instead of succinate buffer.

6.2. Oxidation, Antioxidants, and Other Antioxidation Approaches

Proteins containing methionine, cysteine, cystine, histidine, tryptophan, and tyrosine may be sensitive to oxidative and/or photolytic degradation, depending on the conformation of the protein and resultant exposure of these amino acids to the solvent and environmental conditions such as oxygen, light, high temperature, metal ions, and various free radical initiators. Oxidation of sulfhydryl-containing amino acids (e.g., methionine and cysteine) may lead to disulfide bond formation and loss of biological activity. The free thiol group in a Cys residue of any native biologically active protein not only may oxidize to produce an incorrect disulfide bridge, but also can result in other degradation reactions, such as alkylation, addition to double bonds, and complexation with heavy metals. Atmospheric oxygen has been shown to oxidize Met residues during isolation, purification, and storage (Brot and Weissbach, 1982). The thioether moiety in Met is a weak nucleophile and therefore stays deprotonated even at low pH. However, it has been shown to be selectively oxidized to Met sulfoxide by agents such as hydrogen peroxide under acidic conditions. Other agents that have been used to catalyze this oxidation reaction are periodate, iodine, chloramine-T, dimethylsulfoxide, and *N*-chlorosuccinimide. In several instances these oxidation processes render the protein moiety biologically inactive as has been shown in corticotropin, α- and β-melanotropins, gastrin, calcitonin, corticotropin releasing factor, and parathyroid hormone (Brot and Weissback, 1992; Deadman *et al.*, 1961; Dixon, 1956; Morley *et al.*, 1965; Riniker *et al.*, 1968; Tashjian *et al.*, 1964; Vale *et al.*, 1981). In addition to these examples, several nonhormonal proteins and peptides have been shown to lose activity due to oxidation of the Met residue. In these instances restoration of biological activity concurrently occurs with the reduction of Met sulfoxide to Met. However, there are also several cases where the oxidation of the Met residue does not seem to affect the biological activity of the pharmaceutical agent.

Human growth hormone, chymotrypsin, lysosyme, insulin-like growth factor I, and glucagon are examples of proteins that will degrade by oxidative mechanisms. Hemoglobin, with its oxygen carrying properties dependent on the reduced state of ferrous iron, is very sensitive to oxidation and, as a commercial product in the deoxy state, must contain antioxidants to maintain stability of the heme groups. Kerwin *et al.* (1999) and Cleland *et al.* (1993) list 61 different proteins that can be oxidized with varying degrees of loss of biological activity following oxidation.

There are several choices of antioxidants that can be used in protein formulations (see Table VI). Those which have been used most frequently in protein formulations are ascorbic acid, salts of sulfurous acid (sodium bisulfite, sodium metabisulfite), and thiols such as thioglycerol and thioglycolic acid. Dithriothreitol, reduced glutathione, acetylcysteine, mercaptoethanol, and thioethanolamine are thiols which usually oxidize too readily to be of practical use in pharmaceutical formulations requiring long-term storage.

Precautions must be applied when considering ascorbic acid as an antioxidant in protein formulations. Li *et al.* (1993) found that ascorbate in the presence of Fe^{3+} and oxygen actually induces the oxidation of methionine in small model peptides. Ascorbate is a powerful electron donor and is readily oxidized to dehydroascorbate. It also generates highly reactive oxygen species, such as hydrogen peroxide and peroxyl radicals. These, in turn, will accelerate the oxidation of methionine. Phosphate buffer compared to other buffer systems [e.g., tris(hydroxymethyl)aminomethane (Tris), *N*(2-hydroxyethyl)piperazine-2-(2-ethanesulfonic acid) (HEPES), and 3-(*N*-morpholino)propanesulfonic acid (MOPS)] accelerated the degradation of methionine in the presence of ascorbic acid. The addition of ethylenediamine-

Table VI
Antioxidants and Chelating Agents for Protein Formulations

Agent	Normal % used in formulation
Antioxidants (all water soluble)	
Ascorbic acid (including isoascorbic acid, sodium ascorbate)	0.1–1.0
Sulfurous acid salts (sodium bisulfite, sodium metabisulfite, sodium sulfite)	0.1–0.5
Thioglycerol	0.1–0.5
Thioglycollic acid	0.05–0.2
Cysteine hydrochloride	0.1–0.5
Chelating Agents (all water-soluble)	
Ethylenediaminetetraacetic acid (usually the disodium salt)	0.05–0.1
Citric acid/sodium citrate	0.02–1

tetraacetic acid (EDTA) did not enhance stability, even though ferric ion and other transition metals were components of the formulation, either purposely added or as trace components of the buffer and peptide. This pro-oxidant effect of ascorbate methionine oxidation was reported to be concentration-dependent and occurred most readily at pH 6–7. It is also known that bisulfite can cause stability problems with certain drugs and proteins. For example, bisulfite will rapidly destroy insulin (Brange, 1987).

For protection against oxidation, use of an effective antioxidant is one of several precautions that must be enacted in formulation development which dramatically affects final product manufacture. Other procedures that contribute to protein stability against oxidative degradation include the following:

- Preparation and storage at low temperatures
- Use of chelating agents to eliminate metal catalysis
- Increase of ionic strength (decreases disulfide bond formation; Shahrokh *et al.*, 1994b)
- Elimination of peroxide and metallic contaminants in formulation additives
- Protection from light
- Awareness of possible interaction of light exposure and phosphate buffer in forming free radicals (Fransson and Hagman, 1996)
- Replacement of oxygen with nitrogen or argon during manufacturing
- Removal of oxygen from the headspace of the final container
- Formulation at the lowest pH possible while still maintaining desired protein solubility and hydrolytic stability, because there is

an inverse correlation between oxidative stability and pH (Akers, 1982)

- Use of a container/closure system that allows no oxygen transmission through the package during distribution and storage
- Assurance that phenolic or other oxidizing cleaning agent residues are minimal in the production environment, including the freeze-dry chamber (Kirsch *et al.*, 1993)

Stabilization of human growth hormone against oxidative degradation will be presented as a case history example at the end of this chapter.

Chelating agents (see Table VI) are used in protein formulations to aid in inhibiting free radical formation and resultant oxidation of proteins caused by trace metal ions, such as copper, iron, calcium, manganese, and zinc. Although organic buffer salts, such as sodium citrate, have some capability of binding trace heavy metal contaminants in protein solutions, the major chelating agent used is DSEDTA. The concentration of DSEDTA usually is very small, typically in the range of 0.01–0.05%. DSEDTA tends to dissolve slowly and is usually among the first of formulation ingredients to be dissolved during compounding before adding other ingredients, including the protein.

6.3. Other Protein Chemical Instabilities and Formulation Approaches

At temperatures above ambient temperature, proteins can undergo disulfide cleavage due to β-elimination from cysteine residues. Such a reaction is known to occur at low temperatures and high pH. Volkin and Klibanov (1987) found that the rate of β-elimination is greatly accelerated under alkaline conditions. The free sulfhydryl group that is formed from β-elimination reaction of cysteine can then contribute to other degradation pathways such as disulfide exchange, aggregation, precipitation, and adsorption. In addition to cysteine, other amino acids that undergo degradation via the β-elimination route under alkaline conditions are Cys, Ser, Thr, and Lys. Sen *et al.* (1977) showed that the rate of β-elimination in phosvitin directly correlated with hydroxide concentration, and β-elimination of phosphoserine occurred in the presence of calcium chloride. On the basis of their studies, Lee *et al.* (1977) showed that the antifreeze glycoprotein undergoes β-elimination in proportion to the hydroxide concentration and temperature. Control of pH, therefore, is essential to prevention of β-elimination degradation pathways of susceptible proteins.

Sulfhydryl groups and disulfide bonds play a major role in determining the tertiary structure properties of most proteins. Biological activity can be lost when an interchange of disulfide bonds leads to an altered three-dimensional structure. Various nucleophiles and electrophiles are able to carry out reversible breakage of disulfide linkages. The free thiol group that is generated can then produce incorrect disulfide exchanges. As Srinivasan *et al.* (1990) pointed out, rates of such reactions can be very significant in proteins at elevated temperatures or extremes of pH, and refolding is not usually observed. Liu *et al.* (1991) observed that moisture-induced aggregation of lyophilized proteins such as ovalbumin, bovine serum albumin (BSA), and β-lactoglobulin is caused by incorrect disulfide exchanges in combination with conformational changes. Ryle and Sanger (1955) and Torchinsky (1981, pp. 81–86) were able to successfully prevent disulfide exchanges by using thiol scavengers such as copper ion, *N*-ethylmaleimide, and *p*-mercurybenzoate. Haber and Anfinsen (1962) and Galat *et al.* (1981) observed that proteins with scrambled disulfide bonds can be rearranged to yield native, biologically active species by incubating the protein with mercaptoethanol or cysteine.

With the exception of glycine, all amino acids exhibit chirality at the carbon site to which side chains are attached. Consequently, these amino acids are subject to racemization under basic conditions. These racemizations in proteins can create either a nonmetabolizable enantiomeric form such as D amino acids or introduce peptide bonds at locations inaccessible to proteolytic cleavage. The rate of racemization depends on factors such as inductive and field effects, steric hindrance to solvation, intramolecular solvation, and intramolecular base action. Friedman and Masters (1982) showed that Asp in proteins undergoes racemization at a rate much faster than the free amino acid residue and that the rate of conversion of the L form to D-Asp is more rapid than that of any of the other amino acid side chains. Geiger and Clarke (1987) showed that the racemization of the Asp residue in alrenocorticotropic hormone (ACTH) occurs via a cyclic imide at pH 7.4 and 37°C. Figure 1 graphically summarizes the pH range where the above degradation reactions most likely will occur.

7. PHYSICAL STABILIZATION

Physical instability is rarely encountered in formulations of small drug molecules except for poorly water-soluble compounds. Proteins, because of their ability to adopt higher ordered structures, tend to undergo a number of changes structurally, independent of chemical modifications. Physical instability of proteins is often a greater cause for concern and may be more difficult to control than chemical instability. Virtually all protein structures have hydrophobic regions to some extent, and low solubility in water is regarded as an indication that a material is hydrophobic (Currie and Groves, 1992). Hydrophobic interaction is a major driving force for protein folding, where hydrophobic regions tend to be on the interior of the folded structure. Such exposure will promote aggregation or self-association, possibly leading to physical instability and potential loss of biological activity, because the interaction with the receptor site requires folded structures with the correct conformation.

7.1. Denaturation

Denaturation is unique to proteins and occurs when their native tertiary and frequently secondary structure is disrupted. Denaturation can lead to unfolding, and the unfolded polypeptide chain may undergo further reactions. Such inactivation could be association with surfaces and/or interaction with other protein molecules leading to aggregation and precipitation. Denaturation is of two types: (1) reversible denaturation, caused by temperature or exposure to denaturants (urea, guanidine hydrochloride), where, if the denaturing condition is removed, the protein will regain its native state and recover its activity, and (2) irreversible denaturation, where the protein, once unfolded, will not regain its native form and activity. However, there are several instances where a protein which is "irreversibly denatured" is returned to its native state by the use of denaturant followed by dialysis. One example is T4 lysozyme, where activity can be restored by renaturation with guanidine hydrochloride (Wetzel *et al.*, 1988).

For stabilizing proteins against denaturation in solution, Arakawa and Timasheff (1982, 1984) and Arakawa *et al.* (1993) showed that reversible denaturation can be decreased by the use of additives such as salts that bind to nonspecific binding sites on the proteins. Dahlquist *et al.* (1976) demonstrated increased thermal stability of thermolysin by the binding of ions to specific sites on the protein. Roe *et al.* (1988) reported the ability of $Zn(NO_3)_2$ to significantly increase the thermal stability of superoxide dismutase. Gekko and Timasheff (1981) concluded that the preferential hydration of proteins observed at all conditions in the presence of a glycerol–water mixed solvent system is a prerequisite for stabilizing the native structure of several globular proteins. Pace and Grimsley (1988) found the stability of ribonuclease T_1 to increase in the presence of 0.1 *M* NaCl, $MgCl_2$, and Na_2HPO_4. Through genetic engineering they were then able to introduce appropriate amino acid substitution in ribonuclease T_1. Due to these newly created specific cation/anion-binding sites on the protein, the stability profile of ribonuclease T_1 was enhanced significantly. Pantoliano *et al.* (1988) were able to introduce negatively charged side chains such as Asp in the vicinity of the weakly Ca^{2+} binding site of subtilisin. Such modifications caused an increase in binding affinity for Ca^{2+}, thereby increasing the thermal stability of subtilisin.

7.2. Aggregation

Protein aggregation is caused mainly by hydrophobic interactions resulting from denaturation. When the interior hydrophobic region of a partially or fully unfolded protein is exposed to water, this creates a thermodynamically unfavorable situation due to the fact that a hydrophobic interior is now exposed to a hydrophilic aqueous environment. Consequently, the decrease in entropy from structuring water molecules around the hydrophobic region generates a driving force for the denatured protein to aggregate, mainly through the exposed hydrophobic regions. Thus, solubility of the protein is also compromised (Franks, 1994). In some cases self-association of protein subunits, either native or misfolded, may occur under certain conditions and this may lead to precipitation and loss in activity (Brange and Langkjaer, 1993; Brange *et al.*, 1992a,b; Mitraki and King, 1989; Shahrokh *et al.*, 1994a; Silvestri *et al.*, 1993). The protein antithrombin aggregates during denaturation with guanidine hydrochloride in a manner which proceeds through an intermediate, partially unfolded state. Subsequent aggregation occurs slowly; therefore, it is possible to return antithrombin to its native state by dialyzing the partially unfolded form, which aggregates slowly. However, once aggregation occurs, the native state cannot be reformed by this approach (Fish *et al.*, 1985). Irreversible aggregation due to denaturation can be prevented by the use of surfactants, polyols, or sugars.

7.3. Adsorption

Proteins exhibit a certain degree of surface activity; that is, they adsorb to surfaces due to their innate nature of being amphiphilic polyelectrolytes. Consequently, biological activity may be either reduced or totally lost if such adsorption occurs during manufacturing, storage, or use of the final product. Protein adsorption, besides causing loss of activity, may also be a precursor to aggregation, denaturation, and precipitation phenomena. The process of adsorption depends on protein–protein interactions, time, temperature, pH and ionic strength of the medium, and the nature of the surface. Norde (1995) reviewed the general principles underlying protein adsorption from aqueous solution onto a solid surface. Interactions that determine the overall adsorption process between a protein and a surface include redistribution of charged groups in the interfacial layer, changes in the hydration of the sorbent and the protein surface, and structural rearrangements in the protein molecule. Surface denaturation, which commonly takes place at the liquid/solid and the liquid/air interface, has been shown by Lenk *et al.* (1989) to involve conformational changes such as loss of α helices to β sheets and certain random structures. These structural changes, which are determined by the nature of the interfaces, are similar to those observed with the aggregation phenomenon caused by heat, high pressure, and chemical denaturants. Strategies often used to overcome protein denaturation due to adsorption are include the following:

- Increase protein concentration
- Modify (e.g., siliconize) the surface of the glass containers, providing a resistant barrier to protein–surface interaction
- Decrease the rate of mixing when it is known that shear will affect protein adsorption
- Add excipients such as surfactants that have higher surface activity
- Add macromolecules such as albumin and gelatin [although one must realize the increased concern for using these natural materials because of their potential for endotoxin and/or bovine spongiform encephalopathy (BSE)* contamination].

The literature is replete with problems encountered while delivering insulin, because of its ability to adsorb onto the surfaces of delivery pumps, glass containers, and the inside of the intravenous bags (Brennan *et al.*, 1985; Iwamoto *et al.*, 1982; James *et al.*, 1981; Lougheed *et al.*, 1983; Massey and Sheliga, 1988; Sato, *et al.*, 1984; Twardowski *et al.*, 1983a–c). Insulin adsorption usually is finite once binding sites are covered and such adsorption is usually not clinically significant.

Adsorption to filters and tubing can often be overcome by first saturating the surfaces with excess protein solution, then discarding the filtrate or wash. This approach will waste valuable protein and so may not be a good choice. In some cases, adsorption can be minimized by using certain additives. For example, Oshima (1989) showed that surface denaturation of chymotrypsin can be prevented by the addition of 0.1 *M* NaCl as well as by coating the surface of the container with either lecithin or BSA. Adsorption of urokinase to glass can be prevented by the addition of

*Also known as "mad cow disease," a chronic generative disease affecting the central nervous system of cattle. First diagnosed in Great Britain in 1986, this transmissible disease could be a contaminant in cow-sourced pharmaceutical excipients such as gelatin, bovine serum albumin, polysorbate 80, and others.

0.25% gelatin to the container surface (Patel, 1990). Calcitonin adsorption to glass syringes can be prevented by benzalkonium chloride or benzethenium chloride, presumably due to coating of glass silanol groups by these positively charged antimicrobial preservatives (Kakimoto *et al.*, 1985).

7.4. Precipitation

Precipitation of protein occurs subsequent to denaturation and is a consequence of aggregates combining to form large particles. The mechanism of aggregation leading to precipitation is beyond the scope of this chapter; the reader is referred to Glatz (1992). Brennan *et al.* (1985) established the tendency of insulin to precipitate when loaded into a long-term infusion device. Massey and Sheliga (1988) determined that the presence of large headspace in vials, high concentration of zinc, high pH, and the presence or absence of additives were causative factors for accelerating insulin precipitation in NPH formulations (called "frosting") and concluded that denaturation occurred at the air–water interface, which in turn led to precipitation.

8. FORMULATION APPROACHES FOR SOLVING PHYSICAL STABILITY PROBLEMS

8.1. Surfactants

Surfactants, as their name implies, are surface-active agents, which can exert their effect at solid–liquid, liquid–liquid, and liquid–air interfaces because of their chemical composition, which includes both hydrophilic and hydrophobic groups. These materials reduce the concentration of proteins in dilute solutions at the surface of the solution where they can be adsorbed and/or denatured. Surfactants can bind to hydrophobic interfaces in protein formulations and packaging. Glass, rubber, or plastic adsorption of proteins is well documented (Chawla *et al.*, 1985; Suelter and DeLuca, 1983). Proteins on the surface of water will aggregate, particularly when shaken, because of unfolding and subsequent aggregation of the protein monolayer. Surfactants can denature proteins, but also can stabilize proteins against surface denaturation.

Generally, ionic surfactants tend to denature proteins. Nonionic surfactants usually do not denature proteins even at relatively high concentrations (1% w/v) (Cleland *et al.*, 1993). Most parenterally acceptable nonionic surfactants come from either the polysorbate (sorbitol–polyethylene oxide polymers) or polyether (polyethylene oxide–polyproplyene oxide block co-polymers) groups. Polysorbates 20 and 80 are the only known surfactant stabilizers in marketed protein formulations (see Table IV). However, other surfactants used in protein formulations for clinical studies and/or found in the patent literature include polysorbate 20, Pluronic F68, and other polyoxyethylene ethers (e.g., the "Brij" class) (Wang and Hanson, 1988).

Surfactants are well known to prevent the denaturation and aggregation of insulin (Chawla *et al.*, 1985; Lougheed *et al.*, 1983; Massey and Sheliga, 1988; Sato *et al.*, 1984). However, the choice of surfactant and the final concentration optimal for stabilization are quite dependent on a variety of factors, including other formulation ingredients, for example, sugars, protein concentration, headspace in the container, the type of container, and test methodology.

Recombinant human growth hormone (rhGH) will aggregate readily under mechanical and thermal stress. Aggregation from mechanical stress can be substantially reduced in the presence of surfactants (Katakam *et al.*, 1995). Tween 20 (polysorbate 20) in a molar ratio >4 with rhGH will effectively inhibit aggregation due to weak binding to hydrophobic patches on the protein's surface (Bam *et al.*, 1998). Mechanical stress may cause proteins to be more exposed to air–water interfaces, where denaturation is more likely to occur than in the bulk phase of water. Surfactants will preferentially compete with proteins for accumulation at the air–water interface and keep the protein from undergoing interfacial denaturation resulting from agitation. Pluronic F-68 and Brij 35 will stabilize hGH at their critical micelle concentrations (0.1% and 0.013%, respectively), whereas stabilization with polysorbate 80 requires a concentration of 0.1%, higher than the critical micelle concentration (cmc) value for polysorbate 80 of 0.0013%. The reasons for these differences in stabilizing concentrations are not clear, but may simply reflect differences in interactions between different surfactants and proteins. It is interesting to note that these surfactants do not stabilize hGH against aggregation due to high-temperature stress.

Further substantiation of the important role of surfactants, particularly polysorbate 80, in protecting proteins against surface-induced denaturation during freezing was reported by Chang *et al.* (1996). These investigators found a strong correlation between freeze denaturation (quick freezing of the protein) and surface denaturation (shaking the protein in solution). Proteins that tend to denature under these conditions are protected by the addition of polysorbate 80 (0.1%). Other surfactants—Brij 35, Lubrol-px, Triton X-100, and even the ionic surfactant sodium dodecyl sulfate—also protected the protein from denaturation, although these surfactants have not yet been approved for use in injectable formulations. The authors pointed out that surfactants may protect proteins from denaturation during the freezing step only, and that other stabilizers, for example, sucrose, may be needed to further protect the protein during freeze-drying.

Peroxides are known contaminants of nonionic surfactants. Knepp *et al.* (1996) reported the peroxide levels of polysorbate 80 obtained from different manufacturers using a colorimetric titration method. Levels ranged from less than 1 to more than 27 mEq/kg. They found that peroxide levels increased upon storage at ambient temperatures, probably due to headspace oxygen and/or the container/closure interface allowing ingress of air. Peroxides in polysorbate can result in oxidative degradation of proteins. Obviously, formulators need to screen sources of polysorbate 80 or other polymeric additives used in protein formulations for peroxide contamination and establish peroxide specifications for using the additive. Also, as a precaution, incorporation of an antioxidant can help to overcome the potential for nonionic surfactants to serve as oxidative catalysts for oxygen-sensitive proteins.

Studies on protein–surfactant interactions (Bam *et al.*, 1995; Katakam *et al.*, 1995) have used electron paramagnetic resonance (EPR) spectroscopy to determine the binding stoichiometry of the surfactant to the protein and, thus, the potentially optimal amount of surfactant to use to stabilize the protein against surface denaturation and other physical instability reactions. Protein/surfactant interactions are also discussed in Chapter 3.

8.2. Albumin

Serum albumin is a widely used stabilizer in protein formulations for minimizing protein adsorption to glass and other surfaces (Wang and Hanson, 1988). Albumin competes with other proteins for binding sites on surfaces, but why this is so is not clear. It also is used as a protectant in several lyophilized formulations and has stabilizing effects on other proteins (Wang and Hanson, 1988), yet the mechanism by which it is an effective stabilizer is not understood. Examples of commercial protein formulations containing albumin are found in Table IV.

Because albumin is a natural protein, concerns have been raised about potential contamination of albumin with BSE (Fischer, 1995) if, in fact, bovine serum albumin were to be used rather than human serum albumin. There are also concerns with respect to vial contamination of natural proteins. Epidemiological data collected in the United Kingdom suggest a link between BSE and the feeding to ruminants (e.g. cattle, sheep, goats) of animal feed containing protein derived from transmissible spongiform encephalopathy (TSE)-contaminated tissues. Additionally, the data suggest that exposure to BSE may account for outbreaks of Creutzfeldt–Jakob disease (v-CJD), a central nervous system disease, in humans in the United Kingdom. The formulator should review materials such as enzymes and excipients used in the manufacture of the drug substance and drug product to ensure that there are no concerns with potential TSE contamination. Examples of animal source materials include, in addition to albumin, also gelatin, glycerol, and polysorbate 80. However, it should be noted that synthetic sources of polysorbate 80 are available. Ideally, the use of synthetic versions of these materials would eliminate concerns over potential disease transmission. If animal source materials are used in the manufacture of the product, then assurance must be provided to show what steps are being taken to prevent transmission of BSE. This could include sourcing of material from BSE-free countries and/or using processing such as high temperatures and pressure to achieve inactivation of TSE agents. Several regulatory agencies, such as those of the European Union, the FDA, and the Therapeutic Drugs Administration (Australia), critically reviewed setting guidelines in this area (Federal Register, 1997).

8.3. Sugars and Polyhydric Alcohols

The formulation scientist can select from a wide variety of excipients that can be effective in stabilizing proteins against denaturation due to extremes in temperature. Mannitol (polyol), propylene glycol and polyethylene glycol (polyhydric alcohols), and sucrose, lactose, trehalose, and maltose (dissacharides) are examples. However, at the time of the writing of this chapter, no trehalose-containing protein products had been approved by the FDA. The most effective stabilizers are those that remain amorphous after freeze-drying. Mannitol tends to crystallize during freeze-drying, for example, and thus might not be the best choice as a stabilizer. On the other hand, sugars such as sucrose are usually very effective stabilizers, because sugars tend not to crystallize during freeze-drying. Wang (1992) reviewed the application of these excipients in stabilizing proteins that are subjected to high temperatures ($\sim 60°C$) for short periods of time to pasteurize the protein, particularly those derived from blood. Effects of these solutes in stabilizing proteins during freeze-drying are covered in the following section.

Sucrose has been implicated as the cause of renal failure after the administration of intravenous immunoglobin preparations (Winward and Brophy, 1995). One immunoglobin preparation that did not contain sucrose did not cause acute renal failure, whereas all other preparations with sucrose resulted in this problem. It was pointed out by Winward and Brophy (1995) that acute renal failure was a problem years ago in patients who had received infusions of sucrose when sucrose was used as an osmotic diuretic. It was recommended that sucrose-containing intravenous immunoglobin preparations be avoided in patients with preexisting renal impairment or if the administration of high daily doses is anticipated.

8.4. Cryoprotectants

Proteins are frozen for long-term storage and freeze-dried for increased stability, and can be exposed to freezing conditions inadvertently during distribution of the final product. Some proteins can be deactivated or denatured by freezing. Freezing can produce a "hostile environment" for a protein because of solute concentration, increase in ionic strength, shift in pH, and a loss of hydrophobic interactions that stabilize the native protein (Pikal, 2002). At low temperatures, *freeze* (Franks, 1988) or *cold* (Privalov, 1990) denaturation of proteins can be a problem. Protein instability under very cold conditions may be overcome by the use of certain additives, called cryoprotectants, that protect the protein during freezing.

Arakawa *et al.* (1991, 1993; Arakawa and Timasheff, 1982, 1984) showed that additives that stabilize proteins in solution are preferentially excluded from the surface of native proteins. Based on the use of equilbrium dialysis techniques to calculate interaction parameters between a protein and an additive in water (described by Arakawa *et al.* (1993), and references contained therein), a wide variety of solutes (Table VII) can serve as protein cryoprotectants. These include sugars (sucrose, lactose, and glucose), amino acids (glycine, alanine, and proline), amines (betaine and trimethylamine *N*-oxide), polyols (mannitol and sorbitol), and certain salts (ammonium, sodium, and magnesium sulfate). The presence of these additives in a protein solution creates a thermodynamically unfavorable environment, because the chemical potentials of both the protein and the additive are increased (Arakawa *et al.*, 1993). An entropically unfavorable state occurs when solute molecules are preferentially excluded from contact with the protein. There is a good correlation between preferential exclusion and the effectiveness of an additive to serve as a cryoprotectant. Thus, the native structure of the protein is stabilized, because denaturation of the monomer or dissociation of polymeric oligoproteins would lead to greater area of contact between the protein and the solvent and therefore make a thermodynamically unfavorable situation worse. If a solute binds to the protein, the chemical potential is decreased and the free energy of denaturation decreases, leading to increased stability of the denatured protein and decreased stability of the native protein. Solutes known to bind

Table VII
Common Cryoprotectants and Lyoprotectants for Proteins in Finished Dosage Forms

General Classification	Specific Examples
Sugars	Sucrose, trehalose, lactose, maltose
Amino acids	Glycine, arginine
Polyols	Glycerol, sorbitol, mannitol
Polymers	Polyethylene glycol, dextran, polyvinylpyrrolidone

to proteins in this manner include magnesium chloride, guanidine hydrochloride, and urea. Exceptions are known to exist; for example, polyethylene glycol (PEG) is strongly excluded from the protein surface, but may cause protein destabilization, not by denaturation, but by causing protein molecules to concentrate in smaller volumes and increasing potential for aggregation (Arakawa and Timasheff, 1987; J. C. Lee and Lee, 1987). PEG in dilute solutions of low molecular weights (1000–8000) can provide stable protein solutions by forming a water layer around the protein and separating molecules into soluble monomers (Currie and Groves, 1992). PEG generally is an excellent cryoprotectant (Carpenter and Crowe, 1988).

8.5. Lyoprotectants

Some proteins are not stable during the drying part of the freeze-drying process. During drying, the protein's hydration shell is lost, which can result in irreversible loss of biological activity of some proteins (Liu *et al.*, 1991; Pikal *et al.*, 1991a, b). Many effective cryoprotectants offer no protection during dehydration of proteins. This is because the preferential exclusion of solutes, so important when water is present (either in frozen or liquid solution), is not applicable when water is removed from the system. Two different mechanisms for lyoprotection have been proposed, the water-replacement theory (Arakawa *et al.*, 1993; Carpenter and Crowe 1989) and the vitrification theory (Franks, 1990). The water-replacement theory states that a good stabilizer essentially serves as a water substitute by hydrogen-bonding to the dried protein. Many stabilizers are sugars, which hydrogen-bond to water like proteins hydrogen-bond to water. As water is removed during freeze-drying, the sugar replaces water by forming hydrogen bonds to the surfaces of the protein, and keeps the protein in its native, folded state. The vitrification theory, on the other hand, proposes that lyoprotection by solutes is related to glass formation in both the freeze concentrate and the dried state where the protein and solute, both amorphous glasses, are immobilized together and cannot easily degrade. However, this does not explain why dextran, an amorphous glass-former, does not stabilize and in fact destabilizes human growth hormone (Pikal, 1991) and catalase (Tanaka *et al.*, 1991). More research is needed for reaching a final conclusion on the true mechanism of stabilization of proteins during drying and storage in the dry state.

Common lyoprotectants include mannitol, sugars such as lactose, maltose, trehalose, and sucrose, and amino acids such as glycine, histidine, and arginine. A recent study of lyoprotection of a monoclonal antibody showed histidine buffer at pH 6 to be an effective stabilizer against aggregation, with high concentrations of trehalose and sucrose providing additional protection (Andya *et al.*, 1996). Sucrose was found to be a superior lyoprotectant to trehalose in minimizing conformational changes during the freeze drying of hydrophobic proteins (Kreilgard *et al.*, 1996).

Effectiveness of these stabilizers in the solid state depends at least in part on the following factors:

- The glass transition temperature of the freeze-dried solid (Her and Nail, 1994; Izutsu *et al.*, 1993a; Jang *et al.*, 1995; Pikal, 2002).
- Moisture content (Ahlnek and Zografi, 1990; Bell, *et al.*, 1995; Hageman, 1988; Hsu *et al.*, 1988; Yoshioka *et al.*, 1993).
- The ability of the excipient to remain in the amorphous state (Arakawa *et al.*, 1991, 1993; Costantino *et al.*, 1998; Izutsu *et al.*, 1993b, 1994a).

In fact, these three factors are all interrelated, in that moisture acts as a plasticizer, lowering the glass transition temperature, which increases molecular mobility and promotes crystallization of amorphous solids.

Figure 4 (Izutsu *et al.*, 1993a) provides a good example of the importance of the crystalline versus the amorphous state of a formulation and its ability to stabilize a protein. The open symbols represent amorphous cakes and the closed symbols represent crystalline cakes. As long as the cake remained amorphous, particularly at higher concentrations of mannitol and phosphate buffer, the activity of β-galactosidase was preserved. However, once the mannitol crystallized, activity decreased. Costantino *et al.* (1998) also demonstrated this effect on the stabililization of recombinant human growth hormone, where bulk additives that did not crystallize (lactose, trehalose, cellobiose, methyl α-D-mannopyranoside) during and after freeze-drying were more effective stabilizers against protein aggregation than excipients that did crystallize (mannitol, sorbitol)

To summarize, in pharmaceutical dosage forms, most proteins are too unstable in solution to be marketed as ready-to-use solutions. Therefore, most commercially available protein products as well as proteins in

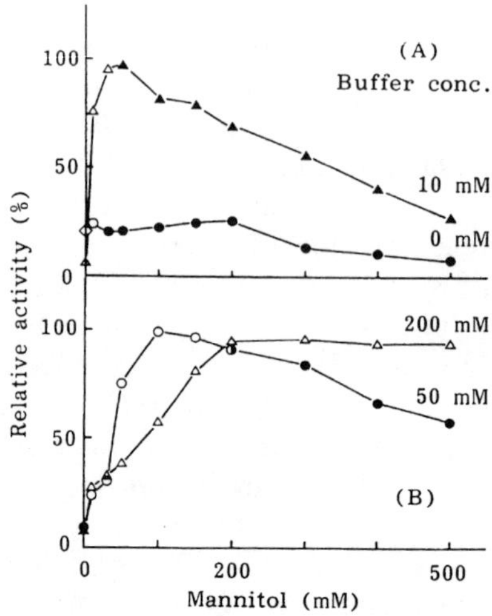

Figure 4. Effect of mannitol on the residual activity of freeze-dried β-galactosidase. Aqueous solutions of the enzyme (2 μg/ml) containing no (◇, ○, and ● in A), 10 m*M* (△ and ▲ in A), 50 m*M* (○ and ● in B), and 200 m*M* (△ and ▲ in B) sodium phosphate buffer solution were freeze-dried. The open and filled symbols represent atmorphous and crystallized cakes, respectively, analyzed by X-ray diffraction crystallization was observed in some samples). The open diamond denotes a collapsed cake which could not be analyzed by X-ray diffraction. The values are means of three separate experimetns. Reprinted with permissed from Izutzu *et al.* (1993b). Copyright 1993 Plenum Press.

development are freeze-dried for long-term stability. Stability problems potentially still exist for proteins in solution (both ready-to-use and after reconstitution) and in the solid state. Chemical instability problems (most frequently hydrolysis and oxidation) can be dealt with from a formulation approach by precise control of pH (buffers), use of antioxidants and chelating agents, storage at low temperatures, and control, of moisture content in the solid state (through processing and packaging). Physical instability problems (aggregation, precipitation, adsorption), occurring both in solution and solid states, can be minimized or prevented by employing additives such as sugars, amino acids, polyols, and surfactants.

9. ADDITIVES FOR ANTIMICROBIAL PRESERVATION

Antimicrobial preservative (AP) agents are required for parenteral products that are intended for multiple-dose use. Many protein products, because of their expense and course of therapy, are packaged as multiple-dose products, with the AP either formulated with the protein or, more commonly, formulated in a special diluent vial to be used to reconstitute a freeze-dried solid. The most common APs used in protein dosage forms are phenol, *m*-cresol, benzyl alcohol, and methyl and propylparaben. Examples of use of these preservatives are listed in Table VIII.

Use of antimicrobial agents requires passing an antimicrobial efficacy test (AET). Unfortunately, the USP and the British and/or European Pharmacopoeia (BP/EP) tests for AET have different acceptance criteria. Table IX summarizes the differences between the two tests. The USP requires a bacteriostatic preservative system, where as the BP/EP requires a bacteriocidal system. For example, the USP requires a three-log reduction in the bacterial challenge by the 14th day after inoculation, while criterion A of the BP/EP test requires the same three-log reduction within 24 hr. This difference in compendial requirements for antimicrobial preservative efficacy has caused many problems in the formulation of protein dosage forms. Passing the BP/EP antimicrobial efficacy test requires the use of relatively high amounts of phenol or cresol or other AP, which may have an impact on the stability of the formulation and could result in sorption of the preservative into the rubber closure. The formulator must keep in mind that increasing the concentration of APs may have a negative impact on protein physical stability (precipitation, aggregation, etc.). Increasing AP levels will increase the hydrophobicity of the formulation and could affect the aqueous solubility of the protein.

It is well known that APs not only protect insulin formulations against inadvertent microbial contamination, but also have a significant effect on stability of Lente formulations and prolong the duration of action of NPH

Table VIII
Antimicrobial Preservative Agents for Protein Products

Type	Concentration (%)	Antimicrobial activity[a] Gram +	Gram −	Fungi	Yeast	PH	Comment
Benzyl alcohol	0.1–3.0	+ + +	+ + +	−	−	3–6	Not effective pH > 7
Cresol	0.1–0.3	+ +	+ +	+ +	+ +	4–10	Most effective
Methylparaben	0.08–01	+ +	+ +	+ +	+ +	3–9	Slowly soluble
Propylparaben	0.001–0.02	+ + +	+ + +	+ + +	+ + +	3–9	Very slowly soluble
Phenol	0.2–0.5	+ +	+ +	+ +	+ +	4–10	Most effective
Thimerosal	0.1–04	+ +	+ +	+ +	+ +	4–8	Japan will not allow

[a] + + +, most effective; + +, moderately effective; −, poor.

Table IX
Comparison of USP 25 and EP 4th Edition Requirements for Antimicrobial Efficacy Testing

Time after Inoculation with microorganisms	Log reduction in microorganism count		
	USP 25	EP Criterion A	EP Criterion B
6 hr	Not required	3 (bacteria)	Not required
24 hr	Not required	No recovery (bacteria)	1 (becteria)
2 days	Not required	No recovery (bacteria)	Not required
7 days	1 (bacteria), no increase (fungi)	No recovery (bacteria), 2 (fungi)	3 (bacteria)
14 days	3 (bacteria), no increase (fungi)	No recovery (bacteria), no increase (fungi)	1 (fungi)
21 days	No increase	No increase	Not required
28 days	No increase	No recovery (bacteria), no increae (fungi)	No increase (bacteria), no increae (fungi)

formulations (Brange, 1987). Phenolic preservatives also have a profound effect on the conformation of insulin in solution (Wollmer *et al.*, 1987) and the self-assembly of the LysPro insulin hexamer (Birnbaum *et al.*, 1997). Preservation of Lente suspension with phenol is not possible without adversely affecting the crystal shape and physical stability of these formulations. NPH formulations, on the other hand, require phenol to obtain the desired crystal and subsequent prolonged duration of action. Methylparaben (methyl 4-hydroxy benzoate) is the preservative of choice for the Lente formulations, because it does not affect the crystal morphology while still providing sufficient antimicrobial protection. However, methylparaben is known not to be as effective an antimicrobial preservative as other preservative systems (Akers *et al.*, 1984; Allwood, 1982) and is absorbed to some extent into the rubber closure of Lente preparations (Brange, 1987).

Antimicrobial preservatives are known to interact with proteins and can cause stability problems such as aggregation. For example, phenolic compounds will cause aggregation of human growth hormone (Kirsch *et al.*, 1993, Maa and Hsu, 1996). Benzyl alcohol, above certain concentrations and depending on other formulation factors, will interact with recombinant human interferon-γ, causing aggregation of the protein (Lam *et al.*, 1997). These examples point out the need for the formulation scientist to understand the importance of potential effects of preservative type, and concentration and other formulation additives on the interaction with proteins in solution.

Table X
D Values against Staphylococcus aureus for Different Antimicrobial Preservative Systems in Insulin Solutions[a]

Antimicrobial preservative system	*D* (hr)
Phenol 0.2% + *m*-cresol 0.3%	0.5
m-Cresol 0.3%	0.6
Phenol 0.5%	0.8
Benzyl alcohol 2.0%	0.8
Phenol 0.2% + *m*-cresol 0.2%	1.3
Methylparaben 0.2% + benzyl alcohol 1.0%	1.4
Chlorobutanol 0.5%	1.8
Phenol 0.2% + *m*-cresol 0.1%	2.2
m-Cresol 0.2%	2.2
Methylparaben 0.2% + propylparaben 0.02%	3.0
Benzyl alcohol 1.0%	4.2
Methylparaben 0.1%	9.5
Methylparaben 0.1% + propylparaben 0.01%	12.3
Phenol 0.2%	16.2

[a]Reprinted with permission from Akers *et al.* (1984). Copyright 1984 American Pharmaceutical Association.

In determining the appropriate AP agent or agents, the model described by Akers *et al.* (1984) might be an appropriate model to follow. The authors used insulin as the protein to be preserved and combined insulin with different types of AP agents either alone or in combination. These formulations were challenged with the five USP antimicrobial efficacy test organisms and *D* values* determined. The *D* value determination allows a single quantitative estimate of the AP effectiveness of a certain agent or combination of agents in a specific formulation against a specific microorganism. An example of the *D*-value data obtained for insulin formulations with different AP systems against *Staphylococcus aureus* is given in Table X with the AP systems listed in order of effectiveness (e.g., 0.2% phenol + 0.3% *m*-cresol was the most effective AP system).

There are instances where a manufacturer, because of concerns regarding aseptic processing and sterility assurance of the product throughout its shelf life, will add an AP agent in the protein formulation even though it is intended only for a single-dose injection. This is very controversial. Regulatory agencies worldwide object to this practice if there

**D* value = time required for a one log reduction in the microbial population due to the effect of the antimicrobial preservative system. The smaller the *D* value, the greater is the effect of the preservative on the microorganism in question.

is opinion that the use of APs in single-dose injections is practiced in order to "cover up" for inadequate aseptic manufacturing practices and controls.

Many countries require that AETs be performed for routine stability protocols and for special stability studies. In addition there may be requests from agencies to do AETs on containers that have been used (i.e., penetrated; partial volume withdrawn) to demonstrate that the product can still kill microorganisms. In mid-1995, the Australian Drug Evaluation Committee (ADEC) passed resolutions that, in light of safety concerns with regard to contamination and cross-contamination, the use of injectable products in multidose packages is discouraged. In order to support the use of a multidose product and the shelf life once a package has been reconstituted or opened for use, preservative efficacy data are required for approval.

10. OTHER ADDITIVES

10.1. Osmolality (Tonicity) Agents

Salts or nonelectrolytes (e.g., glycerin) are added to protein formulations in order to achieve an isotonic solution. Nonelectrolytes often are preferred over salts as tonicity adjusters, because of the potential problems salts can cause with respect to protein stability (Pikal, 1990). Generally, solutions containing proteins administered intravenously do not have to be precisely isotonic, because of immediate effects from dilution by the blood. This cannot necessarily be the case for proteins administered by intramolecular and subcutaneous routes, because dilution of the injected product occurs less readily. Intrathecal and epidural injections into the cerebrospinal fluid require very precise specifications for the product to be isotonic and at physiological pH. This is because extremes in osmolality and/or pH can damage or destroy cells, and cerebrospinal fluid cells cannot be reproduced or replaced (Cradock *et al.*, 1977).

10.2. Bulking Agents for Freeze-Drying

These materials not only serve to give mass and pharmaceutical elegance to freeze-dried solids, but also may stabilize the protein during and after freeze-drying. Mannitol, the most commonly used bulking agent for freeze-dried formulations, tends to crystallize during freeze-drying (Pikal *et al.* 1991b) and will dominate X-ray diffraction patterns obtained from protein formulations containing mannitol (Oliyai *et al.*, 1994), provided that mannitol is at a sufficiently high concentration relative to other solutes. Sugars such as lactose and sucrose have been used as bulking agents in freeze-dried protein dosage forms. An example is the use of lactose in glucagon freeze-dried formulations.

10.3. Suspending Agents

Formulations for injectable suspensions generally require a suspending agent to enable the insoluble drug to be easily resuspended and allow the withdrawal of a uniform dose. Although there are not many protein suspensions commercially available, suspension dosage forms of insulin are the most widely used formulations of insulin. NPH insulin employs protamine as a complexing agent, which forms a suspension of insulin by stoichiometrically binding to insulin molecules (Brange, 1987). Lente insulin suspensions employ high levels of zinc ion to suspend insulin in solution. There are many other suspending or complexing agents known to form stable suspensions of insulin (Brange, 1987).

Bovine somatotropin (BST) is another example of a protein that is available as a suspension for once-a-month injection in cattle. However, unlike insulin, this suspension uses a vegetable oil (e.g., sesame oil) as the liquid phase.

10.4. Solubilizing Agents

Drug solubility may be improved by the use of solubilizing agents such as glycerin, propylene glycol, polyethylene glycol (300 or 400), and ethanol. Cyclodextrin derivatives have received considerable attention in the literature for their effective solubilization of a wide variety of molecules (Brewster *et al.*, 1989, 1991; Hora *et al.*, 1991, 1992; Rajewski and Stella, 1996; Stratton, 1991; Thompson, 1997). Cyclodextrins are crystalline, cyclic oligosaccharides derived from starch and have six (alpha), seven (beta) or eight (gamma) glucose units. Cyclodextrins are "bucket-like" toroid molecules having a rigid structure with a central cavity (Nash, 1994). The internal surface of the cavity is hydrophobic and the outer surface of the torus is hydrophilic. Thus, hydrophobic molecules can form an inclusion complex within the internal cavity and be rendered soluble in an aqueous solution. The cyclodextrin derivative having the most promise as an excipient in parenteral formulations is 2-hydroxypropyl-β-cyclodextrin. This derivative has been shown to increase solubility and stability of several proteins (Brewster *et al.*, 1989, 1991; Hora *et al.*, 1991, 1992; Simkins, 1991; Stratton, 1991). Relatively high cyclodextrin concentrations (10–20%) are required to stabilize these proteins. A relatively new and very promising cyclodextrin is an anionic β-derivative with a sodium sulfonate salt separated from the hydrophobic cavity by a butyl ether spacer group, Captisol™ (CyDex, Overland Park, KS). Captisol has been shown to increase drug stability (Jarvinen *et al.*, 1994) and solubility (Badawy *et al.*, 1995; Gorecka *et al.*, 1996; Okimoto *et al.*, 1995) of parenteral drugs including insoluble peptides (Johnson *et al.*, 1994). Thompson (1997) has written an excellent review of the safety evaluation and regulatory status of Captisol and other cyclodextrins. Safety of these excipients by the parenteral route of administration is still in question.

There is always a desire for newer excipients that can aid in solubilization, stabilization, and preservation of parenteral formulations. However, new excipients must undergo significant safety and regulatory hurdles. Expensive toxicological studies must be performed to convince drug manufacturers and regulatory authorities of long-term safety of the new excipient. Investment in these studies frequently presents a barrier to further study of new excipients.

11. PACKAGING

The packaging system is vital to successful marketing and use of a parenteral dosage form. All parenteral packaging components must be characterized according to appropriate compendial (USP, EP, Japanese pharmacopeia[JP]) requirements for glass, rubber, and/or plastic. These requirements include testing for reactivity with water and other solvents and safety evaluation of whatever is extracted into these solvents. Effect of sterilization methods (dry heat for glass; steam heat under pressure for rubber closures; gas or gamma radiation for plastic) on the integrity and potential leachables from packaging components must be determined. The International Conference on Harmonization (ICH) (Federal Register, 1994) stability guidelines require definitive stability studies to be performed in the final container/closure system, which has to have been thoroughly tested with the product formulation and found not to have any major physical–chemical stability issues.

All manufacturers of packaging materials are required to have a drug master file (DMF) on their components used in a registered product filed with the FDA (U.S. Food and Drug Administration, 1994). A DMF for packaging components is a submission to the FDA that contains information about the materials used for packaging for one or more human drug products. It allows detailed information about the packaging component to be referenced while maintaining confidentiality of the component or formulation. The pharmaceutical industry is not allowed to access these files. Sometimes this is unfortunate, because the packaging manufacturer may change a component of the container or closure formulation that is not divulged to the drug product manufacturer. Therefore, it is very important to determine lot-to-lot variability of packaging materials and to use good practices in purchased material quality control to assure that the package system originally selected always meets the desired specifications.

Because of their tendency to adsorb to surfaces, proteins can be difficult to stabilize in the appropriate package. As previously discussed, proteins are known to adsorb to glass, rubber and plastic surfaces, with insulin being the most studied protein for its adsorptive properties (Kraegen *et al.*, 1975; Mitrano and Newton, 1982; Petty *et al.*, 1974; Seres, 1990; Twardowski *et al.*, 1983a–c). Protein formulations may also react adversely with extractables from rubber and plastic formulations, although we could not find any published information on this possibility.

11.1. Glass

The USP categorizes glass into three types (see Table XI), differentiated by the amount of alkali leachables titrated from the glass when subjected to a high-temperature stress condition (USP, 1995, pp. 1781–1782). Type I borosilicate glass is the most resistant, least reactive (fewest potential leachables), highest quality, and most expensive glass available for injectable products and is normally the type of glass used for ready-to-use injectable solutions. For protein formulations, particularly those marketed as solution products, Type I glass should be used because of these properties. Type III soda-lime glass is of a lower quality as far as alkaline leachables are concerned, but normally is quite acceptable for solid state products such as freeze-dried preparations. Type II soda lime glass compositionally is the same as Type III glass, except that Type II glass receives a surface coating of sulfur dioxide which forms a sodium sulfate film, which can be easily washed off. This treatment renders Type II glass more resistant to alkali leachables, although it is not as pure as Type I glass. Type II glass is used for large-volume parenteral solutions.

Glass is prepared either by a process known as blow molding or by a process where the final glass container is formed from glass tubing. Blow molding processes usually are used to form glass for large vials (usually greater than 50 ml) and bottles, whereas tubing glass processes are used to make ampoules, smaller vials (<50 ml), syringes, and cartridges.

Table XI
Types of Glass Used in Protein Dosage Forms[a]

Type	General description	Type of USP test	Limits: Size (ml)	Limits: Amount of 0.02 *N* acid (ml)
I	Highly resistant, borosilicate glass Very little alkali leachables Used for most protein products Most expensive type of glass	Powdered glass	All	1.0
II	Treated (sulfur dioxide) soda-lime glass	Water attack	100 or less	0.7
	Sodium sulfate forms on glass because of reaction between sulfuric acid and sodium ions Used for large-volume bottles and acidic-to-neutral small-volume solutions		>100	0.2
III	Soda-lime glass About 10 times less resistant to alkali leachables than type I glass Used for dry powder small-volume products Cannot be terminally sterilized with final product	Powdered glass	All	8.5

[a]See *United States Pharmacopeia*, 24th ed. (2000), pp. 1930–1931.

Glass particulates are known to be a major source of particulate matter in parenteral solutions. Regardless of the quality of glass, the reputation of the manufacturer, the method of manufacture, or the method of cleaning, glass particulates are unavoidable. Glass particulates are generated most commonly during sterilization and depyrogenation of the glass containers (Whyte, 1983). Barium sulfate is a known Type I glass particulate contaminant in parenteral solutions formed from the leaching of the barium oxide component of glass and sulfate ions, the sulfate coming from either a sulfate salt or a bisulfite antioxidant excipient (Boddapatti *et al.*, 1980). Procedures for the isolation and identification of glass extractables are available (Parenteral Drug Association, 1986).

Proteins are packaged not only in glass vials, but also in glass cartridges and, potentially, in glass syringes. Normally, glass vials are not siliconized, but glass cartridges and syringes must be siliconized in order for the rubber-tip plunger rod to be moved easily through the lumen of the glass barrel. Studies must be done to assure that there is little or no interaction between the silicone on the glass and the protein or other formulation ingredients. For cartridge manufacture, the silicone is added before heat sterilization of the glass container, with the high temperature (usually above 250°C). Some believe that this will "chemically bond" the silicone to the glass, although personal communications with silicone oil manufacturers indicate that such bonding is physical, not chemical. In any case, there is always the possibility of excessive silicone not being completely bonded on the glass and thus the potential for "free" silicone to interact with the product.

11.2. Rubber

Rubber is used as the closure of glass vials and bottles, at the tip of the plunger in syringes, and the septum in glass cartridges, and can be a component of intravenous administration sets. Types of elastomeric material acceptable for parenteral products are listed in Table XII along with some general properties and other comments. Natural and butyl rubber are the two most commonly used elastomers for rubber closures of parenteral vials, bottles, and other containers. Elastomers are classified as saturated or unsaturated, with the degree of unsaturation affecting elastomer physical and chemical properties. Rubber closures and other rubber devices are composed of a wide variety of ingredients, examples of which are given in Table XIII.

Rubber closures are most commonly formed in steel molds under extremely high pressures (tons per square inch) and vulcanized (cured) at temperatures greater than 200°C. Mold surfaces where the molten rubber is placed are sprayed with an aerosolized release agent. This process is not perfect; gaps in the spray pattern will leave open areas where the rubber will stick tightly to the mold. When the rubber sheet is removed from the mold, tiny tears can occur in the rubber, which, in turn, can be the source of leachables and particulates. It is not known how prevalent this problem may be. However, the reality of this type of manufacturing problem causes some manufacturers to coat rubber closures with Teflon, silicone, nonsiliconized coatings such as HelvoPure (Helvoet Company), or Purcoat (West Company), or other polymeric material in order to minimize particulate and other potential incompatibilities with the product and improve the machinability of the rubber closure process. However, rubber closure coatings increase the cost of the closure, may not form as tight a container/ closure seal (unpublished observations; they can especially be a problem for

Table XII
Types and Selected Physical and Chemical Properties of Rubber Elastomers[a,b]

	Butyl/ halobutyl	Natural	Neoprene	NBR	Silicone	Fluoro elastomers	Urethane	EFDM	Polybutadiene
Chemical name	Isobutylene–isoprene copolymer	*Cis*-1,4,poly isoprene	Polychloro-prene	Butadiene–acrylonitrile copolymer	Poly-dimethyl alloxane	Fluro-rubber	Polyester isocyanate	Etylene propylene diene monomer	*Cis*-polybutadiene
Moisture vapor resistance	Excellent	Good	Fair	Fair	Poor	Good	Poor	Fair	Fair
Gas transmission resistance	Excellent	Good	Fair	Good	Poor	Good	Poor	Fair	Fair
Coring	Fair	Excellent	Good	Fair	Poor	ND	Excellent	Fair	Fair
Compression recovery	Poor	Excellent	Good	Good	Poor	Good	Excellent	Good	Good
Shelf life	Good	Fair	Good	Fair	Excellent	Excellent	Excellent	Excellent	Fair
Automatic handling ease	Poor	Good	Fair	Good	Fair	ND	Fair	Fair	Good
Heat resistance	Excellent	Good	Good	Good	Excellent	Excellent	Poor	Very Good	Good
Resistance to									
Water	Excellent	Good	Fair	Good	Excellent	Good	Poor	Good	Good
Animal oil	Excellent	Poor	Good	Excellent	Good	Excellent	Excellent	Fair	Fair to Poor
Vegetable oil	Excellent	Poor	Good	Excellent	Excellent	Excellent	Excellent	Fair	Fair to Poor
Mineral oil	Poor	Poor	Good	Excellent	Fair	Excellent	Excellent	Poor	Poor
Aliphatic solvents	Poor	Poor	Good	Poor	Poor	Excellent	Excellent	Good	Poor
Aromatic solvents	Good	Good	Poor	Good	Poor	Excellent	Poor	Fair	Poor
Chlorinated solvents	Poor	Poor	Poor	Poor	Poor	Excellent	Good	Poor	Poor
Dilute Acid	Good	Good	Good	Good	Fair	Fair	Poor	Good	Fair
Alkali, dilute	Good	Good	Good	Good	Good	Good	Poor	Good	Fair
Abrasion resistance	Fair	Good	Fair	Good	Fair	Good	Excellent	Good	Fair
Resillence	Poor	Excellent	Good	Good	Good	Fair	Good	Good	Good
Ozone resistance	Excellent	Poor	Good	Fair	Excellent	Excellent	Good	Good	Fair
Radiation resistance	Fair to Poor	Good	Good	Good	Fair to Good	Fair to Good	Fair	Fair	Poor

[a]Reproduced with permission from Smith and Nash (1992). Copyright 1992 Marcel Dekker, Inc.
[b]Ratings expressed in this guide are average values for typical rubber formulations using these base elastomers. Ratings can vary significantly with specific rubber

Table XIII
Additives in Rubber and Plastic formulations

Rubber	Plastic
Curing (vulcanizing) agent: forms crosslinks to shape rubber, e.g., sulfur, zinc oxide, phenolic resins, peroxide	Lubricant: improves processibility, e.g., stearic acid, paraffin waxes, polyethylene waxes
Accelerator: increases curing rate, e.g., 2-mercaptobenzothiazole, amines, thiols, ureas	Stabilizer: retards degradation, e.g., epoxy, organotions, mixed metals
Activator: increases efficiency of accelerator, e.g., zinc oxide, stearic acid	Plasticizer: enhances flexibility, resiliency, melt flow, e.g., phthalates
Antioxidant: resists aging, e.g., phenol, diethiocarbamates, paraffin waxes	Antioxidant: prevents oxidative degradation, e.g., butylated hydroxy toluene thioesters
Plasticizer: aids shaping process	Antistatic agents: minimizes surface static charge, e.g., quaternary ammonium compounds
Filler: modifies hardness, e.g., carbon black, chromium oxide, iron oxide	Slip agents: minimizes coefficient of friction
Lubricant: paraffin, mineral oil, fatty oil	Dye, pigment
Pigment	

vacuum-sealed or gas-flushed vials), and may themselves be sources of particulates (DeLuca and Knapp, 1993).

Qualifying a rubber closure formulation for use as the primary closure for a protein product is an aspect of product development that should not be taken lightly. Physical–chemical properties of the elastomer itself, such as water vapor transmission, hardness, and oxygen transmission, must be known. Usually butyl rubber derivatives have properties more suitable for labile proteins, for example, low water vapor transmission coefficient. Rubber leachate toxicities must be determined according to compendial tests. Besides generation of particulates and coring, other mechanical properties that need to be known include seal integrity, hardness, penetrability or insertion force (force required to penetrate the rubber with a needle), and resealability (ability of the rubber to reseal after needle puncture, vitally important for multidose vial products). Examples of protocols for such tests can be found in a bulletin published by the Parenteral Drug Association (1982).

Formulation scientists must conduct comprehensive preformulation screening studies in order to select the most compatible rubber closure formulation for the protein product being developed. Screening studies should include accelerated stability studies with various closures, including one or more coated closures, at high temperatures, exposure to light, and positioning of vials in inverted positions. By using stability-indicating assays both for the protein and, if present, antimicrobial preservative agent, primary and alternative rubber formulations can be selected from these preformulation studies for long-term stability studies.

Some rubber formulations are known both to contain components that *leach* into a pharmaceutical product (e.g., zinc, 2-mercaptobenzothiazole) and to adsorb components from the product formulation (*sorption*). *Permeation* of water vapor and other gases through rubber materials can sometimes readily occur. Rubber may also be a primary source of particulate matter and may be "cored" (i.e., a piece or pieces of rubber material gorged out by the needle) easily if the method of penetrating the rubber with a needle is not done carefully. These potential problems when using rubber closures are discussed in more detail below.

11.2.1. LEACHING

There are several types of substances which can leach from rubber closures into pharmaceutical solutions. Metal ions, such as zinc from cure activators (Boyett and Avis, 1976) and aluminum from fillers (Milano *et al.*, 1982), have been found to be common extractables. Phosphate buffer salts used in parenteral formulations can complex zinc and aluminum extractables and form insoluble precipitates. Other potential extractables from rubber formulations include other metals (aluminum, calcium, magnesium, cadmium, lead, mercury), sulfur from the curing agents, accelerators such as mercaptobenzothiazole (Danielson *et al.*, 1984), carbon black, lubricants (stearic acid), antioxidants, and various plasticizers. Latex leachables, although predominately originating from natural rubber latex gloves, catheters, and condoms, may also leach from natural rubber elastomeric closures (MacCracken, 1996; Towse *et al.*, 1995; Vassallo *et al.*, 1995), although the level of extractables and the associated health risk remains to be determined. Standard operating procedures are available for formulation development scientists to extract, isolate, and identify extractable materials from rubber closures (Parenteral Drug Association, 1980). To minimize leaching, butyl and halobutyl rubber formulations are the best choice, although natural rubber imparts desirable mechanical properties such as resealability after multiple needle punctures. It is important to keep in mind that any type of elastomeric formulation used in pharmaceutical packaging systems has the potential for producing leachable materials.

Stability studies of the final product in the final container/closure system must include data from samples stored in the inverted position in order to allow maximum contact of the product with the rubber closure to be sure that either product adsorption or rubber leachables will not occur to a significant extent. This also simulates what may happen during shipping and long-term storage of a commercial product when the container may fall over and the solution come into contact with the rubber surface for a long time.

11.2.2. SORPTION

Formulation components, particularly the active drug substance and the antimicrobial preservative, can be adsorbed (surface adsorption), even absorbed (migrate through the rubber) by rubber materials. Sorption depends on a number of factors, including physical and chemical characteristics of the drug (molecular weight, concentration, partition coefficient), drug product (composition, pH, surface tension, volatility), and rubber material (hydrophobicity, surface roughness, surface charge), temperature, extent of product–rubber surface contact on storage, and composition of formulation ingredients. The authors' personal experiences indicate that antimicrobial preservatives are the most difficult components of the product formulation, including the protein itself, to stabilize against rubber sorption. Phenol, *m*-cresol, and benzyl alcohol are especially troublesome. Prevention of some loss of antimicrobial preservatives over product shelf life is almost impossible. Therefore, the formulator must determine how much preservative can be lost over the shelf life of the product yet still pass the appropriate compendial preservative efficacy test(s).

11.2.3. PERMEATION

Selection of rubber closures depends on the water vapor and oxygen transmission coefficients of these materials. Freeze-dried protein solids must be protected from exposure to water vapor in order to maintain protein stability. Rubber formulations known to have low water vapor transmission coefficients (butyl and halobutyl) are the best choices for freeze-dried vials. Specific data on water vapor transmission of a particular rubber formulation can be obtained from the manufacturer. Rubber formulations that minimize water vapor transmission can still be sources of water adsorbed from the closure washing process or the closure sterilization process. Obviously, it is imperative to dry rubber closures as thoroughly as possible in order to avoid moisture problems originating from the washing and sterilization process. Coating rubber closures with a hydrophobic surface like Teflon may aid in preventing absorbed water from migrating from the rubber closure into the product, although Teflon itself also is porous. Another, less costly, alternative is to use rubber formulations (see Table XII) that do not readily absorb water that with time might be desorbed into the product.

11.2.4. PARTICULATE MATTER

Rubber closures can be a source of particulate matter in one or more of several ways. If closures are not cleaned properly, foreign particulates will be carried into the final packaged product and can be loosened from the closure surface into the solid or liquid product. Specially engineered rubber closure washing machines must be used, which are designed to wash and rinse closures with minimal abrasion and directing drainage of solutions away from clean closures (Huber, Capsolut, DCI Smeja Pharma-Technik). Particulates can be generated from rubber closures in the form of leachable materials that may interact with components of the protein solution to form insoluble precipitates. Finally, particulates can occur from rubber closures when needles are inserted inappropriately causing rubber pieces to be gouged out and to fall into the product solution. Practitioners are trained to insert needles with the "bevel edge up" in order to minimize the potential for coring.

11.2.5. LATEX TOXICITY

Natural rubber latex has been under regulatory scrutiny recently because of increased concern regarding latex hypersensitivity reactions in humans (Lang, 1996; Slater, 1994; Wilkinson and Beck, 1996). The *Pharmacopeial Forum* (1996) proposed a ban on natural rubber in elastomeric closures effective November 15, 1999 (it was later rescinded). The source of the allergic reaction is certain proteins in the rubber tree, *Hevea brasiliensis*. Items such as rubber gloves, catheters, and other thin-walled items, which are the primary sources for latex rubber exposure to humans, are prepared from a liquid form of natural rubber ("latex") by a dipping process. Elastomeric closures are made from a "hard rubber" by a molding process. It is believed that the latex protein(s) causing the allergic reactions are more readily extractable from items made using the dipping process. Since elastomeric closures are not prepared in this manner, there is no evidence that latex reactions are originating from natural rubber closure formulations. However, scientifically sound studies on the health risk associated with natural rubber latex extractables from pharmaceutical closures are needed.

11.3. Plastic

Plastic or polymeric materials are used in many different aspects of manufacture and use of protein dosage forms. Process tubing, syringes, the container system (e.g., plastic bottles or bags), filters, and other devices are all composed of polymeric materials. Plastic surfaces have the potential to adsorb protein or leach out impurities when the protein solution and the polymeric material are in contact with the surface. Plastic impurities may be unpolymerized monomers of the plastic material, many of which have known human toxicities (e.g., vinyl chloride, acrylonitrile, styrene, and benzene) (Liebe, 1995). Like rubber closures, plastics contain many additives, such as lubricants, stabilizers, plasticizers, antioxidants, antistatic agents, slip agents, and dyes, any or all of which potentially can form leachable material in the protein product it holds.

With the advances in form-fill-seal technology, where sheets of plastic are sterilized and formed into the final container, and the container is filled with product and sealed, all within the same apparatus, future protein products may be packaged in plastic containers more routinely than in glass containers.

Insulin binding to plastic is well known. Hirsch *et al.* (1977, 1981) were the first to report on the clinical significance of insulin adsorption to the internal surface of poly(vinyl chloride) (PVC) intravenous tubing. Insulin also binds to polyolefin infusion containers, although the binding is not as great as to PVC surfaces (Hirsch *et al.*, 1981). Insulin and other proteins will bind to some extent to almost any type of plastic surface; therefore, formulation scientists must take precautions to ensure that such binding can be prevented or minimized. For process tubing protein adsorption can be minimized by the following procedures:

- Prerinsing the tubing with the formulation to saturate potential sites
- Using tubing that has minimal binding properties
- Using tubing only once, to avoid the development of areas that can serve as "seeds" for protein binding and particulates if tubing is used again and not cleaned properly

Usually it is not worth the cost of cleaning validation to reuse tubing. For plastic containers containing proteins, protein adsorption can be minimized by adding albumin or nonionic surfactants to compete for the binding sites or using a plastic container material that has lower surface hydrophobicity.

Table XIV summarizes the types of plastic packaging materials currently available along with comments about their advantages and disadvantages.

Table XIV
Plastic Materials Used with Proteins in Processing and Packaging

Material	Advantage	Disadvantage	Application
Polyvinylchloride	Inexpensive	Plasticizer migration	In-process tubing, finished product bags
Polyethylene	Inert, no additives	Fragile	Disposable syringes, IV infusion bags
Polypropylene	Rigid	Too hard	Disposable syringes, IV infusion bottles
Ethyl vinyl acetate	Inexpensive	Poor sterilization props	Infusion bags, plastic tubing
Polyurethane	No migration, strong	Darkens easily	Short-term implants
Silicone	Resilent	Poor bonding	Pump tubing, in-process tubing

Table XII, already referenced for rubber closures, also provides information about typical formulation ingredients of plastic materials.

Formulation scientists must understand how their products will be used commercially. If the product is not going to be injected directly from its primary package, but instead be (a) diluted or admixed with intravenous diluents in glass or plastic containers, or (b) infused through plastic intravenous administration sets, filtered with in-line intravenous filters, and/or placed into plastic syringes, then data must be generated for understanding the compatibility of the protein formulation when in contact with these different products and surfaces. Preparation of dilute protein solutions for infusion may require an agent such as human serum albumin to minimize protein binding to the surface of the large container.

Sterilization of plastic materials most commonly is accomplished by either ethylene oxide gas or gamma irradiation. Tests must be done on sterilized polymeric material to determine potential leachables. For example, gamma irradiation of ethylene vinyl acetate will produce acetic acid leachate, which could affect the final pH of the administered product. Gamma irradiation of plastic material also forms free radicals, which may cause discoloration of certain plastics and affect cosmetic appearance as well as ability to see evidence of particulate matter.

11.4. Silicone

Silicone generally is required to lubricate the rubber closure to provide surface lubricity such that closures or stoppers will not "clump" during and after autoclaving and will "flow" in high-speed stoppering equipment. Silicone coating may also facilitate the proper fit of closures into the necks of vials, thus increasing the integrity of the container/closure interface. Insufficient surface lubricity can cause "clumping" or "twinning" of stoppers resulting in a shutdown of the filling line. Not only can this be costly, it can also jeopardize assurance of sterility, because of human intervention on the filling line to remove poorly siliconized or nonsiliconized stoppers. Glass vials of suspension products sometimes may be siliconized to facilitate product drainage and help to assure uniformity of the withdrawn dose. Silicone also is used in cartridge delivery systems where a rubber plunger must be pushed easily inside the glass barrel to accurately deliver the right volume of solution and dose of protein. Quantities of silicone applied to rubber closures are considered extremely low with respect to potential chemical and biological incompatibilities of this relatively inert material (Riffkin, 1968).

Silicone represents a hydrophobic surface that can cause problems with some proteins. Glass surfaces coated with silicone, originally in the expectation of decreasing protein adsorption to glass, in fact were found to increase the adsorption of a variety of proteins, including insulin, globulin, and lysozyme (Mizutani, 1981). Only albumin did not bind to the siliconized glass surface as much as it did with nonsiliconized glass, an observation leading to the use of albumin in minimizing adsorption of other proteins to glass.

11.5. Delivery Devices

Convenience of use drives technology development for protein pharmaceutical products in two important ways: development of more convenient product presentations (e.g., insulin cartridges and pens) and development of sustained-release formulations (e.g., microspheres). Cartridges and delivery pens are becoming a primary administration system for injection of insulin (Fig. 5).

Delivery devices greatly aid in the convenience, ease, and accuracy of administration of drugs by injection. However, there are many issues that must be resolved concerning the compatibility of the drug product and the device, the reproducible use of the device (unless it is a disposable system), and manufacturing quality concerns whereby the device needs meet to strict performance criteria lot after lot. For example, with insulin device delivery systems, stability of insulin in the device must be demonstrated, because the device normally is stored and used under

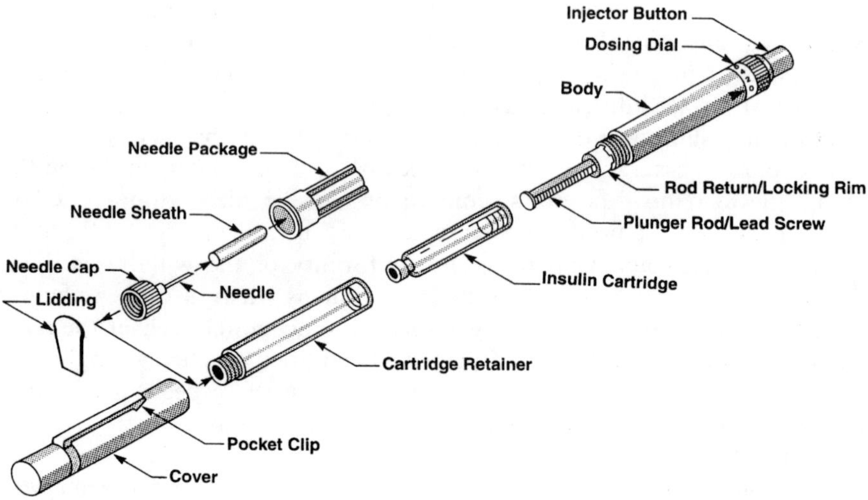

Figure 5. A cartridge-pen delivery system for insulin and other protein formulations.

ambient temperature conditions, whereas insulin in vials is stored refrigerated. Repeated use of the same device with a large number of insulin cartridges must assure that dose accuracy and delivery does not change over time. Compatibility of the needle system of the device with the cartridge rubber septum must be demonstrated. Finally, if suspensions are administered by these devices, procedures must be in place and data available to validate that resuspension and dose homogeneity are achieved.

11.6. Container/Closure Integrity

Increased concerns in recent years have focused on the integrity of the container/closure interface of sterile products in order to maintain sterility of the product over its shelf life. A number of methods are available for detecting and quantitating leakage in sterile product packaging systems (Chrai *et al.*, 1994; Guazzo, 1994; Morton *et al.*, 1989). A brief description and advantages and limitations of each of these methods are listed in Table XV. The microbial challenge test continues to be the most commonly used leak test for container/closure integrity. Newer methods, such as the application of mass spectrometry for leak detection of sterile packaging systems, are generating significant interest from quality assurance experts in industry and from regulatory agencies. Kirsch (1997a,b,c) reported excellent correlation between leakage rates (0.00000001–0.05 cm^3/sec of helium as measured by mass spectroscopy) and the microbial challenge test results. The critical leak rate, where samples started failing the microbial challenge test, was approximately 0.000003 cm^3/sec, corresponding to a pinhole leak in the range of 0.1–0.3 μm.

12. INTERFACE BETWEEN FORMULATION DEVELOPMENT AND MANUFACTURING

All unit operations—compounding, mixing, filtration, sterilization, filling, stoppering, freeze-drying, and in-process storage—can affect protein quality and stability. In fact, the purpose of using some of the stabilizers discussed previously is to protect the protein not only during shelf life, but also during its manufacture. Cryo- and lyoprotectants are examples of stabilizers used to protect proteins from degradation during freeze-drying procedures. Surface-active agents and albumin help to minimize binding of proteins to surfaces during manufacture. It has already been emphasized in this chapter that one cannot separate formulation from processing to obtain a quality-finished product having all the current good manufacturing practice (cGMP) values (safety, identity, strength, purity, and quality).

12.1. Front-End Analysis of Manufacturing Methods

Early inclusion of personnel representing manufacturing technical services and quality control/quality assurance (QC/QA) are desirable for development of a robust manufacturing process, methods for in-process and final product quality control, and effective technology transfer to manufacturing. For proteins, development issues which may be encountered during processing (chemical, physical, and microbiological) should be addressed in the design and evaluation of the biopharmaceutical manufacturing process. For example, few proteins can withstand the rigors of terminal sterilization; thus, effective aseptic filtration sterilization procedures must be developed. Factors such as viscosity can be critical in filtration, especially if the product must be maintained at subambient temperatures. In-process excursions of temperature, pH, and agitation also need to be considered. Furthermore, exposure to trace reactive substances

Table XV
Container/Closure Integrity Methods[a]

Test	Basic principle	Advantage	Disadvantage
Visual inspection	Look for leaks	Simple Inexpensive	Insensitive Operator-dependent Qualitative
Bubble test	Submerge package in liquid, pressurize and/or temperature cycling to accelerate leakage, improvement in sensitivity	Simple Inexpensive Location of leaks can be observed	Relatively insensitive Operator-dependent Wets package seal Qualitative
Pressure vacuum decay	Change in pressure or vacuum is measured inside the package or outside in a sealed package chamber	Relevant to package performance Quantitative Clean, nondestructive	Sensitive to temperature and, atmospheric pressure
Dye tests	Movement of dye across a seal is visually or instrumentally detected	Widely accepted in the industry Easy Inexpensive No special equipment required for visual detection	Qualitative Destructive Slow
Chemical tracer tests	Solution containing a tracer chemical is applied to one side of package seal; pressure or vacuum is applied as a driving force; chemical leakage detected by appropriate instrument	More sensitive and quantitative than dye test Correlates to liquid leakage Operator independent	Destructive
Microbial challenge	Containers are media-filled and the seal is either challenged directly with microorganisms or is allowed to sit in ambient storage environment; presence of microbial growth is visually confirmed	FDA-condoned No special equipment required	Insensitive Media fills are expensive due to production down time and labor costs

Table XV
(*continued*)

Test	Basic principle	Advantage	Disadvantage
			Ambient storage challenge requires large storage area
Weight change	Container is filled with liquid or dessicant, sealed, stored at various stress conditions, and reweighed over time	Easy Directly relates to closure performance Quantitative Inexpensive	Time-consuming Leak location not detected
Helium mass spectrometry	Helium is place either inside or outside of the container; vacuum is applied to seal interface and migrating helium is detected by mass spectrometry	Inert gas Extremely sensitive test Rapid test time Quantitative	May confuse helium diffusion with leakage
Gas detection tests	Test tracer gas is placed on one side of container seal; inert carrier gas is passed along opposite seal side; tracer gas is detected either by a coulombic detector (oxygen) or by a photoelectric sensor (water or carbon dioxide); instruments designed to pierce containers and test package headspace for oxygen or carbon dioxide are another type of gas detection method	Directly relates to package performance Does not pick up false leaks as helium detection can	Slow Often fixture-dependent
Vacuum-ionization test	High-voltage, high-frequency field is applied to vials sealed under vacuum; the field causes residual gas to glow; glow intensity is function of vacuum level	On-line nondestructive test Rapid	Some protein decomposition has been reported
Electrical conductivity/capacitance tests	High frequency, high voltage is applied to seal; increase in conductivity correlated to presence of liquid along the seal	Both permit 100% inspection Not operator-dependent Clean Nondestructive	On-line capacitance test not yet available

[a]See Guazzo (1994).

such as sanitizing agents (formaldehyde, hydrogen peroxide, phenolic compounds) has been shown to cause protein degradation (Kirsch *et al.*, 1993). Quality of excipient raw materials can have a significant impact on protein product stability. A recent example was provided by Dubost *et al.* (1996), who found that reducing sugar impurities in mannitol were responsible for the oxidative degradation of a cyclic heptapeptide. Physical or chemical stability can be affected at several points during the process (freezing step or primary drying of freeze-drying process), as demonstrated for proteins such as hGH (Pikal *et al.*, 1991a,b).

"Seamless" manufacturing of proteins is a relatively new technology that should be considered for proteins that cannot be crystallized and where the amorphous form is exceedingly difficult to handle. In seamless manufacturing, the protein "bulk solution" is not stored, but is processed continuously from the bulk production facility directly to the finished dosage form manufacturing area. In situations where continuous processing cannot be done, proteins may be isolated in bulk in frozen granules by a process called "cryogenic granulation" (Ryan *et al.*, 1995; Schmidt and Akers, 1997). Cryogenic granulation or cryogranulation is a process whereby a solution containing a protein is poured onto a stream of liquid nitrogen, where the droplets rapidly freeze and are carried through a series of descending helical trays and deposited as small, discrete frozen pellets.

Selection of equipment will be dictated by the formulation, the critical product attributes, the primary packaging, the production scale, and the site of manufacture. For example, for those drug substances which are heat-labile, it may be necessary to control the temperature during manufacture. Certain proteins and/or excipients may also require special equipment considerations such as glass versus 316 stainless steel. If oxidation is a problem, equipment will be needed to allow for manufacture in an oxygen-free environment and, possibly, protection from light. Mixing of proteins during the compounding stages must be accomplished without generating foam; therefore, special mixing apparatus must be designed to mix thoroughly without excessive foaming.

During this front-end analysis, a preliminary control strategy should be developed. The strategy encompasses the critical product quality attributes and the means by which these attributes can be assured during manufacturing, shipping, distribution, and product use. The control system which is ultimately in place will typically have the following key elements:

1. Release and expiry specifications (or retest intervals) for drug substance, excipients, drug product, and packaging components.

Specifications need to reflect USP, EP, and JP monographs. Preliminary specifications should be based on previous experience (laboratory and clinical scale) and should be set to control the product, but be sufficiently flexible to allow for uncertainty with a process under development.

2. Methods required to evaluate and characterize the protein active pharmaceutical ingredient and the formulation for potency, purity, impurities, and safety. The rationale behind selection of each method should be based on scientific principles. The methods used depend on the complexity of the protein.
3. In-process controls, according to which, samples are routinely collected during the manufacture of the clinical materials. For example, samples may be collected as a function of mixing time, before and after sterile filtration, and at different holding times in the process. Each process step should be evaluated with respect to criteria for forward processing (CFP), proven acceptance ranges (PAR), and the methods for analysis.
4. Instructions for handling product during shipping and distribution. Studies will need to be conducted to evaluate sensitivity to vibration, special handling, aberrant temperature conditions such as freezing or elevated temperatures, and high relative humidities for solid-state products.

12.2. Critical Control Parameters for Protein Product Manufacture

In the development of a protein as a commercial product, process screening studies are essential for assessing parameters which may impact the chemical, physical, and microbiological stability of the product as well as overall manufacturability. Each unit operation in the manufacturing process should be evaluated in order to identify critical process parameters (CPP). Optimization studies should then be conducted to establish acceptance criteria (both target levels and ranges). A thorough understanding of the process and how it will perform over a broad range of conditions, including interaction effects among the variables, is critical in order to develop a robust process. The process, furthermore, has to be validated such that the product can be manufactured on a routine and consistent basis without the need for reprocessing or the occurrence of recurring manufacturing deviations. Data to support worst-case limits should be generated in the lab during the development phase. The development scientist must keep in mind that the scale of the process may influence results. For example, it is commonly assumed that development of the freeze-drying cycle changes greatly on going from lab scale to pilot scale to production scale. Ideally, experiments can be performed at pilot scale using equipment representative of the production process. In addition, the sequence of steps in the process, acceptable environmental conditions (temperature, humidity, air, etc.) for the unit operation, and acceptable excipient ranges should be evaluated.

For solutions, the following process conditions may impact product quality:

1. *Agitation* (rate, duration): High rates may result in undesired foaming, denaturation, aggregation, or oxidation. Kim *et al.* (1994) reported on the effect of high shear force on the α helix-to-β sheet conversion of insulinotropin resulting in a reduced solubility of the protein. Thus, agitation or mixing rates will affect the conformation and solubility of this protein, thus requiring special control of these rates.
2. *Compounding sequence:* Order of addition of excipients and drug substance must be determined. Normally, excipients are added and dissolved before drug substance is added (Harwood *et al.*, 1993). This allows sufficient time for excipients to dissolve while minimizing time in solution for unstable drug substances.
3. *pH adjustments:* Prolonged exposure to acid or basic pHs can cause protein degradation.
4. *Filtration:* A membrane is selected which offers low protein binding as confirmed by protein assays on the solutions before and after filtration to detect any losses (Hawker and Hawker, 1975). This is discussed in detail in Chapter 5.
5. *Filtration and filling:* The temperature may need to be controlled to prevent chemical degradation during filling. Rates of filtration and filling should be considered to prevent shearing of the protein, although this possibility is remote. Effects of filtration and filling rates were studied for human growth hormone (Hsu *et al.*, 1988; Pikal *et al.*, 1991b). Shear forces encountered during processing were found not to cause aggregation of human growth hormone. However, aggregation of proteins during filtration and filling can be caused by interactions of the protein with polymeric hydrophobic surfaces, such as those composing sterilizing filters and process tubing for filling equipment, suspected to be the cause of human growth hormone aggregation during processing (Hsu *et al.*, 1988).

Insulin aggregation and fibrillation can be caused by interaction of the protein in solution with plastics (Thurow and Geisen, 1984) (see previous discussion on insulin binding to plastic surfaces).

6. *Freeze-drying:* Denaturation of a protein solution may occur because of pH shifts or ionic strength changes during freezing (Orii and Morita, 1977). The freezing rate, concentration, endpoint pH, and time and temperature of holding prior to lyophilization may affect the chemical and physical stability of the product (see Chapter 6).
7. *Environmental conditions:* Sparging of nitrogen into solution to remove dissolved gases or use of a nitrogen overlay to replace the air headspace during filling to retard the oxidation of oxygen-labile formulation may be required (Brown and Leeson, 1969).
8. *Materials compatibility:* During the manufacture, the drug is exposed to various materials such as stainless steel, filters, tubing, and pump diaphragms. It is important to ensure that the formulation components are compatible with these materials.
9. *Time and temperature:* Throughout manufacture, critical holding times and temperatures need to be established, not only for protein stability protection purposes, but also to assure microbiological control, particularly to prevent endotoxin contamination.

See Chapter 7 on quality assurance and quality control for additional details on this topic.

12.3. Clinical Trial Supplies

As the transfer of a process from laboratory scale into the clinical phase occurs, the development scientist needs to be involved during the early production of clinical trial supplies as well as during any subsequent formulation and process changes. The development scientist can obtain valuable insight into the conditions of manufacturing as well as provide key input to the manufacturing groups about the rationale for formulation and process decisions. These lots also provide an opportunity to collect information under larger scale manufacturing conditions than in laboratory-scale experiments. In-process testing should be routinely conducted during the manufacture of clinical supplies. The data are used to evaluate the process and lead to process improvements, which ultimately support the transfer into the manufacturing phase at full scale.

13. QUALITY CONSIDERATIONS DURING FORMULATION DEVELOPMENT

13.1. Early Product Assessment

All known physical and chemical properties of a protein should be viewed in light of how these properties will affect final quality of the finished dosage form. The development scientist should build quality into the process so that it can be validated and data exist to support "worst-case" limits (e.g., extremes in pH, excipient levels, time/temperature limits). Process design should be evaluated to assure control of quality parameters.

13.2. Documentation

An essential element of validation is establishing thorough documentation that a process will consistently meet its predetermined specifications and quality attributes. Ongoing documentation is essential during formulation development. Development history reports are required during preapproval inspections and are generally a high-level overview describing the history of the drug product from preliminary studies to the commercial formulation and the process submitted in the regulatory document. There may be one report or several to describe the various steps of development (e.g., fermentation, granule isolation, drug substance, drug product, methods development). It is as important to document what did not work as what worked and why decisions were made. Where necessary, bioequivalence of lots used in clinical trials should be demonstrated. A method history report should contain information on regulatory commitments during the development process and give the history of each method. Rationale for specifications, reference standards, and cleaning methods should be provided. For product development reports, lot rejections, deviations, and resolution of the issues need to be covered. Typical contents of a parenteral drug product development history report include the following:

- Preformulation data
- Selection of excipients supported by preformulation studies
- Antimicrobial characteristics of the product
- Rationale and basis for packaging component choices
- Description of container/closure integrity studies

- Overview of the manufacturing process with a description of each unit operation, acceptance criteria for all critical steps of the process, and a history of results for each step
- Scale-up studies
- Stability overview, including product use studies

Development history reports include, but are not limited to, method histories, formulation, primary packaging, process development, and control strategy.

Technical reports prepared on an ongoing basis aid in compiling the development history reports and are key in the information transfer to the manufacturing sites. Typical information contained in these periodic reports are the rationale for decisions, description of what worked and what did not work, and information for solving manufacturing problems.

13.3. Stability Studies

Stability testing of protein and peptide dosage forms should follow the ICH guidelines (U.S. FDA, 1996; ICH, 1995). Table XVI summarizes the requirements of this guideline. Three batches or more of the final product in the final container/closure system that represents the finished product manufacturing scale must be put on stability testing. The batches should use different lots of bulk material. A minimum of 6 months of data at desired storage conditions must be available at the time of submission (less than 6 months of data for products that will have less than 6 months dating). Product expiration dating is based on real-time data, not extrapolated from accelerated stability studies. If different volumes and/ or strengths of the same formulation are to be tested, a matrix system or bracketing may be permitted. Details of matrixing and bracketing are found in the guideline.

Stability testing must use methods that are stability-indicating and validated. Methods should monitor changes in potency, purity, and other product characteristics* as a function of time and storage conditions as defined in the stability protocol. Data on final product in containers maintained in an inverted or horizontal position and all different container/ closure combinations must be obtained. Multiple dose containers must have data to support stability during simulated use, for example, repeated

*Visual appearance, visible particulates in solutions. pH, moisture level of powders, sterility testing or container/closure integrity, degradation, if any, of additives.

Table XVI
Summary of Guideline for Stability Testing of Biotechnological and Biological Products[a]

Selection of batches	Guidance for stability studies for regulatory submissions
A. Drug substance	At least three batches representative of production scale A minimum of 6 months of stability data at time of submission (unless storage period will be less than 6 months) Pilot-plant-scale batch data acceptable at submission as long as there is commitment to place first three manufacturing-scale batches into long-term stability program after approval Storage containers should represent actual holding containers to be used during manufacture; containers of reduced size acceptable, provided they are constructed of same material and use same type of container/closure system
B. Intermediates	Identify intermediates and generate in-house data and process limits to assure final product stability
C. Drug product	At least three batches of final container product representative of final product at production scale Different batches of bulk material should be studied in final product A minimum of 6 months of data at time of submission (unless storage period will be less than 6 months) Product expiration dating based on real-time/real-temperature data Continuing stability updates should occur during review process Quality of final product must be representative of quality of material used in clinical studies Pilot-plant-scale batch data acceptable at submission as long as there is commitment to place first three manufacturing scale batches into long-term stability program after approval
D. Sample selection	Matrix system and/or bracketing is acceptable when product consists of different fill volumes, units, or mass

Stability-indicating profile	Guidance
A. Protocol	Include detailed protocol for assessment of stability of both bulk drug substance and drug product to support proposed storage conditions and expiration dating periods; include statistical methods, specifications, test intervals
B. Potency	Potency must be compared to an appropriate reference standard Perform at appropriate intervals as defined in the protocol and report in units of biological activity calibrated against some recognized standard
C. Purity/molecular characteristics	Purity should be assessed by more than one method Limits of acceptable degradation should be documented and justified, taking into account levels observed in material used in preclinical and clinical studies Usual methods include electrophoresis, high-resolution chromatrography, and peptide mapping

Table XVI
(continued)

Stability-indicating profile	Guidance
D. Other product characteristics	Visual appearance, visible particulates in ready-to-use or reconstituted solutions, pH, moisture level of solid products Sterility testing or alternatives (e.g., container/closure integrity) should be performed at a minimum initially and at the end of the proposed shelf life Additives such as stabilizers and preservatives that may degrade during the dating period of the product and might adversely affect product quality need to be monitored during stability
Storage conditions	**Guidance**
A. Temperature	Storage conditions for real-time/real-temperature stability studies may be confined to the proposed storage temperature
B. Humidity	As long as humidity-protected containers are used, stability tests at different relative humidities can be omitted
C. Accelerated and stress conditions	Studies should be conducted on the drug substance and drug product under accelerated and stress conditions to provide useful support data for establishing the expiration data, be used for support of change in formulation or scaleup, assist in analytical method validation, or generate information that may help elucidate degradation profile of the drug; other benefits include data on accidental exposure to stress conditions, e.g., during transportation, determine what test parameters are best indicators of product stability, and give further insight into degradation patterns not seen under normal storage conditions
D. Light	Appropriate regulatory authories should be consulted on a case-by-case basis
E. Container/closure	Stability studies should include samples maintained in inverted or horizontal position for product to have maximum contact with closure Data should be supplied for all different container/closure combinations that will be marketed Data must be generated that demonstrate that the closure used for multiple-dose application is capable of withstanding repeated insertions and withdrawals so that product retains full potency, purity, and quality
F. Stability after reconstitution of freeze-dried product	Stability of product after reconstitution should be demonstrated for the conditions and maximum storage period specified
Testing frequency	
Shelf life $\leqslant$ 1 year	Monthly testing for first 3 months; 3-month intervals thereafter
Shelf life $\geqslant$ 1 year	Every 3 months during first year of storage, every 6 months during second year; annually thereafter

[a]Summarized from FDA (1996). Also published in *Federal Register* on July 10, 1996 (61 FR 36466).

punctures of the closure, effect of volume change due to repeated withdrawals on potency, purity, and quality.

Stability testing intervals usually are every 3 months during the first year of storage, every 6 months during the second year, and annually thereafter.

Although accelerated stability testing is not acceptable for direct assignment and support of expiration dates for finished protein products, it does have a role in developmental stability studies. ICH stability guidelines emphasize the importance of conducting stability studies under accelerated conditions, because such studies usually provide useful supporting data for establishing the expiration date. Additionally, accelerated stability studies provide information for short-term assessment of proposed formulation, packaging, or manufacturing changes, evaluate equivalence of materials from different bulk process methods, assist in the validation of analytical methods for the stability program, and can generate information which may help elucidate degradation profiles of the protein alone and in the final formulation.

Accelerated stability studies also may be useful in determining whether accidental exposures to stress conditions during transportation of the final product are harmful to the product. The guidelines generally define accelerated stability test conditions at temperatures at least 15°C above the designated long-term storage temperature for the product.

Typically, kinetic experiments are conducted at elevated temperatures. Estimates of rate constants at lower temperatures are then obtained by extrapolation of an Arrhenius plot. An assumption of this approach is that the kinetics does not change as a function of temperature. For proteins, it is important to determine the appropriate temperature range for these stability studies (Yoshioka *et al.*, 1994). Ideally, it is desirable to obtain an approximately linear relationship to permit prediction of shelf life of the drug product. However, in cases where the results cannot be extrapolated, these data are still useful for determining the effects of short-term temperature excursions on the product and for specifying any special storage precautions that may be encountered during shipping and distribution.

Shnek *et al.* (1998) described two automated physical stress tests for evaluating physical stability of insulin suspensions and solutions. One is a moderate stress test combining temperature cycling (25°C to 37°C) and resuspenion by agitation, and the other is a more considerable stress test combining high temperature (37°C) and extreme agitation (4 h daily). These tests provide more relevant physical stability predictions for using suspensions and solutions in pen-cartridge devices and determining effects of stressful conditions these protein products could encounter during shipping, distribution, and patient use.

13.4. Stability Studies Supporting Distribution of Protein Products

Formulation development typically focuses on protein dosage form stability under well-controlled conditions. However, studies must be done to demonstrate that the final protein formulation in the final package produced by a validated process will retain its stability and other quality properties during distribution of the product throughout the world. These studies can be accomplished both by simulation in the laboratory and actual distribution of the product. During distribution, even if the product is to be held in controlled (e.g., refrigerated) conditions, aberrant situations can occur, such as due to transportation breakdown, mechanical failure, dropping or otherwise mishandling packages, and even overt violation of required handling procedures. Data should be available to aid in knowing what these extreme situations will do to the quality of the product. The effect of shear (i.e., agitation, mixing, and other mechanical forces experienced during processing and handling) and temperature extremes on protein stability during simulated and actual conditions should be evaluated.

14. EXAMPLES OF FORMULATION PROBLEMS

14.1. Aggregation

Protein aggregation, as already emphasized in this chapter, is a major problem in developing stable protein solutions.* Acidic fibroblast growth factor (aFGF) is an example of a protein that aggregates readily when exposed to temperatures above refrigeration (Tsai *et al.*, 1993). To solve this problem, the authors described a four-part approach:

1. Use a rapid screening procedure, based on turbidimetric measurement, to identify solution conditions, polyanions, and common excipients that stabilize aFGF against heat-induced aggregation.
2. Combine the most promising agents with aFGF and use circular dichroism and differential scanning calorimetry to quantitate stabilization effects against aggregation.
3. Formulate the most promising agents with aFGF in solution and monitor protein stability at room temperature.

*For more thorough treatment of protein formulation problems, the reader is referred to the two books edited by Wang and Pearlman (1993; Pearlman and Wang, 1996) on case histories involving stability and characterization of protein and peptide drugs.

4. Test *in vivo* efficacy of the best formulations of aFGF using a wound healing animal model.

This approach enabled the authors to learn that in addition to the well-known protection of aFGF by heparin, a surprisingly wide variety of polyanions (e.g., ATP, dextran sulfate, pentosan polysulfate, sulodexide, phytic acid, inositol hexasulfate, inorganic phosphate salts) stabilize aFGF against aggregation by increasing the temperature at which the protein unfolds by 15–30°C. Stabilization by polyanionic compounds occurs because these compounds enhance the structural integrity by binding to the protein. Common excipients also were studied for their stabilizing effects. Many were effective (e.g., glycine, glycerol, sucrose, dextrose, trehalose), but at relatively high concentrations. Their final formulation choices, all of which were equally capable of accelerating wound healing, contained either heparin, inositol hexasulfate, or sulfated β-cyclodextrin (with the latter two formulations also containing 1% hydroxyethyl cellulose in phosphate buffer with 0.2 m*M* EDTA). This is an excellent paper from which to learn approaches for solving protein aggregation problems.

14.2. Oxidation and Deamidation

Human growth hormone is a protein that both in aqueous solution and in the solid state can undergo chemical decomposition both by oxidation (methionine 14 to methionine sulfoxide) and deamidation (asparagine at position 149) (Becker *et al.*, 1987). This protein is also prone to aggregation. Pikal *et al.* (1991a, b) studied how to overcome these stability problems. Variables that they studied included pH, levels of salts, type of lyoprotectant excipient, residual water content, and oxygen level in the vial headspace. They found that a combination of glycine and mannitol enhanced stability, largely because glycine in this combination can remain amorphous. In the amorphous state glycine serves as a lyoprotectant and as a "sink" for residual moisture, both effects serving to stabilize the protein against deamidation and aggregation. The level of phosphate buffer is important as an aid in the minimization of deamidation, as a pH value on either side of pH 7 increased chemical decomposition. Increases in levels of sodium chloride increased the rate of decomposition.

Oxidation of human growth hormone is due to methionine oxidation. Oxidation occurred even when the air in the vial headspace was replaced with nitrogen (headspace $<1\%$ oxygen). Phosphate-buffered formulations containing glycine and mannitol as stabilizing excipients, although found to be overall the most stable of all hGH formulations studied (Pikal *et al.*, 1991a), were also found to be the most sensitive to changes in the oxygen level of the vial headspace. Because these formulations are amorphous due to the effect of glycine in the presence of mannitol, any dissolution of headspace oxygen in the amorphous phase can significantly affect protein oxidation. It was assumed by the authors that most of the oxygen supply in the formulated vial with nitrogen headspace came from oxygen "trapped" in the amorphous phase of the product during freeze-drying, the trapped oxygen originally being dissolved in the solution filled into the vial.

14.3. Binding to Glass

Early identification of the properties of surface adsorption to solid surfaces such as glass is critical to the development of packaging and delivery systems for proteins. As discussed in earlier sections, various formulation additives can be added to minimize or inhibit adsorption. Johnston (1996) examined the effects of solvent additives which would minimize the degree of adsorption to various surfaces for a model protein, recombinant human granulocyte colony stimulating factor (rGH-CSF). For glass vials, he examined solutions containing the following additives: 0.5% w/w Pluronic® F-127, 0.05% and 0.5% w/w Pluronic® F-68, 0.5% v/v glycerin USP, and 0.005%, 0.05%, and 0.5% Tween 20. A concentration of

Table XVII
Adsorption of Bovine Serum Albumin (BSA) and Bovine Immunoglobulin G (IgG) to Commercial 0.2-μm Microfiltration Capsules[a]

Membrane type and capsule manufacturer	Protein concentration μg/ml; (wt%)	BSA adsorption μg/cm²	IgG adsorption μg/cm²
Polyvinylidene difluoride, millipore	50 (0.0050)	—	1
	250 (0.025)	2	3
	1000 (0.10)	11	—
	5000 (0.50)	56	—
Nylon, Pall	50 (0.0050)	6	108
	250 (0.025)	31	126
	1000 (0.10)	31	—
	5000 (0.50)	102	—
Cellulose acetate, sartorius	50 (0.0050)	—	1
	250 (0.025)	5	5
	1000 (0.10)	17	—
	5000 (0.50)	56	—

[a]Reproduced with permission from Brose and Waibel (1996). Copyright 1996 Aster Publishing.

2 ng/ml rGH-CSF in D5W was used as a control. Each solvent additive was examined in triplicate at each concentration tested, and solutions were sampled at times 0, 0.5, 1, 2, and 3 hr and assayed for rGH-CSF. A rapid adsorption was observed up to 1 hr, after which time a slower rate approaching steady state was observed, similar to adsorption reported for insulins and other proteins. Furthermore, viscometry was applied to estimate that the thickness of the adsorbed layer to glass was about 1 μm. Tween 20 at a concentration of 0.5% showed the most potential for inhibiting surface adsorption to parenteral glass vials.

14.4. Binding to Filter Surfaces

Proteins are known to bind to filter surfaces. Brose and Waibel (1996) reported on the adsorption of BSA and sheep IgG to three types of commercial filter surfaces, polyvinylidene difluoride (PVDF, Millipore), nylon (Ultipor N66, Pall), and cellulose acetate (CA, Sartorius). They filtered 5-mg/ml solutions of proteins at pressures <1 psi and measured protein concentration in the filtrate as a function of filtrate mass. The mass of adsorbed protein was determined by measuring the difference between theoretical (5 mg/ml = 0.5 wt%) and actual protein weight percent. This mass was divided by the membrane surface area (500 cm^2) to obtain the amount of protein adsorbed per unit surface area. They found that both proteins adsorb to all filters, but the extent of adsorption is markedly affected by protein concentration and the type of filter (see Table XVII). Nylon filters adsorbed the most protein. BSA adsorption was linearly related to its concentration, with adsorption to nylon being twice as high as its adsorption to PVDF and CA membranes. IgG adsorption was linear, but very low to PVDF and CA filters, whereas its adsorption to nylon was extensive and nonlinear (it leveled off at higher concentrations). Other authors (Brose and Waibel, 1996) have also reported on protein adsorption on filter surfaces, so the phenomenon will occur, but the question is how much and how can the adsorption be minimized? The choice of filter, the concentration and type of protein being filtered, and the filtration conditions (pressure, rate, size of filter, etc.) all will affect the extent of protein adsorption to filters.

REFERENCES

Absolom, D. R., Zingg, W., and Neumann, A. W., 1987, Protein adsorption to polymer particles: Role of surface properties, *J. Biomed. Mater. Res.* **21:**161–171.

Ahlnek, C., and Zografi, G., 1990, The molecular basis of moisture effects on the physical and chemical stability of drug in the solid state, *Int J. Pharm.,* **62:**87–95.

Akers, M. J., 1982, Antioxidants in pharmaceutical products, *J. Parenteral Sci. Technol.* **36:**222–228.

Akers, M. J., Boand, A., and Binkley, D., 1984, Preformulation method for parenteral preservative efficacy evaluation, *J. Pharm. Sci.* **73:**903–907.

Allwood, M. C., 1982, The effectiveness of preservatives in insulin injections, *Pharm. J.* **229:**340–344.

Andya, J., Wu, S., Hsu, C., and Shire, S. J., 1996, The effect of sugars and buffer excipients on the stabilization of a lyophilized formulation of an anti-IgE humanized monoclonal antibody, *Pharm. Res.* **13:**S-78.

Anik, S. T., and Hwang, J. Y., 1983, Adsorption of D-Nal(2)6LHRH, a decapeptide, onto glass and other surfaces, *Int. J. Pharm.* **16:**181–190.

Arakawa, T., and Timasheff, S. N., 1982, Preferential interaction of protein with salts in concentrated solution, *Biochemistry* **21**, 6545–6552.

Arakawa, T., and Timasheff, S. N., 1984, Mechanism of protein salting in and salting out by divalent cation salts: Balance between hydration and salt binding, *Biochemistry* **23:**5913–5923.

Arakawa, T., and Timasheff, S. N., 1985, Mechanism of polyethylene glycol interactions with proteins, *Biochemistry* **24:**6756–6762.

Arakawa, T, Carpenter, J. F., Kita, Y. A., and J Crowe, J. H., 1990, The basis for toxicity of certain cryoprotectants: A hypothesis, *Cryobiology* **27:**401–415.

Arakawa, T., Kita, Y., and Carpenter, J. F., 1991, Protein–solvent interactions in pharmaceutical formulations, *Pharm. Res.* **8:**285–291.

Arakawa, T., Prestrelski, S. J., Kenney, W. C., and Carpenter, J. F., 1993, Factors affecting short-term and long-term stabilities of proteins, *Adv. Drug Deliv. Rev.* **10:**1–29.

Badawy, S. I. F., Ghorab, M. M., and Adeyeye, C. M., 1995, Study of the complexation between danazol and hydrophilic cyclodextrin derivatives, *Pharm. Res.,* **12:**S204.

Baker, R. W. (ed.), 1980, *Controlled Release of Bioactive Materials*, Academic Press, New York.

Bam, N. B., Cleland, J. L., Yang, J., Manning, M. C., Carpenter, J. F., Kelley, R. F., and Randolph, T. W., 1998, Tween protects recombinant human growth hormone against agitation-induced damage via hydrophobic interactions, *J. Pharm. Sci.* **87:**1554–1559.

Bam, N. B., Randolph, T. W., and Cleland, J. L., 1995, Stability of protein formulations: Investigation of surfactant effects by a novel EPR spectroscopic technique, *Pharm. Res.* **12:**2–11.

Becker, G. W., Tackitt, P. M., Bromer, W. W., Lefeber, D. S., and Riggin, R. M., 1987, Isolation and characterization of a sulfoxide and a desamido derivative of a dimeric form of biosynthetic human growth hormone, *Biotechnol. Appl. Biochem.* **9:**478–487.

Bell, L. N., Hageman, M. J., and Muroaka, L. M., 1995, Thermally induced denaturation of lyophilized bovine somatotropin and lysozyme as impacted by moisture and excipients, *J. Pharm. Sci.* **84:**707–711.

Birnbaum, D. T., Kilcomons, M. A., DeFelippis, M. R., and Beals, J. M., 1997, Assembly and dissociation of human insulin and Lys pro-insulin hexamers: A comparison study, *Pharm. Res.* **14**:25–36.

Boddapati, S., Butler, L. D., Im, S., and DeLuca, P. P., 1980, Identification of subvisible crystals of barium sulfate in parenteral solutions, *J. Pharm. Sci.* **69**:608–611.

Boyett, J. B., and Avis, K. E., 1976, Extraction rates of marker compounds from rubber closures for parenteral use. II. Mechanism of extraction and evaluation of select extraction parameters, *Bull. Parenteral Drug Assoc.* **30**:169–177.

Brange, J., 1987, *Galenics of Insulin*, Springer-Verlag, New York.

Brange, J., 1992, Chemical stability of insulin. 3. Influence of excipients, formulation, and pH, *Acta Pharm. Nord.* **4**:149–158.

Brange, J., and Langkjaer, L., 1993, Insulin structure and stability, in: *Stability and Characterization of Protein and Peptide Drugs: Case Histories* (Y. J. Wang and R. Pearlman, eds.), Plenum Press, New York, pp. 315–350.

Brange, J., Langkjaer, L, Havelund, S., and Volund, A., 1992a, Chemical stability of insulin. 1. Hydrolytic degradation during storage of pharmaceutical preparations, *Pharm. Res.* **9**:715–726.

Brange, J., Havelund, S., and Hougaard, P., 1992b, Chemical stability of insulin. 2. Formation of higher molecular weight transformation products during storage of pharmaceutical preparations, *Pharm. Res.* **9**:727–734.

Brennan, J. R., Gebhart, S. S. P., and Balckard, W. G., 1985, Pump-induced insulin aggregation, *Diabetes* **34**:353–359.

Brewster, M. E., Simpkins, J. W., Hora, M. S., Stern, W. C., and Bodor, N., 1989, The potential use of cyclodextrins in parenteral formulations, *J. Parenteral Sci. Technol.* **43**:231–240.

Brewster, M. E., Hora, M. S., Simpkins, J. W., and Bodor, N., 1991, Use of 2-hydroxypropyl-β-cyclodextrin as a solubilizing and stabilizing excipient for protein drugs, *Pharm. Res.* 8:792–795.

Brose, D. J. and Waibel, P., 1996, Adsorption of proteins in commercial microfiltration capsule, *Pharm. Technol.* **20**:48–52.

Brot, N., and Weissbach, H., 1982, The biochemistry of methionine residues in proteins, *Trends Biochem. Sci.* **7**:137–139.

Brown, M. and Leeson, L. J., 1969, Protection of oxygen-sensitive pharmaceuticals, *J. Pharm. Sci.* **58**:242–245.

Capasso, S., Mazzarella, L., and Zagari, A., 1991, Deamidation via cyclic imide of asparaginyl peptides: Dependence on salts, buffers, and organic solvents, *Peptide Res.* **4**:234–238.

Carpenter, J. F., and Crowe, J. H., 1988, The mechanism of cryoprotection of proteins by solutes, *Cryobiology* **25**:244–255.

Carpenter, J. F., and Crowe, J. H., 1989, An infrared spectroscopic study of the interactions of carbohydrates with dried protein, *Biochemistry* **28**:3916–3922.

Carpenter, J. F., Prestrelski, S. J., and Arakawa, T., 1993, Separation of freezing and drying-induced denaturation of lyophilized proteins using stress-specific stabilization, *Arch Biochem Biophys.* **303**:456–464.

Chang, B. S., Kendrick, B. S., and Carpenter, J. F., 1996, Surface-induced denaturation of proteins during freezing and its inhibition by surfactants, *J. Pharm. Sci.* **85:**1325–1330.

Chawla, A. S., Hinberg, I., Blais, P., and Johnson, D., 1985, Aggregation of insulin, containing surfactants, in contact with different materials, *Diabetes* **34:** 420–424.

Chrai, S., Heffernan, G., and Myers, T., 1994, Glass vial container–closure integrity testing—an overview, *Pharm. Technol.* **18:**162–175.

Christensen, P., Johansson, A., and Nielsen, V., 1978, Quantitation of protein adsorbance to glass and plastics: Investigation of a new tube with low adherence, *J. Immunol. Meth.* **23:**23–28.

Clarke, S., Stephenson, R. C., and Lowenson, J. D., 1992, Lability of asparagine and aspartic acid residues in proteins and peptides. Spontaneous deamidation and isomerization reactions, In: *Stability of Protein Pharmaceuticals, Part A: Chemical and Physical Pathways of Protein Degradation* (T. J. Ahern and M. C. Manning, eds.), Plenum Press, New York, pp. 2–29.

Cleland, J. L., Powell, M. F., and Shire, S. J., 1993, The development of stable protein formulations: A close look at protein aggregation, deamindation and oxidation, *Crit. Rev. Ther. Drug Carrier Syst.* **10:**307–377.

Costantino, H. R., Carrasquillo, K. G., Cordero, R. A., Mumenthaler, M., Hsu, C. C., and Griebenow, K., 1998, Effect of excipients on the stability and structure of lyophilized recombinant human growth hormone, *J. Pharm. Sci.* **87:**1412–1420.

Cradock, J. C., Kleinman, L. M., and Davignon, J. P., 1977, Intrathecal injections—A review of pharmaceutical factors, *Bull. Parenteral Drug Assoc.* **31:** 237–247.

Currie, B. L., and Groves, M. J., 1992, Protein stability and degradation mechanisms, in: *Pharmaceutical Biotechnology* (M. E. Klegerman and M. J. Groves, eds.), Interpharm Press, Englewood, CO, pp. 213–216.

Dahlquist, F. W., Long, J. W., and Bigbee, W. L., 1976, Role of calcium in thermal stability of thermolysin, *Biochemistry,* **15:**1103–1111.

Danielson, J. W., Oxborrow, G. S., and Placencia, A. M., 1984, Quantitative determination of chemicals leached from rubber stoppers into parenteral solutions, *J. Parenteral Sci. Technol.* **38:**90–93.

Davis, S. S., Illum, L., and Tomlinson, E., 1986, *Delivery Systems for Peptide Drugs*, Plenum Press, New York.

Deadman, M. L., Farmer, T. H., and Morris, C. J. O. R., 1961, Studies on pituitary adrenocorticotrophin, *Biochem. J.* **78:**348–352.

DeLuca, P. P. and Knapp, J. Z., 1993, Particulate matter, in: *Pharmaceutical Dosage Forms, Parenteral Medications, Volume 3,* 2nd ed. (K. E. Avis, H. A. Lieberman, and L. Lachman, eds.), Marcel Dekker, New York, pp. 117–230.

Dixon, H. B. F., 1956, The melanophore stimulating activity of corticotrophin, *Biochim. Biophys. Acta.* **19:**392.

Dong, D. E., Andrade, J. D., and Coleman, D. L., 1987, Adsorption of low density lipoproteins onto selected biomedical polymers, *J. Biomed. Mater. Res.* **21**, 683–700.

Dubost, D. C., Kaufman, M. J., Zimmerman, J. A., Bogusky, M. J., Coddington, A. B., and Pitzenberger, S. M., 1996, Characterization of a solid state reaction product from a lyophilized formulation of a cyclic heptapeptide. A novel example of an excipient-induced oxidation, *Pharm. Res.* **13:**1811–1814.

Federal Register, 1994, Stability testing of new drugs and products, *Fed. Reg.* **59**(183):48754–48759.

Federal Register, 1997, Proposed Rule 21CFR Part 589, Substances prohibited from use in animal food or feed, animal proteins prohibited in ruminant feed, *Fed. Reg.* **62**(2):551–583.

Fischer, J., 1995, Bovine spongiform encephalopathy, *Deutsche Apotheker Zeitung,* **135:**35–47.

Fish, W. W., Danielsson, A., Nordling, L., Miller, S. H., Lam, C. F. and Bjork I., 1985, Denaturation behavior of antithrombin in guanidinium chloride: Irreversibility of unfolding caused by aggregation, *Biochemistry* **24:**1510.

Franks, F., 1990, Freeze drying: From empiricism to predictability, *Cryo-Lett*, 11:93.

Franks, F., 1994, Long term stabilization of biologicals, *Biotech.* **12:**253–256.

Franks, F., Hatley, R. H. M., and Friedman, H. L., 1988, The thermodynamics of protein stability: Cold destabilization as a general phenomenon, *Biophys. Chem.* **31:**307–315.

Franks, F., Hatley, R. H. M., and Mathias, S. F. 1991, Materials science and the production of shelf-stable biologicals, *BioPharm* **4:**38–42.

Fransson, J., and Hagman, A., 1996, Oxidation of human insulin-like growth factor I in formulation studies. 2. Effects of oxygen, visible light, and phosphate on methionine oxidation in aqueous solution and evaluation of possible mechanisms, *Pharm Res.* **13:**1476–1481.

Friedman, M., and Masters, P. M., 1982, Kinetics of racemization of amino acid residues in casein, *J. Food Sci.* **47:**760–765.

Galat, A., Creighton, T., Lord, R. C., and Blout, E. R., 1981, Circular dichroism, Raman spectroscopy, and gel filtration of trapped folding intermediates of ribonuclease, *Biochemistry* **20:**594–601.

Geiger, T., and Clarke, S., 1987, Deamidation, isomerization, and racemization of asparginyl and aspartyl residues in peptides, *J. Biol. Chem.* **262:**785–794.

Gekko, K., and Timasheff, S. N., 1981, Mechanism of protein stabilization by glycerol: Preferred hydration in glycerol–water mixtures, *Biochemistry* **20:** 4667–4676.

Glatz, C. E., 1992, Modeling of aggregation–precipitation phenomena, in: *Stability of Protein Pharmaceuticals, Part A: Chemical and Physical Pathways of Protein Degradation,* (T.J. Ahern, and M. C. Manning, eds.), Plenum Press, New York, pp. 135–166.

Gorecka, B. A., Sanzgiri, Y. D., Bindra, D. S., and Stella, V. J., 1995, Effect of SBE4-β-CD, a sulfobutylether β-cyclodextrin, on the stability and solubility of O^6-benzyguanine (NSC-637037) in aqueous solutions, *Int. J. Pharm.* **125:** 55–61.

Gu, L. C., Erdos, E. A., Chiang, H. -S., Calderwood, T., Tsai, K., Visor, G. C., Duffy, J., Hsu, W. -C., and Foster, L. C., 1991, Stability of interleukin 1b (IL-1b) in aqueous solution: Analytical methods, kinetics, products, and solution formulation implications, *Pharm. Res.* **8:**485–490.

Guazzo, D., 1994, Package integrity testing, in: *Parenteral Quality Control: Sterility, Pyrogen, Particulate Matter and Package Integrity Testing* (M. J. Akers, ed.), Marcel Dekker, New York, pp. 258–262.

Haber, E., and Anfinsen, C. B., 1962, Side-chain interactions governing the pairing of half-cystine residues in ribnuclease, *J. Biol. Chem.* **237**, 1839.

Hageman, M., 1988, The role of moisture in protein stability, *Drug Dev. Ind. Pharm.* **14**, 2047–2070.

Haley, E. E., Corcoran, B. J., Dorer, R. D., and Buchanan, D. L., 1966, β-Aspartyl peptides in enzymatic hydrolysates of proteins, *Biochemistry* **5:**3229–3235.

Hanson, M. A., and Rouan, S. K. E., 1992, Introduction to formulation of protein pharmaceuticals, in: *Stability of Protein Pharmaceuticals, Part B: In Vivo Pathways of Degradation and Strategies for Protein Stabilization* (T. J. Ahern and M. C. Manning, eds.), Plenum Press, New York, pp. 209–233.

Harwood, R. J., Portnoff, J. B., and Sunbery, E. W. 1993, The processing of small volume parenterals and related sterile products, in: *Pharmaceutical Dosage Forms: Parenteral Medications, Vol. 2*, 2nd ed. K. E. Avis, H. A. Lieberman, and L. Lachaman, eds.), Marcel Dekker, New York, pp. 36–37.

Hawker, R. J. and Hawker, L. M., 1975, Protein losses during sterilizing by filtration, *Lab. Practice* **24:**805–807.

Her, L. M. and Nail, S. L., 1994, Measurement of glass transition temperatures of freeze-concentrated solutes by differential scanning calorimetry, *Pharm. Res.* **11:**54–59.

Hillery, A. M., Lloyd, A. W., and Swarbrick, J., eds., 2001, *Drug Delivery and Targeting for Pharmacists and Pharmaceutical Scientists*, Taylor and Francis, London.

Hirsch, J. I., Fratkin, M. J., Wood, J. H., and Thomas, R. B., 1977, Clinical significance of insulin adsorption by polyvinylchloride infusion systems, *Am. J. Hosp. Pharm.* **34:**583–588.

Hirsch, J. I., Wood, J. H., and Thomas, R. B., 1981, Insulin adsorption to polyolefin infusion bottles and polyvinylchloride administration sets, *Am. J. Hosp. Pharm.* **38:**995–997.

Hoffman, A. S., 1974, Principles governing biomolecule interactions at foreign interfaces, *J. Biomed. Mater. Res. Symp.* **5**(Part 1):77–83.

Hora, M. S., Rana, R. K., Wilcox, C. L, Katre, N. V., Hirtzer, P., Wolfe, S., and Thomson, J. W., 1991, Development of a lyophilized formulation of interleukin-2, *Dev. Biol. Stand.* **74:**295–303.

Hora, M. S., Rana, R. K., and Smith, F. W., 1992, Lyophilized formulations of recombinant tumor necrosis factor, *Pharm. Res.* **9:**33–36.

Hsu, C. C., Pearlman, R., and Curley, J. C., 1988, Some factors causing protein denaturation and aggregate formation in pharmaceutical processing, *Pharm. Res. Suppl.* **5:**S34.

Hsu, C. C., Ward, C. A., Pearlman, R., Nguyen, H. M., Yeung, D. A., and Curley, J. G., 1991, Determining the optimum residual moisture in lyophilized protein pharmaceuticals, *Dev. Biol. Stand.* **74:**255–271.

International Conference on Harmonization, 1995, *Quality of Biotechnology Products: Stability Testing of Biotechnological/Biological Products*, Annex to the Tripartite ICH Guideline for the Stability Testing of New Drug Substances and Products, ICH, Geneva.

Iwamoto, G. K., Van Wagenen, R. A., and Andrade, J. D., 1982, Insulin adsorption: Intrinsic tyrosine interfacial fluorescence, *J. Colloid Interface Sci.* **86:**581–585.

Izutsu, K., Yoshioka, S., and Terao, T., 1993a, Stabilization of β-galactosidase by amphiphilic additives during freeze drying, *Int. J. Pharm.* **90:**187–194.

Izutsu, K., Yoshioka, S., and Terao, T., 1993b, Decreased protein-stabilizing effects of cryoprotectants due to crystallization, *Pharm. Res.* **10:**1232–1237.

Izutsu, K., Yoshioka, S., and Terao, T., 1994a, Effect of mannitol crystallinity on the stabilization of enzymes during freeze drying, *Chem Pharm. Bull.* **42:**5–8.

Izutsu, K., Yoshioka, S., and Kojima, S., 1994b, Physical stability and protein stability of freeze dried cakes during storage at elevated temperatures, *Pharm. Res.* **11:**995–999.

James, D. E., Jenkins, A. B., Kraegan, E. W., and Chisholm, D. J., 1981, Insulin precipitation in artificial infusion devices, *Diabetologia* **21:**554–557.

Jang, J. W., Kitamura, S., and Guillory, J. K., 1995, The effect of excipients on glass transition temperatures for FK906 in the frozen and lyophilized states, *PDA J. Pharm. Sci. Technol.* **49:**166–174.

Jarvinen, K., Jarvinen, T., Thompson, D. O., and Stella, V. J., 1994, The effect of a modified β-cyclodextrin, SBE4-β-CD on the aqueous stability and ocular absorption of pilocarpine, *Curr. Eye Res.* **13:**897–905.

Johnson, M. D., Hoesterey, B. L., and Anderson, B. D., 1994, Solubilization of a tripeptide HIV protease inhibitor using a combination of ionization and complexation with chemically modified cyclodextrins, *J. Pharm. Sci.* **83:** 1142–1146.

Johnston, T. P., 1996, Adsorption of recombinant human granulocyte colony stimulating factor (rGH-CSF) to polyvinyl chloride, polypropylene, and glass: Effect of solvent additives, *PDA J. Pharm. Sci. Technol.* **50:**238–245.

Kakimoto, F., Asakawa, N., Ishibashi, Y., and Miyake, Y., 1985, Prevention of adsorption of peptide hormones, Eisai Co. Ltd, Japan Patent JP61221125-A2861001.

Katakam, M., Bell, L. N., and Banga, A. K., 1995, Effect of surfactants on the physical stability of recombinant human growth hormone, *J. Pharm. Sci.* **84:**713–716.

Kerwin, B. A., Akers, M. J., Apostol, I., Moore-Einsel, C., Etter, J. E., Hess, E., Lippincott, J., Levine, J., Mathews, A. J., Revilla-Sharp, P., Schubert, R., and Looker, D. L., 1999, Acute and long-term stability studies of deoxy hemoglobin and characterization of ascorbate-induced modifications, *J. Pharm. Sci.* **88:**79–88.

Kim, Y., Rose, C. A., Liu, Y., Ozaki, Y., Datta, G., and Ut, A. T., 1994, RT-IR and near-infared FT-raman studies of the secondary structure of insulinotropin in the solid state: α-Helix to β-sheet conversion induced by phenol and/or high shear force, *J. Pharm. Sci.* **83:**1175–1180.

Kirsch, L. E., Riggin, R. M., Gearhart, D. A., Lefeber, D. S., and Lytle, D. L., 1993, In-process protein degradation by exposure to trace amounts of sanitizing agents, *J. Parenteral Sci. Technol.* **47:**155–160.

Kirsch, L. E., Nguyen, L., Moeckly, C. S., 1997a, Pharmaceutical container/closure integrity I: Mass spectrometry-based helium leak rate detection for rubber-stoppered glass vials, *PDA J. Pharm, Sci. Tech.* **51:**187–194.

Kirsch, L. E., Nguyen, L., Moeckly, C. S., Gerth, R., 1997b, Pharmaceutical container/closure integrity II: The relationship between microbial ingress and helium leak rates in rubber-stoppered glass vials, *PDA J. Pharm, Sci. Tech.* **51:**195–202.

Kirsch, L. E., Nguyen, L., Gerth, R., 1997c, Pharmaceutical container/closure integrity III: Validation of the helium leak rate for rigid pharmaceutical containers, *PDA J. Pharm. Sci. Tech.* **51:**203–207.

Knepp, V. M., Whatley, J. L., Muchnik, A., and Calderwood, T. S., 1996, Identification of antioxidants for prevention of peroxide-mediated oxidation of recombinant human ciliary neurotrophic factor and recombinant human nerve growth factor, *PDA J. Pharm. Sci. Technol.* **50:**163–171.

Kraegen, E. W., Lazarus, L., Meler, H., Campbell, L., and Chia, Y. O., 1975, Carrier solutions for low level intravenous insulin infusion, *Br. Med. J.* **1975:** 464–466.

Kreilgard, L., Frokjaer, S., Flink, J. M., Randolph, T. W., and Carpenter, J. F., 1996, The effects of sugars, polymers, and surfactants on conformational stability of a hydrophobic protein during freeze-drying, *Pharm. Res.* **13:**S-79.

Lam, X. M., Costantino, H. R., Overcashier, D. E., Nguyen, T. H., and Hsu, C. C., 1996, Replacing succinate with glycolate buffer improves the stability of lyophilized interferon-gamma, *Int. J. Pharm.* **142:**85–95.

Lam, X. M., Patapoff, T. W., and Nguyen, T. H., 1997, The effect of benzyl alcohol on recombinant human interferon-γ, *Pharm. Res.* **14:**725–729.

Lang, L. A., 1996, A review of latex hypersensitivity, *Toxic Substances Mechanisms* **15:**1–11.

Lee, H. S., Osuga, D. T., Nashaf, A. S., Ahmed, A. I., Whitaker, J. R., and Feeney, R. E., 1977, Effects of alkali on glycoproteins: β-Elimination and nucleophilic addition reactions of substitued threonyl residues of antifreeze glycoprotein, *J. Agr. Food Chem.* **25:**1153–1158.

Lee, J. C. and Lee, L. L. Y., 1987, Thermal stability of proteins in the presence of polyethylene glycols, *Biochemistry* **26:**7813–7819.

Lee, V. H. L. (ed.), 1997, *Peptide and Protein Drug Delivery*, 2nd ed., Marcel Dekker, New York.

Lenk, J. R., Ratner, B. D., Gendreau, R. M., and Chittur, K. K., 1989, IR spectral changes of bovine serum albumin upon surface adsorption, *J. Biomed. Mater. Res.* **23:**549–569.

Li, S., Schoneich, C., Wilson, G. S., and Borchardt, R. T, 1993, Chemical pathways of peptide degradation. V. Ascorbic acid promotes rather than inhibits the oxidation of methionine to methionine sulfoxide in small model peptides, *Pharm. Res.* **10:**1572–1579.

Liebe, D. C., 1995, Pharmaceutical packaging, in: *Encyclopedia of Pharmaceutical Technology, Vol. 12* (J. Swarbrick and J. C. Boylan, eds.), Marcel Dekker, New York, pp. 16–28.

Liu, W. R., Langer, R., and Klibanov, A. M., 1991, Moisture-induced aggregation of lyophilized proteins in the solid state, *Biotechnol. Bioeng.* **37:**177–182.

Lougheed, W. D., Albisser, A. M., Martindale, H. M., Chow, J. C., and Clement, J. R., 1983, Physical stability of insulin formulations, *Diabetes* **32:**424–432.

Maa, Y. F., and Hsu, C. C., 1996, *Int. J. Pharm.* **140:**155–168.

MacCracken, J., 1996, Latex allergy in diabetic patients: A call for latex-free insulin tops, *Diabetes Care* **19:**184.

Manning, M. C., Patel, K., and Borchardt, R. T., 1991, Stability of protein pharmaceuticals, *Pharm. Res.* **6:**903–918.

Massey, E. H. and Sheliga, T. A., 1988, Human insulin (HI) isophane solution (NPH) with improved physical stability, *Pharm. Res.* **5:**S34.

Milano, E. A., Waraszkiewicz, S. M., and Dirubio, R., 1982, Extraction of soluble aluminum from chlorobutyl rubber closures, *J. Parenteral Sci. Technol.* **36:**116–120.

Mitraki, A., and King, J., 1989, Protein folding intermediates and inclusion body formation, *Biotechnology.* **7:**690–696.

Mitrano, F. P. and Newton, D. W., 1982, Factors affecting insulin adherence to Type I glass bottles, *Am. J. Hosp. Pharm.* **39:**1491–1495.

Mizutani, T., 1981, Estimation of protein and drug adsorption onto silicone–coated glass surfaces, *J. Pharm. Sci.* **70:**493–496.

Mizutani, T., and Mizutani, A., 1975, Prevention of adsorption of protein on controlled-pore glass with amino acid buffer, *J. Chromatogr.* **111:**214–216.

Mizutani, T., and Mizutani, A., 1978, Adsorption of a protein on glass surfaces in detergent solutions, *J. Non-cryst. Solids* **27:**437–439.

Morley, J. S., Tracey, H. J., and Gregory, R. A., 1965, Structure–function relationships in the active C-terminal tetrapeptide sequence of gastrin, *Nature* **207:**1356–1359.

Morton, D. K., Lordi, N. G., Troutman, L. H., and Ambriosio, T. J., 1989, Quantitative and mechanistic measurements of container/closure integrity. Bubble, liquid, and microbial leakage tests, *J. Parenteral Sci. Technol.* **43:** 104–108.

Murase, M., and Franks, F., 1989, Salt precipitation during the freeze concentration of phosphate buffer solution, *Biophys. Chem.* **34:**293–300.

Nash, R. A., 1994, Cyclodextrins, in: *Handbook of Pharmaceutical Excipients*, 2nd ed. (A. Wade and P. J. Weller, eds.), American Pharmaceutical Association, Washington, DC, pp. 145–148.

Niebergall, P. J., 1990, Ionic solutions and electrolytic equilibria, in: *Remington's Pharmaceutical Sciences*, 18th ed., Mack Publishing Co., Philadelphia, PA, pp. 213–230.

Norde, W., 1995, Adsorption of proteins at solid–liquid interfaces, *Cells Materials* **5:**97–112.

Ogino, J., Noguchi, K., and Terato, K., 1979, Adsorption of secretin on glass surfaces, *Chem. Pharm. Bull.* **27:**3160–3163.

Okimoto, K., Rajewski, R. A., Jona, J. A., and Stella, V. J., 1995, The interaction of charged and uncharged drugs with a neutral (HP-β-CD) and anionically charged (SBE7-β-CD) cyclodextrin, *Pharm. Res.* **12:**S205.

Oliyai, C., Patel, J. P., Carr, L., and Borchardt, R. T., 1994, Chemical pathways of peptide degradation. VII. Solid state chemical instability of an aspartyl residue in a model hexapeptide, *Pharm. Res.* **11**, 901–908.

Orii, Y., and Morita, M., 1977, Measurement of the pH of frozen buffer solutions by using pH indicators, *J. Biochem.* **81:**163–168.

Oshima, G., 1989, Solid surface-catalyzed inactivation of bovine alpha-chymotrypsin in dilute solution, *Int. J. Biol. Macromol.* **11:**43–48.

Pace, C. N. and Grimsley, G. R., 1988, Ribonuclease T1 is stabilized by cation and anion binding, *Biochemistry* **27:**8311–8315.

Pantoliano, M. W., Whitlow, M., Wood, J. F., Rollence, M. L., Finzel, B. C., Gilliand, G. L., Poulos, G., and Bryant, P. N., 1988, The engineering of binding affinity at metal ion binding sites for the stabilization of proteins: Subtilisin as a test case, *Biochemistry* **27:**8311–8316.

Parenteral Drug Association, 1980, Extractables from Elastomeric Closures: Analytical Procedures for Function Group, Characterization, and Identification, Methods Bulletin No. 1, PDA, Bethesda, MD.

Parenteral Drug Association, 1982, Elastomeric Closures: Evaluation of Significant Performance and Identity Characteristics, Technical Methods Bulletin 2, PDA, Bethesda, MD.

Parenteral Drug Association, 1986, Glass: Isolation and identification of extractables from USP grade glass, *J. Parenteral Sci. Technol.* **40**, S3–S11.

Patel, J. P., 1990, Urokinase: Stability studies in solution and lyophilized formulations, *Drug Dev. Ind. Pharm.* **16:**2613–2626.

Pearlman, R., and Wang, Y. J., 1996, *Formulation, Characterization, and Stability of Protein Drugs*, Plenum Press, New York.

Petty, C., Cunningham, N., and Major, M. C., 1974, Insulin adsorption by glass infusion bottles, polyvinylchloride infusion containers, and intravenous tubing, *Anesthesiology* **40:**400–404.

Pharmacopeial Forum, 1996, The United States Pharmacopeial Convention, Inc. No. 31, Sept.–Oct.

Pikal, M. J., 1990, Freeze drying of proteins. II. Formulation selection, *BioPharm.* **3:**26–30.

Pikal, M. J., 2002, Freeze drying of proteins, in: *Peptide and Protein Delivery*, 2nd ed. (V. H. L. Lee, ed.), Marcel Dekker, New York (in press).

Pikal, M. J., Dellerman, K. M., Roy, M. L., and Riggin, R. M., 1991a, The effects of formulation variables on the stability of freeze dried human growth hormone, *Pharm. Res.* **8:**427–436.

Pikal, M. J., Dellerman, K. M., and Roy, M. L., 1991b, Formulation and stability of freeze dried proteins: Effects of moisture and oxygen on the stability of freeze dried formulations of human growth hormone, *Dev. Biol. Stand.* **74:**21–37.

Privalov, P. L., 1990, Cold denaturation of proteins, *Crit. Rev. Biochem. Mol. Biol.* **25:**281–305.

Rajewski, R. A. and Stella, V. J., 1996, Pharmaceutical applications of cyclodextrins. 2. *In vivo* drug delivery, *J. Pharm. Sci.* **85:**1142–1169.

Riffkin, C., 1968, Panel discussion: Siliconization of parenteral packaging components, *Bull. Parenteral Drug Assoc.* **22:**66–75.

Riniker, V. B., Neher, R., Maier, R., Kahnt, F. W., Byfield, P. G. H.,Gundmundsson, T. V., Galante, L., and MacIntyre, I., 1968, Menschliches calcitonin I: Isolierung und charakterisierung, *Helv. Chim. Acta* **51:**1738–1743.

Robinson, A. B., and Rudd, C. J., 1974, Deamidation of glutaminyl and asparaginyl residues in peptides and proteins, in: (B. L. Horecker and E. R. Stadtman, eds.), *Current Topics in Cellular Regulation* , **Vol. 8**, Academic Press, New York, pp. 247–295.

Roe, J. A., Butler, A., Schaller, D. M., and Valentine, J. S., 1988, Differential scanning calorimetry of Cu, Zn-superoxide dismutase, the apoprotein, and its Zn-substituted derivatives, *Biochemistry* **27:**950–995.

Ryan, W., Collier, P., Satyagal, V., Sachdev, R., and Sumodojo, F., 1995, Characterization of a cryogenic pelletizer for preserving delicate biologicals, *BioPharm* **8:**32–38.

Ryle, A. P., and Sanger, F., 1955, Disulfide interchange reactions, *Biochem J.* **60:**535–539.

Sato, S., Ebert, C. D., and Kim, S. W., 1984, Prevention of insulin self-association and surface adsorption, *J. Pharm. Sci.* **72:**228–232.

Schmidt, D. J., and Akers, M. J., 1997, Cryogranulation—A potential new final process for bulk drug substances, *BioPharm* **10:**28–32.

Sen, L. C., Gonzalez-Flores, E., Feeney, R. E., and Whitaker, J. R., 1977, Reactions of phosphoproteins in alkaline solutions, *J. Agr. Food Chem.* **25:**632–638.

Senior, J., and Radomsky, M., eds., 2000, *Sustained-Release Injectable Products*, Interpharm Press, Englewood, Co.

Seres, D. S., 1990, Insulin adsorption to parenteral infusion systems: Case report and review of the literature, *Nutr. Clin. Pract.* **5:**111–117.

Shahrokh, Z., Eberlein, G., and Wang, Y. J., 1994a, Probing the conformation of protein (bFGF) precipitates by fluorescence spectroscopy, *J. Pharm. Biomed. Anal.* **12:**1035–1041.

Shahrokh, Z., Sluzky, V., Stratton, P. R. and Eberlein, G. A., 1994b, Disulfide-linked oligomerization of basic fibroblast growth factor: Effects of sulfated compounds, in: *ACS Symposium Series on Protein Formulation and Delivery* (J. Cleland and R. Langer, eds.), American Chemical Society Publishing, pp. 85–99.

Shnek, D. R., Hostettler, D. L., Bell, M. A., Olinger, J. M., and Frank, B. H., 1998, Physical stress testing of insulin suspensions and solutions, *J. Pharm. Sci.* **87:**1459–1465.

Silvestri, S., Lu, M. Y., and Johnson, H., 1993, Kinetics and mechanism of peptide aggregation I: Aggregation of a cholecystokinin analog, *J. Pharm. Sci.* **92:**689–693.

Simkins, J. W., 1991, Solubilization of ovine growth hormone with 2-hydroxypropyl-β-cyclodextrin, *J. Parenteral Sci. Technol.* **45:**266–269.

Slater, J. E., 1994, Latex allergy, *J. Allergy Clin. Immunol.* **94:**139–149.

Smith, E. J., and Nash, R. J. (1992). Elastomeric closures for parenterals, in: *Pharmaceutical Dosage Forms: Parenteral Medications, Vol. 1,* 2nd ed., (K. R. Avis, H. A. Lieberman, and L. Lachman, eds.), Marcel Dekker, New York, pp. 445–512.

Srinivasan, N., Sowdhamini, R., Ramakrishnan, C., and Balaram, P., 1990, Conformation of disulfide bridges in proteins, *Int. J. Peptide Protein Res.* **36:**197–155.

Stratton, C. E., 1991, Cyclodextrins and biological macromolecules, *BioPharm.* **4:**44–49.

Suelter, C. H., and DeLuca, P. P., 1983, How to prevent losses of protein by adsorption to glass and plastic, *Anal. Biochem.* **135:**112–119.

Szkudlarek, B. A., Garcia, G. A. and Rodriguez-Hornedo, N., 1996, The effect of freezing on the stability of lactate dehydrogenase in phosphate buffer solutions in the absence and presence of additives, *Pharm. Res.* **13:**S85.

Tanaka, R., Takeda, T., and Miyajima, R., 1991, Cryoprotective effect of saccharides on denaturation of catalase during freeze drying, *Chem. Pharm. Biol.* **39:**1091–1094.

Tashjian, A. H., Ontjes, D. A., and Munson, P. L., 1964, Alkylation and oxidation of methionine in bovine parathyroid hormone: Effects on hormonal activity and antigenicity, *Biochemistry* **3:**1175–1182.

Thompson, D. O., 1997, Cyclodextrins-enabling excipients: Their present and future use in pharmaceuticals, *Crit. Rev. Ther. Drug Carrier Syst.* **14:**1–104.

Thurow, H., and Geisen, K., 1984, The stabilisation of dissolved proteins against denaturation at hydrophobic interfaces, *Diabetologica* **27:**212–218.

Torchinsky, Y. M., 1981, *Sulfur in Proteins*, Pergamon Press, New York.

Towse, A., O'Brien, M., Twarog, F. J., Braimon, J., and Moses, A. C., 1995, Local reaction secondary to insulin injection, *Diabetes Care* **18:**1195–1197.

Tsai, P. K., Volkin, D. B., Dabora, J. M., Thompson, K. C., Bruner, M. W., Gress, J. O., Matuszewska, B., Keogan, M., Bondi, J. V., and Middaugh, C. R., 1993, Formulation design of acidic fibroblast growth factor, *Pharm. Res.* **10:**649–659.

Twardowski, Z. J., Nolph, K. D., McGary, T. J., Moore, H. L., Collin, P., Ausman, R. K., and Slimack, W. S., 1983a, Insulin binding to plastic bags: A methodologic study, *Am. J. Hosp. Pharm.* **40:**575–578.

Twardowski, Z. J., Nolph, K. D., McGary, T. J., and Moore, H. L., 1983b, Nature of insulin binding to plastic bags, *Am. J. Hosp. Pharm.* **40:**579–582.

Twardowski, Z. J., Nolph, K. D., McGary, T. J., and Moore, H. L., 1983c, *Am. J. Hosp. Pharm.* **40:**583–586.

Tyler-Cross, R., and Schirch, V., 1991, Effects of amino acid sequence, buffers, and ionic strength on the rate and mechanism of deamidation of asparagine residues in small peptides, *J. Biol. Chem.* **266:**22549–22556.

U.S. Food and Drug Administration, 1994, *Guideline for Submitting Documentation for Packaging for Human Drugs and Biologics*, U.S. FDA, Rockville, MD.

U.S. Food and Drug Administration, 1996, *Guideline for Industry: Quality of Biotechnological Products: Stability Testing of Biotechnology/Biological Products*, U.S. FDA, Rockville, MD.

USP, 1995, Biological Tests Protocol, Sections 381, 387, and 388, *United States Pharmacopeia*, 23rd ed.

Vale, W., Sepiss, J., Rivier, C., and Rivier, J., 1981, Characterization of a 41-residue ovine hypothalamic peptide that stimulates secretion of corticotrophin and β-endorphin, *Science* **213:**1394–1397.

Van den Berg, L., and Rose, D., 1959, Effect of freezing on the pH and composition of sodium and potassium phosphate solutions: The reciprocal system KH_2PO_4–Na_2PO_4–H_2O, *Arch. Biochem. Biophys.* **81:**319–329.

Van den Oetelaar, P. J. M, Mentink, I. M., and Brinks, G. J., 1989, *Drug Dev. Ind. Pharm.* **15:**97–106.

Vassallo, S., *et al.*, 1995, Allergic reaction to latex from stopper of a medication vial, *Anesthesia Analgesia* **80:**1057–1058.

Volkin, D. B. and Klibanov, A. M., 1987, Thermal destruction processes in proteins involving cystine residues, *J. Biol. Chem.* **262:**2945–2950.

Wang, Y. J., 1992, Parenteral products of peptides and proteins, in: *Pharmaceutical Dosage Forms: Parenteral Medications*, Vol. 1 (K. E. Avis, H. A. Lieberman, and L. Lachman, eds.), Marcel Dekker, New York, pp. 283–320.

Wang, Y. J. and Chien, Y. W., 1984, Sterile Pharmaceutical Packaging: Compatibility and Stability, Technical Report No. 5 Parenteral Drug Association, Philadelphia, PA.

Wang, Y. J. and Hanson, M. A., 1988, Parenteral Formulations of Proteins and Peptides: Stability and Stabilizers, Technical Report Vol. 42, No. 10, 1988 Supplement, Parenteral Drug Association, Phildelphia, PA.

Wang, Y. J., and Pearlman, R., 1993, *Stability and Characterization of Protein and Peptide Drugs*, Plenum Press, New York.

Wetzel, R., Perry, L. J., Baase, W. A., and Becktel, W. J., 1988, Disulfide bonds and thermal stability of T4 lysozyme, *Proc. Natl. Acad. Sci. USA* **85:**401–405.

Whitaker, J. R., and Feeney, R. E., 1982, Chemical and physical modification of proteins by the hydroxide ion, *CRC Crit. Rev. Food Sci. Nutr.* **19:**173–212.

Whyte, W., 1983, A multicentered investigation of clean air requirements for terminally sterilized pharmaceuticals, *J. Parenteral Sci. Technol.* **37:** 138–144.

Wilkinson, S. M., and Beck, M. H., 1996, Allergic contact dermatitis from latex rubber, *Br. J. Dermatol.* **134:**P910–P914.

Winward, D. B., and Brophy, M. T., 1995, Acute renal failure after administration of intravenus immunoglobulin: The literature and case report, *Pharmacotherapy*, **15:**765–772.

Wollmer, A., Rannefeld, B., Johnasen, B. R., Hejnaes, K. R., Balschmidt, P., and Hansen, F. B., 1987, Phenol–promoted structural transformation of insulin in solution, *Biol. Chem. Hoppe-Seyler* **368:**903–911.

Yoshioka, S., Aso, Y., Izutsu, K., and Terao, T., 1993, Stability of β-glactosidase, a model protein drug, is related to water mobility as measured by ^{17}O nuclear magnetic resonance (NMR), *Pharm. Res.* **10:**103–108.

Yoshioka, S., Aso, Y., Izutu, K., and Terao, T., 1994. Application of accelerated testing to shelf-life prediction of commercial protein preparations, *J. Pharm. Sci.* **83:**454–456.

3

Aseptic Processing of Protein Pharmaceuticals

Michael Townsend

1. INTRODUCTION

The objective of this chapter is to present a planning guide for designing and implementing aseptic processing for drug product manufacturing, for both clinical supplies and the commercial market, from the perspective of the unique characteristics of protein-containing drug products. The discussion will be restricted to solution, suspension, and lyophilized dosage forms in final containers made of glass or plastic. It will present a conceptual framework for the development of aseptic processes at the lab bench, subsequent scale-up to clinical batches, and, finally, validation of the fully scaled-up and well-defined processes for commercial manufacturing of the drug product.

2. ASEPTIC PROCESSING OF SOLUTION, LYOPHILIZED, AND SUSPENSION DOSAGE FORMS

The processes necessary to transform a pharmacologically active protein or peptide into a safe, efficacious, and reliable parenteral drug

---

Michael Townsend • Isis Pharmaceuticals, Inc., Carlsbad, California 92008.
Development and Manufacture of Protein Pharmaceuticals, edited by Nail and Akers. Kluwer Academic/Plenum Publishers, New York, 2002.

product must first be adequately developed and validated to reproducibly yield product that meets specifications. These processes must then be carried out in controlled areas on suitably qualified equipment by trained personnel using standard operating procedures and manufacturing instructions to document the operation. Most of the processes in sterile pharmaceutical manufacturing must be done aseptically so that the final product can be guaranteed to be nonpyrogenic and have a sterility assurance level (SAL) of 1×10^{-3} or less by the end of solution or suspension filling, or lyophilization, if applicable.

The SAL of an aseptic filling process refers to the probability of a nonsterile vial being produced during the aseptic operations. It is an empirically derived value, which is determined by the results of media fills. A media fill entails the compounding, sterile filtration, and filling of microbiological growth media as simulated product into vials (to include the lyophilization step, if applicable), followed by incubation of the filled vials to detect the presence of microbiological contamination. At a minimum, it is recommended that at least 3000 vials be filled for each test. The ratio of the number of contaminated vials to the total number filled gives the SAL (i.e., to achieve an SAL of 1×10^{-3} no more than 3 vials may contain microbial growth for every 3000 vials filled). Such media fills must be repeated on a regular schedule to revalidate the aseptic operations. In addition to media fills, other microbiological control measures must be employed, such as periodic environmental filter inspections, routine microbiological monitoring of the manufacturing personnel and environment, and routine air quality monitoring in the sterile core (see Section 2.1.4) (USP, 2000a, pp. 2099–2106; 2000b, p. 2147).

The U.S. Food and Drug Administration (FDA) has proposed that terminal sterilization should be employed as an adjunct to aseptic processing to further increase a parenteral product's SAL (U.S. Food and Drug Administration [FDA], 1991). However, most protein or peptide drug products do not possess sufficient heat stability to allow for the use of typical high-temperature sterilizing methods (autoclaving or dry heat), and the use of ionizing radiation (gamma, E-beam, or X-ray) can also result in unacceptable levels of physical and chemical degradation (Yamamoto, 1992). Therefore, aseptic processing is of fundamental importance in preventing microbial contamination of a protein-containing formulation during drug product manufacturing.

The term "aseptic processing" implies a number of fundamental requirements: (1) All product contact equipment and final packaging components must be sterilized using validated cycles, (2) the processes must be carried out in environmentally controlled facilities under Class 100 high-efficiency particulate air (HEPA)-filtered,* laminar flow air, (3) the operators must be properly trained in aseptic processing, and (4) the operators must be gowned in attire to prevent the shedding of viable and nonviable particulates into the work area. In a restrictive sense, aseptic processing begins with the sterile filtration of the formulated bulk solutions and continues through to the sealing of the closures in the primary containers. In a broader sense, though, all of the parenteral product manufacturing facilities, utilities, and support functions that supply raw materials, equipment, components, and formulated bulk solutions to the these downstream operations fall under the "aseptic processing" umbrella. These functions play a critical role in achieving the desired SAL and preventing pyrogenic contamination of the final product containers produced in a parenteral drug product manufacturing operation.

2.1. Current Aseptic Processing

Descriptions of current practices in aseptic processing are available in a number of guidelines, articles and books (Center for Drug Evaluation and Research, 1987; (Akers, 1988; Akers *et al.*, 1988; Center for Drug Evaluation and Research, 1987; Fry, 1987; Hofmann, 1993; Trappler, 1993), so this section will not devote significant space to the subject. However, many of the requirements necessary for successful aseptic processing will be reviewed here so as to furnish the reader with a baseline of information.

2.1.1. FACILITIES

The facilities that house an aseptic processing area must be logically planned so as to facilitate the unidirectional flow of components, equipment, bulk product, and filled/sealed product containers through the area. An example of such a facility is shown in Fig. 1. The design should provide an increasing level of cleanliness as the process moves toward the final aseptic filling, lyophilization, and sealing area(s). The heightened levels of cleanliness are achieved by increasing air quality levels, controlling access to the areas (airlocks and gowning areas), and increasing the degree to which the personnel are gowned.

The manufacturing areas should be easily cleaned (i.e., seamless ceilings and walls, and nonporous floors with coved molding) and large enough to allow for the proper orientation of permanent equipment (e.g., filling lines,

*A HEPA filter is an extended-medium dry-type filter in a rigid frame having a minimum particle collection efficiency of 99.97% for particles having diameters >0.3 μm.

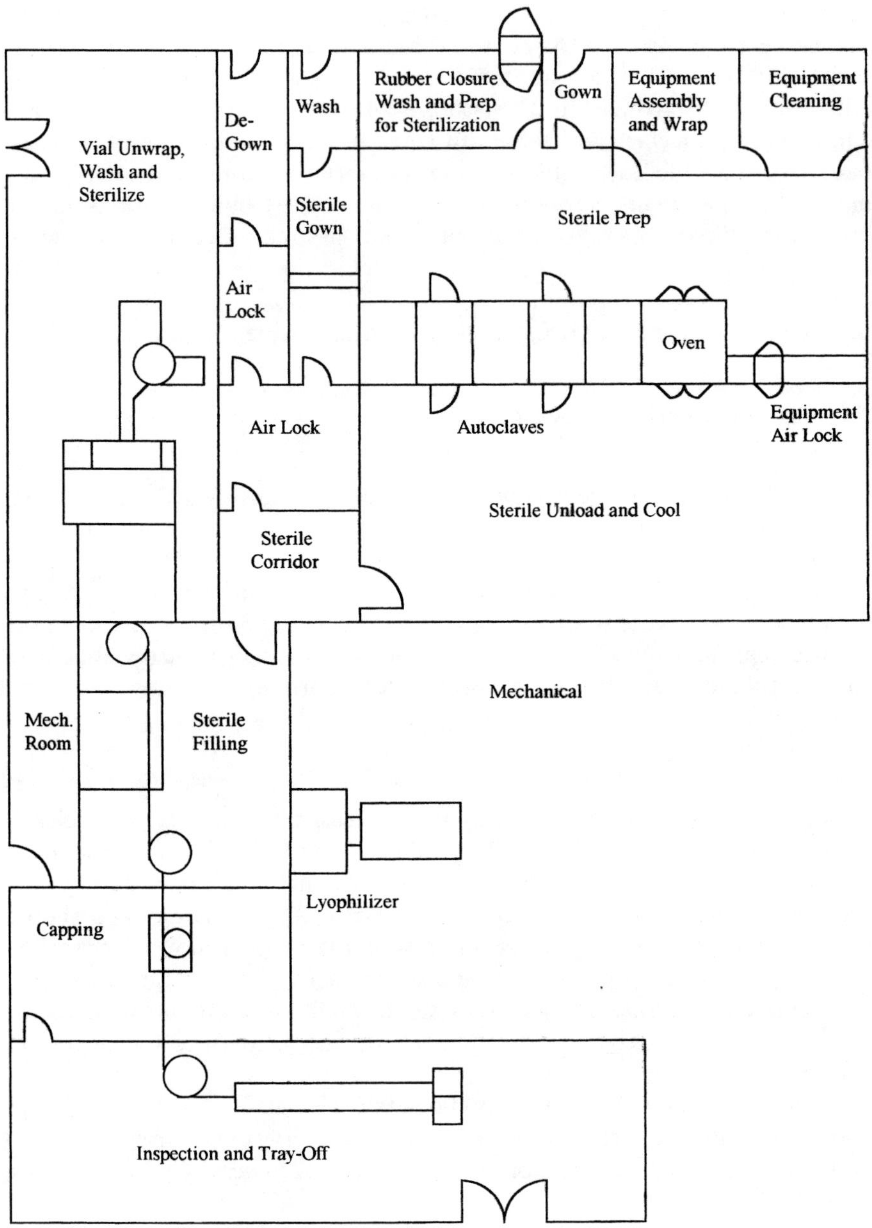
Rubber Closure
Wash and Prep
for Sterilization
Gown
Equipment
Assembly
and Wrap
Equipment
Cleaning
Wash
De-
Gown
Vial Unwrap,
Wash and
Sterilize
Sterile
Gown
Sterile Prep
Air
Lock
Oven
Air Lock
Autoclaves
Equipment
Air Lock
Sterile Unload and Cool
Sterile
Corridor
Mechanical
Mech.
Room
Sterile
Filling
Lyophilizer
Capping
Inspection and Tray-Off

lyophilizers) and unencumbered movement of personnel and portable equipment (e.g., tanks, pumps, tray racks, etc.) around these permanent pieces of equipment. The materials of construction should be non- or low-particle shedding and impervious to aggressive cleaning and disinfecting solutions.

2.1.2. UTILITIES

Water, compressed gases, clean steam, and heat, ventilation, and air conditioning (HVAC) services must all meet the current good manufacturing practice (cGMP) regulations for parenteral drug product processing (U.S. FDA, 1996). These include descriptions of the types of materials used for construction, required operating specifications, equipment design, and requirements for controlling and testing for microbiological and extraneous chemical contamination.

The quality of water to be used in each of the various processes within the manufacturing area is dictated by cGMP regulations:

- Water for injection (WFI), defined as distilled or reverse osmosis water kept at 80°C in a recirculating loop or batched on a 24-hr schedule, must be used as the final rinse water for all product contact equipment and final product containers and closures. It must also be the water used as a raw material for compounding formulated bulk drug product.

Figure 1. A modern parenteral processing area consists of zones of different degrees of cleanliness. This is designed to act as a barrier to the movement of viable and nonviable contamination from noncontrolled areas into the critical areas where sterile products are filled, and lyophilized, if applicable. The levels of cleanliness are specified in U.S. Federal Standard 209D (Clean Room and Work Station Requirements, Controlled Environment, 15 June 1988), and are defined by the number of particles greater than or equal to specified diameters allowed per cubic foot of air. A diagram of a parenteral facility is shown. The "sterile filling room" is the most critical area, where sterile products and sterilized equipment/components are manipulated while being exposed to the environment. An air cleanliness level of at least Class 100 (i.e. NMT 100 particles $\geqslant 0.5$ µm, NMT 300 particles $\geqslant 0.3$ µm, and NMT 750 particles $\geqslant 0.2$ µm) must be maintained in this area. Rooms in which sterile products are not exposed to the environment, but through which sterilized components must move, like the sterile corridor, and sterile unload and cool areas, are designed to meet less stringent air cleanliness levels (generally Class 10,000). Separate in-going and out-going air locks and gown/degown rooms act as buffers to further protect the cleanest areas. HEPA-filtered air supply, high air flows, and pressure differentials moving from the cleanest to less clean areas, as well as special room finishes, all aid in achieving the environment necessary for parenteral manufacturing.

- Demineralized water, generated either by ion exchange methods and/or reverse osmosis, may be used as process wash or rinse water prior to a final rinse by WFI, when applicable.

Compressed gases used in the aseptic operation that could potentially contact sterile drug product or equipment to be used in its processing must be free of particulate contamination, both viable and nonviable, and must be free of extraneous chemical contamination, for example, compressor oil.

Clean steam, generated from WFI, must be used in the autoclaves or sterilize-in-place (SIP) systems that sterilize product contact equipment.

HVAC systems that support the aseptic manufacturing areas must be designed to prevent microbiological contamination (especially fungal) and maintain low, controlled relative humidity levels during the extremes of outside weather conditions.

2.1.3. PARENTERAL MANUFACTURING

A process flow chart for the manufacturing of a typical lyophilized drug product is shown in Fig. 2. For comparison, a flow chart for a suspension drug product is shown in Fig. 3.

2.1.3.1. Preparation of Product Contact Equipment

Preparation of product contact equipment is identical for all three of the dosage forms. Dirty equipment is disassembled and certain items, such as silicone transfer tubing, are generally discarded. Potentially disposable items such as transfer tubing are often viewed as single-use components, because of the difficulty and cost of developing and validating suitable cleaning procedures. Stainless steel and reusable polymeric parts, for example, gaskets and pump diaphragms, are cleaned in a suitable washing device, such as a high-agitation water recirculation bath, a cabinet spray washer, or an ultrasonic bath, using a validated cycle. After the final WFI rinse, the equipment is dried with filtered compressed air and stored in closed stainless steel containers in a Class 10,000 room. When the clean equipment is needed to manufacture a batch of a protein product it is assembled in a Class 10,000 room, wrapped in particle-free paper or placed in a stainless steel–covered box, and then sterilized in a double door autoclave opening into the aseptic processing suite. Compounding and receiving tanks may be cleaned manually or by an automated clean-in-place (CIP) system. Cleaned tanks

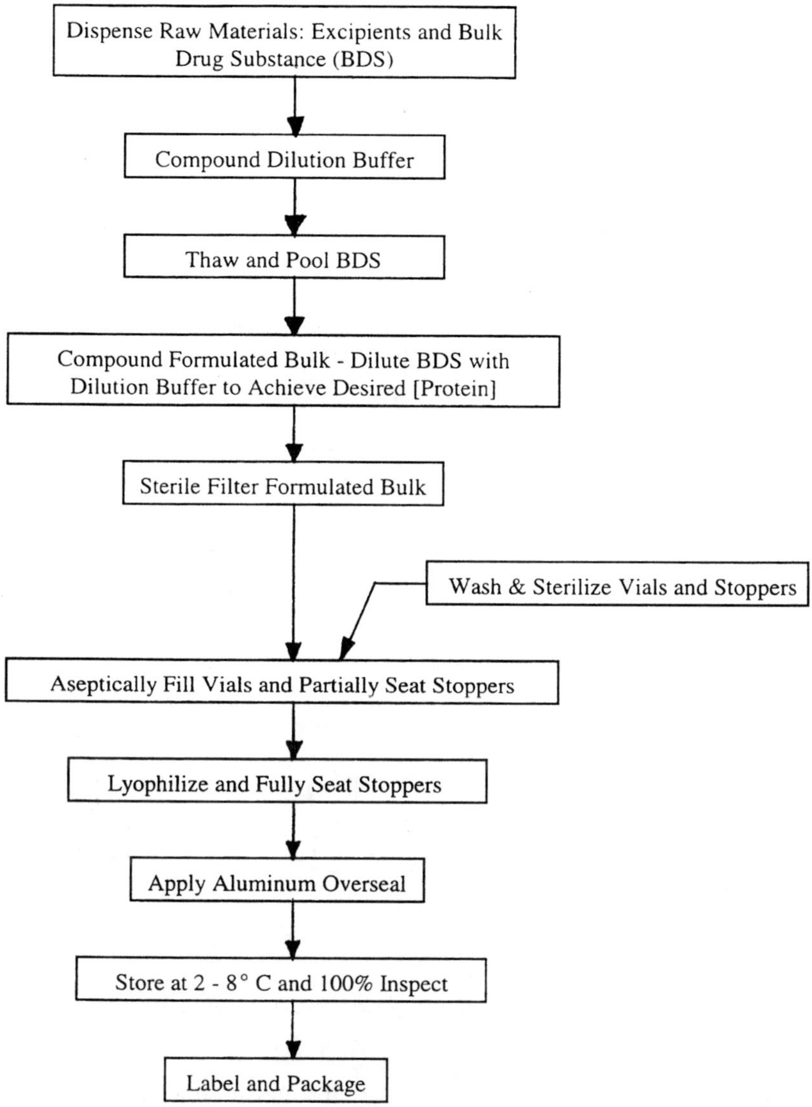

Figure 2. Flow chart of lyophilized drug product manufacturing process when the API is supplied as a frozen solution.

intended as sterile reservoirs for filter-sterilized, formulated bulk drug product may be autoclave sterilized or sterilized-in-place using high-pressure, saturated steam delivered directly into the tank.

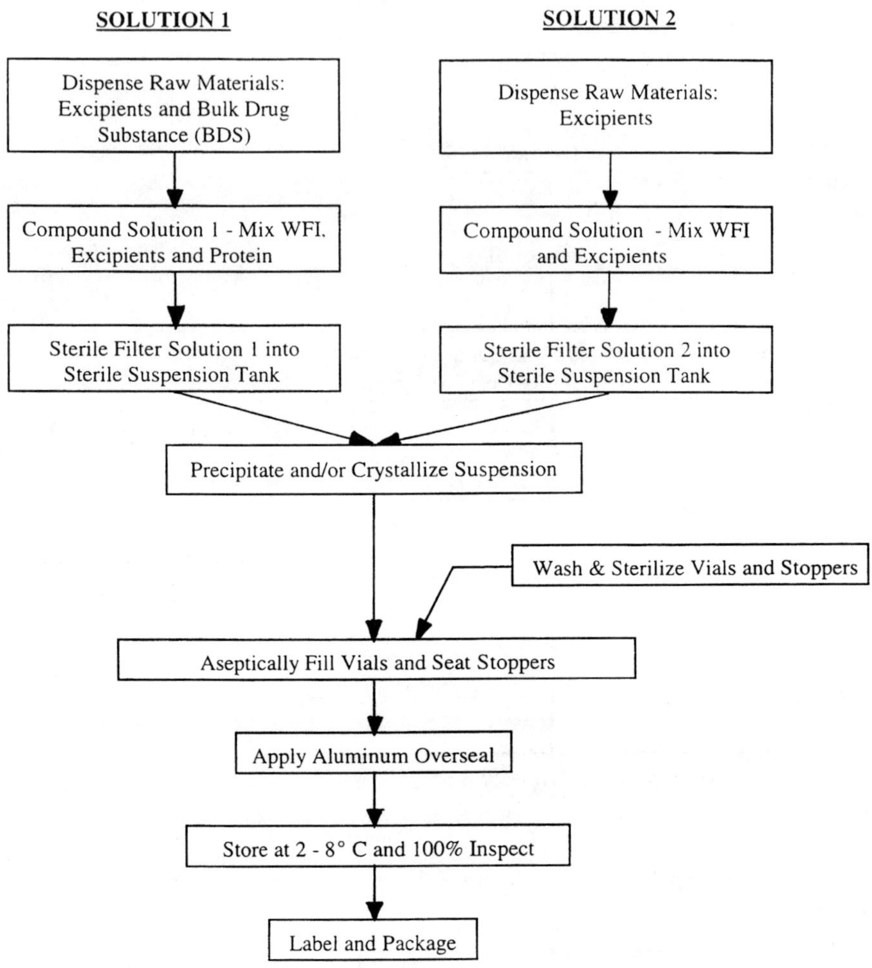

Figure 3. Flow chart of a suspension drug product manufacturing process.

Sterilization and depyrogenation of product contact equipment and packaging components must be accomplished prior to their entry into the aseptic processing area for use in the filling operation. Fundamental aspects of thermal sterilization processes are covered in Chapter 4 of this volume, and there are numerous, more detailed documents describing development and validation of sterilization processes (Parenteral Drug Association [PDA], 1978, 1981, 1996). Depyrogenation can be achieved by thermal destruction or by the removal of pyrogenic materials (bacterial endotoxins) by washing and/or rinsing. Validation of depyrogenation is accomplished by demonstrating that a process can reduce a spiked bacterial endotoxin challenge of at least 1000 USP units to no more than 1/1000 of the starting quantity, that is, a three-log cycle reduction (USP, 2000c, pp. 2144–2145). Many of the component parts of product contact equipment cannot withstand the high temperatures necessary for depyrogenation, which for practical purposes must be greater than 170°C (Nakata, 1994, Tjusi and Harrison, 1978). Therefore, depyrogenation is accomplished by a combination of appropriate cleaning and a final WFI rinse. Subsequent sterilization is carried out using saturated steam, either in an autoclave or via an SIP system.

2.1.3.2. Component Preparation

Preparation of packaging components is also identical for all three of the dosage forms.

(a) Glass containers (vials, syringes, or ampoules) are washed in either a fully automated or semiautomatic washer that employs multiple water and air injections at differing temperatures to help clean the vials by thermally shocking them. The removal of glass spicules (small glass particles, sometimes partially fused to the vial wall, left in the containers as a result of the vial-making process) from the interior of the glass containers is best accomplished by using an ultrasonic bath prior to the water and air blasts. The final WFI rinse occurs immediately before a siliconization step, if it is required for syringe barrels or as an antiadsorptive coating.

(b) The glass containers may be batch-sterilized and depyrogenated in hot air convection ovens at temperatures of 200–250°C, or in a continuous-process, tunnel oven at approximately 300°C through which the containers pass on a mesh, metal belt. In both cases, the ovens are supplied with HEPA-filtered, Class 100 air. Heating of the glass containers in a tunnel oven occurs either via direct radiation from infrared heating elements or convection from heated air that is forced through high-temperature HEPA filters located immediately over the glass containers on the moving belt.

(c) Rubber closures are prepared by washing in a suitable washing device, for example, a sparging overflow kettle, or a fluidized bed washer. In many cases, rinsing with heated WFI along with air sparging to generate agitation will optimally clean rubber closures. The use of detergents or other cleaning agents during washing to facilitate removal of leachables, manufacturing residues, or particulate matter must be carefully evaluated during the development of the closure cleaning cycle. Many of the newer rubber formulations currently used for closures contain little or no undefined filler material that could be a source of leachables. However, if cleaning agents must be employed, there is an added burden on the rinse portion of the cycle to assure the reduction of these potential extractables (i.e., the cleaning agents) to levels that would not adversely affect the stability of the drug product formulation.

(d) If siliconization of rubber stoppers is required, and it is in most cases, in order to achieve filling line machinability and prevent clumping during autoclaving, it may be carried out in the same vessel in which the washing took place or the closures may be transferred to another piece of equipment for this operation. Silicone may be added as a dispersion in water or as an emulsion, which contains emulsifying agents. Closures are then placed in either stainless steel trays with lids or in combination Tyvek/heat-resistant plastic autoclave bags and sterilized in a double-door autoclave opening into the aseptic processing suite. As with the product contact equipment, depyrogenation of closures is accomplished using optimized washing techniques and a final WFI rinse.

2.1.3.3. Compounding

The manufacturing differences for the three dosage forms (solution, lyophilized, and suspension) first appear at the compounding stage. For a solution or lyophilized product, the bulk protein, either as a dried powder or as thawed frozen protein concentrate, is mixed with the excipient solution. The resulting solution is sterile-filtered into a sterilized receiving vessel prior to filling. Parenteral suspension dosage forms are produced in two ways: (1) by aseptically dispersing a sterile powder of known particle-size distribution into a sterile liquid medium in which it is insoluble or (2) by combining two sterile solutions which bring about the formation of an amorphous or crystalline solid phase *in situ* (M.J. Akers *et al.*, 1987). The sterile powder referred to in method 1 may be formed by bulk lyophilization and milling to achieve a uniform particle distribution, by precipitation or recrystallization, or by spray drying. An oil-based suspension of recombinant bovine somatotropin is an example of a protein suspension prepared by this method. For method 2, multiple solutions must be compounded, sterile-filtered, and then aseptically mixed to generate the suspension solid phase. Examples of protein product suspensions manufactured in such a manner are insulin zinc suspensions and isophane zinc insulin suspensions.

For a suspension generated by combining two filter-sterilized solutions (method 2 above), if the solid phase is to be crystalline, a hold time may be required to allow for crystallization to take place. Crystallization may occur *de novo* or through an intermediate amorphous precipitate, which generates the microenvironment necessary for crystallization to be initiated and then propagated. Examples of critical parameters in this step may be the temperatures of the solutions, the agitation rates, and the hold times. A terminating solution may be added to the crystalline-growing medium to end the crystal growth process and stabilize the crystals once the desired particle size has been achieved.

2.1.3.4. Filling

Filling of solution and lyophilized dosage forms can be carried out on virtually any of the currently available filling machines, for example, positive-displacement syringe filler, rolling-diaphragm pump filler, or time-pressure filler. The different types must be evaluated for their compatibility with the protein product in question. Filling rates, needle sizes, pump sizes, the need for bottom-up filling, and the manner of supplying an inert gas for an oxygen-sensitive product must all be part of the process development work. In order to assure elegant dried cakes, lyophilized formulations must be filled in such a manner as to avoid splashing of product onto the walls or neck area of the vial or foam formation during filling.

For oxygen-sensitive products, an inert gas such as nitrogen may be introduced to replace the air in the headspace of compounding and sterile holding tanks as well as the final product containers. On the filling line this may be accomplished by flushing the vial with nitrogen from a delivery needle prior to, during, and/or after solution filling. The nitrogen must be sterile-filtered prior to its introduction into the tanks or vials to prevent viable and nonviable particulate contamination. The gas pressure, flow rate, and purge time must be optimized during process development so as to meet the oxygen tension specifications necessary to prevent oxidative degradation of the product. Closure insertion and capping immediately after filling/nitrogen purge should prevent air contamination.

Filling of suspension dosage forms requires that the solid phase be maintained homogeneously distributed throughout the suspending medium during the entire operation. This can be accomplished by using a suitable sterile holding tank mixer, for example, a VIBRO-Mixer or a magnetically coupled mixer (see Section 3.1.1.4 for a description of these mixers), and a recirculation system which transports the bulk suspension formulation from the holding tank to a manifold, from which the filling pumps are fed, and then back to the holding tank. A peristaltic pump can be used to power the recirculation system, because of its suitability for a closed, aseptic operation. The recirculation system not only must transport the formulation to the filling machine, but also maintain the homogeneity of the suspension during its transit. Mixing during recirculation can be achieved by the maintenance of turbulent flow in the system. The diameter of the tubing (or piping) and the pump speed will affect the amount of turbulence or mixing to which the suspension is subjected during recirculation. The recirculation system must be free of dead ends and irregular curved or kinked paths at which suspension particles could collect leading to heterogeneity. The sedimentation properties of the suspension particles in the formulation will determine the degree of mixing turbulence required to maintain homogeneity from the holding tank into the vial.

The design of the recirculation system must take into account the viscosity of the suspension formulation, the particle size and density of the suspension, the morphology of the particles (globular, needle-like, etc.), and the fragility of the suspended particles. All mixing and recirculation parameters must be optimized to balance the requirement for maintenance of suspension homogeneity with the necessity to avoid damaging or altering the suspension particle size or morphology. All mixing and recirculation systems impart varying degrees of stress to the formulation, mostly in the form of shear stress to the suspension particles. Long exposures may result in significant abrasion or breakage of the suspension particles, potentially altering their dissolution characteristics. Therefore, the recirculation system design should be optimized to avoid suspension particle damage for the maximum time period during which the suspension might be exposed during a filling operation.

2.1.3.5. Lyophilization

Lyophilization is an important unit operation for production of either final pharmaceutical product or a process intermediate, where it may be necessary to concentrate the protein or to store it in a relatively non-growth-promotive state. Aseptic processing considerations apply primarily to production of finished, sterile lyophilized products, and is the focus of this discussion.

Lyophilization takes place in chambers incorporating sanitary design principles. Current FDA guidelines require sterilization of the chamber/condenser after each batch of final pharmaceutical product. Newer lyophilizers have CIP and SIP capabilities, which greatly increases the assurance of achieving reproducible levels of cleanliness and desired probabilities of sterility.

A lyophilizer chamber and condenser leak test should be performed prior to each cycle to assure that the sterility of the product load will not be compromised due to an adventitious leak. This can be done with a vacuum hold test, which can be run and evaluated by an automated lyophilizer controller. If all of the volatile materials have been removed from the chamber before the test, there should be minimal pressure rise observed during the vacuum hold period (Trappler, 1993).

Lyophilization trays that must be transported outside of HEPA-filtered, laminar flow protection before loading into the freeze-dryer chamber should be equipped with covers to protect the partially stoppered vials from particulate contamination. Tray covers are removed immediately before loading the trays into the chamber. The entrances to the lyophilization chambers should be protected by HEPA-filtered, laminar flow air to protect open vials during the loading phase. Trays of product can be loaded into the lyophilizer either manually or using automated tray loaders, usually onto shelves that have been preset to the initial temperature for the cycle, for example, approximately + 5°C. It is desirable to remove tray bottoms after loading to achieve better and more uniform heat transfer from the shelves to the vials. For bottomless trays, rectangular rings usually remain around the vials from each tray to maintain tray identity and to aid in replacing the tray bottom for unloading following cycle completion.

In order to monitor the temperature profile of the product in the vials during lyophilization, small thermocouples are usually placed in vials prior to the initiation of the cycle. Probed vials, generally at least three, are placed in trays at different locations within the lyophilizer load, for example, on the top, middle, and bottom shelves of the dryer. Thermocouples should be located in the same position within each vial—in the center of the vial and touching the bottom—because this is the location where the last ice sublimes during drying. Special thermocouple holders are available to facilitate precise placement (see Fig. 2 in Chapter 6 of this volume). The thermocouple wires should be of large enough gauge (small enough diameter) so as not to add significant mass or volume to the filled solution, which may alter its freezing or drying characteristics.

Placement of thermocouples in the vials, by necessity a manual operation, poses a risk to sterility assurance, because of the proximity of the operator and his or her activities to other open vials in the load. This raises the issue of where the probed vials should be located in a tray and also at what position on the lyophilizer shelf. The thermocouple wire, initially sterile before being handled by the operator while being located in the vial, can pose a potential hazard of both viable and nonviable particulate contamination to open vials over which it may be placed. Because of lateral heat transfer from the walls and door of the chamber and the resulting "edge effect" causing faster drying of vials at the edge of the shelves, the most representative location for thermocoupled vials is in the center of a tray at the center of a shelf. However, some manufacturers believe that the risk to the product in terms of decreased sterility assurance outweighs the benefit of monitoring the most representative vials. Therefore, many manufacturers choose to place probed vials at the front edges of trays located closest to the lyophilizer door, where the risk to product sterility is minimized. The probed vials are then discarded after the completion of lyophilization.

In addition to the trade-off between process monitoring considerations and sterility assurance aspects of the process, formulation scientists and process development personnel should recognize that temperature data from thermocoupled vials are biased relative to the rest of the batch. This arises because the thermocouple acts as a seed site for heterogeneous nucleation of ice crystallization. As a result, monitored vials undergo less supercooling, freeze slower, have a larger average ice crystallite size, and dry faster than nonmonitored vials. This bias must be taken into account when determining freezing, primary drying, and secondary drying times for initial cycle development as well as during scale-up. These points underscore the need for improved process monitoring technology that does not depend upon monitoring of individual vials. This point is discussed further in Chapter 6.

The vacuum level in the chamber and condenser should be controlled automatically during the cycle by the addition of either dry, sterile-filtered air or nitrogen. A capacitance manometer should be used to measure the vacuum level for both monitoring and controlling purposes. Vacuum level in combination with the shelf temperature and the inherent heat and mass transfer characteristics of the frozen and partially dried cake affect the product temperature of the frozen solution in the vials. The product temperature must be maintained below thermal transition events above which meltback or collapse of the cake may occur, for example, eutectic temperature (T_e) or collapse temperature (T_g'). These thermal transitions are characterized during formulation and lyophilization cycle development using a variety of techniques, such as thermal analysis and lyomicroscopy. The cycles developed to accommodate these critical parameters must be checked out at both lab and commercial scale to evaluate the ability of a scaled-up lyophilizer to achieve adequate temperature and vacuum control to prevent collapse.

After the completion of the lyophilization cycle, stoppering of the vials is done while the vials are still in the chamber and after the required headspace environment and pressure have been established. Products that are sensitive to oxygen could have dry, filtered nitrogen introduced into the chamber or a high level of vacuum could be maintained, for example, 100- to 200-μm Hg vacuum. In most cases stoppers are seated in the vials with at least a partial vacuum remaining in the vial, for example, approximately 1 psig chamber pressure, so as to assist in maintaining container/closure seal integrity during subsequent processing and storage. Stoppering takes place by compressing the shelves together with a mechanism that is completely within the lyophilizer chamber so as to avoid the potential for external contamination of the load. Properly siliconized stoppers are a necessity to assure that they will easily slide into the vial neck and seat fully in the vials, and then will not pop out of the vial after the shelf pressure is removed.

2.1.3.6. Capping

Assuring that the contents of the vial remains sterile requires the application of an aluminum overseal to keep the stopper properly seated in the vial and to maintain container/closure seal integrity. This operation, commonly termed "capping," should be done in a protected environment. Until the vials are capped they are still vulnerable to inadvertent contamination, because there is nothing in place to firmly compress the stopper flange onto the vial crown to form a secure seal. Capping is generally done in a room immediately adjacent to the aseptic filling room with an HEPA-filtered, laminar flow hood covering the capping machine and the conveyor transporting the vials out of the aseptic area and into the capper. Cappers are known to generate microscopic aluminum particles during their operation, which is the main reason that this operation is performed outside of the aseptic area. Container/closure seal integrity depends on the application of adequate seal compression force on the stopper. This important parameter can be monitored by utilizing a device to automatically measure the forces applied during the sealing process on each vial. This attachment to the capper, such as a seal force monitor (e.g., Genesis Machinery Products, Inc., Exton, PA), will monitor sealing forces and reject vials to which insufficient compression force has been applied during the sealing process.

2.1.3.7. Inspection

All parenteral product lots are 100% inspected after the completion of the capping operation and before labeling takes place. For solution products the objectives of the inspection are to determine (1) if the container/closure/seal system is defective in some way, for example, a vial missing a closure, cracked glass container, or inappropriately crimped aluminum overseal, and (2) if the vial contents contain particles of foreign matter. Evaluation of solution color and clarity are also part of an inspector's charge during the inspection process. Product containers that are defective are removed from the batch and rejected. A tabulation of the numbers and types of defects is maintained for each batch of product inspected. This information is critical in determining the final disposition of the batch by the quality control group.

For protein and peptide solution products, the inherent aggregation characteristics of the molecules may result in a decrease in solution clarity and the appearance of opalescence. The *European Pharmacopoeia* addresses the issue of small, uniformly suspended particles in a product container by requiring an objective evaluation of solution clarity measured in reference to standards (European Pharmacopoeia, 1997). Eckhardt *et al.* (1994) describe the development of a spectrophotometric assay that can be correlated with the opalescence reference standards, effectively eliminating the need for their repeated preparation. Such an analytical tool may aid an inspector in estimating the degree of solution opalescence or haziness during an acceptability evaluation.

Inspection of lyophilized products includes the same container/closure defect evaluation as described above, but the examination of the vial contents focuses on the consistency and appearance of the lyophilized cake. Lyophilization should yield a cake that retains the same volume and shape as the starting solution with no evidence of collapse, melting, or foaming. Partial collapse is often observed in freeze-dried cakes of protein formulations, at least in part because of low collapse temperatures of some protective solutes (see Chapter 6). If protein stability is significantly lower in a partially collapsed matrix, then collapse becomes more than a pharmaceutical elegance issue: Areas of local collapse which go undetected could result in subpotent product vials prior to the expiration date. For this reason, it is good practice to carry out special stability studies on vials culled from scale-up or production batches which contain partially collapsed cakes. The stability of the protein in these cakes is then compared with that of normal cakes. If the stability in a partially collapsed matrix is not significantly different from that of noncollapsed cakes, then these data can support the position that collapse is a cosmetic defect, with more liberal inspection acceptance criteria.

Inspection of a suspension product focuses primarily on defects in the container/closure system. This is because the product suspension particles prevent detection of contaminating particles or fibers, unless they are visually distinguishable from the product suspension particles.

Visual inspection of solution product containers may be conducted by trained and qualified inspectors, automated equipment that incorporates an image analysis system for detecting defects, or a combination of both. Human inspections must be carried out in specially lighted enclosures having both a light and a dark background against which the product containers are viewed. Groves (1993) presented a comprehensive description of the procedures, operator training, and equipment that is used for visual inspection of parenteral product containers.

Automated equipment must be calibrated using particles of known sizes seeded into the product solution, and then the detection process in the product containers must be validated. Knapp and Abramson (1990) published a performance-based validation procedure for automated particle inspection systems that is applicable to a wide variety of machines.

Swirling or other types of agitation are used in both human and machine inspections to suspend contaminating particles so they can be detected. For protein or peptide products, agitation employed during the inspection process may have to be minimized, because many proteins and peptides are sensitive to interfacial denaturation that may be induced by such movements (see Section 3.1.1).

Inspection of lyophilized and suspension product containers can only be conducted by trained and qualified human inspectors. Automated systems do not have the capability, at this time, of evaluating the quality of lyophilized cakes. Suspension products are opaque, or at best transluscent, which would also preclude an automated inspection process.

2.1.4. MEDIA FILLS AND ENVIRONMENTAL MONITORING

Media fills, air filter certifications, and routine environmental monitoring of the aseptic processing area and personnel are all part of the evaluation plan for assuring that aseptic operations meet predetermined specifications and yield products having a high level of sterility assurance. Media fills are an integral part of the validation (and revalidation) of aseptic filling operations. Suitable microbial growth media, for example, tryptic soy broth, is used as a surrogate for the drug product formulation, so that any microorganisms that may contaminate a vial during processing would be detected by the appearance of microbial growth and turbidity in the vial contents after appropriate incubation times and temperatures. Aseptic filling processes must demonstrate an SAL of 1×10^{-3} or lower.

Media fills that simulate the inclusion of lyophilization in the manufacturing process must also attain this same SAL. FDA guidelines indicate that only certain aspects of the lyophilization operation should be validated and thus mimicked in a media fill (U.S. FDA, 1993a). These include the transportation of the filled and partially stoppered vials to the lyophilizer, the loading of the trays of vials into the lyophilizer, and the full insertion of the stoppers into the vials, which occurs at the end of a cycle.

The guidelines suggest that neither freezing nor drying should be carried out after loading the media vials, because those steps may invalidate the results by killing some contaminating microbes or by drying the growth medium and making it less supportive of microbial growth. The sterilization of the lyophilization chamber and the gas backfill system must be validated separately, and is not the objective of the media fills.

Maintaining a suitable aseptic environment in which sterile drug products can be manufactured requires a well-defined plan for routinely evaluating the viable and nonviable particulate levels within the aseptic processing area (Munson and Sorensen, 1990). This includes the use of air particle monitors to quantitate total airborne particulates and either active or passive microbial monitoring methods. Microbial monitoring of the air is conducted via passive settling plates which contain suitable agar growth media and by devices that actively draw air in and capture viable microorganisms. The three types of active devices are liquid impingement, membrane filtration, and agar impaction. Surfaces within the aseptic processing area are also sampled to evaluate microbial levels. Microbial counts per unit surface area are obtained by using contact agar plates or swabbing a specified area as defined by a template.

Monitoring the aseptic operator's gowns and gloves for microbial contamination using agar contact plates is also routinely performed to assure that proper gowning techniques and aseptic procedures are being used in the aseptic area. Personnel monitoring is critical because it is well known that the major source of viable particulates in an aseptic processing area is of human origin.

2.2. The Future of Aseptic Processing

It is well accepted now that the next quantum leap in parenteral processing will be achieved by the optimization and commercialization of barrier or isolation technology as a substitute for conventional clean room operations (Haas, 1995). Several manufacturers are currently marketing their versions of such a technology (e.g., Bosch/TL Systems, LaCahlene, Despatch). There are potentially two main benefits to be gained from its use: (1) an economic benefit, because the expensive design, construction, and operation of typical clean room suites is not required for the use of this technology and (2) a potential for greater sterility assurance, because humans are effectively removed from the environment in which all aseptic operations are performed.

One objective of barrier technology is to enclose the aseptic operations so as to isolate them from humans, because particles shed from humans, both viable and nonviable, are the major source of contamination in a clean room. At the same time, a barrier system should still allow for indirect human intervention for machine setup, remedying machine interruptions, and repair. This objective is being addressed in several ways by different suppliers of barrier technology systems. All of them share one common characteristic, though, minimization and enclosure of the controlled, ultraclean environment in which aseptic processing is carried out. All of the designs employ a system for controlled entry where equipment and supplies being moved into or out of the enclosure must be either chemically or thermally sterilized so that the sterility assurance within the barrier can be maintained to levels $<1\times10^{-3}$. In fact, some suppliers of barrier systems are claiming sterility assurance levels as high as 1×10^{-5}. Indirect human access is achieved by the use of plastic half-body ports or rubber glove ports. The equipment and surfaces within the barrier are chemically sterilized and the barrier is supplied with HEPA or ultra low penetration air (ULPA)-filtered* air to minimize particulate contamination of the filled drug product. The residual levels of the chemical sterilants used in both the main enclosure and the airlocks in these barrier systems must be evaluated for their potential to adversely affect the protein and peptide products that may be processed within them. This topic will be discussed in more detail in Section 3.6.

A barrier system could be designed that isolates virtually the entire parenteral manufacturing operation from beginning to end. Such a system would require detailed design and engineering to overcome the difficulties of materiel movement between different pieces of processing equipment with only minimal human assistance. This would be exceptionally challenging for a lyophilized product, where the traying of vials prior to loading of the lyophilizer would require sophisticated robotic tray handlers and loaders. In addition, to maintain the operational readiness and high production output expected of commercial facilities, such a design would have to address the difficulties of equipment changeovers in a limited access enclosure between filling operations.

CIP and SIP systems could be employed to clean and sterilize much of the equipment within the barrier. For use with macromolecular drug products there are cautions that must be heeded in the development of such systems, which will be discussed in Section 3.5.

Another approach might be to bring already cleaned, assembled, and wrapped changeover equipment into the barrier system through a direct connection to a specially designed autoclave. A double-door autoclave with its outlet linked directly to the barrier would avoid the need for time-

*An ULPA filter is an extended-medium dry-type filter in a rigid frame having a minimum particle collection efficiency of 99.999% for particle diameters >0.12 μm.

consuming movement of equipment through airlocks, which requires a chemical sterilization step. The dissipation of heat emanating from the autoclave door during and after the cycles would be engineered into the design. Clean, siliconized stoppers could be prepared outside the barrier and placed into stainless steel containers or sealed Tyvek/plastic bags, which could be sterilized in this autoclave and unloaded directly into the barrier for use in a filling order. Specially designed docking ports through which materiel, such as sterilized stoppers, could be passed into the barrier are another option to be considered in the design. Vials would be washed automatically and then passed onto a belt to carry them through a tunnel oven for sterilization and depyrogenation. The exit from the tunnel oven would open directly into the barrier system and feed the filler/stoppering machine. Such a system has been designed and is shown in Fig. 4.

The interior surfaces of a barrier system are sterilized on a routine basis using chemical sterilants, most of which are vapor-phase systems using agents such as hydrogen peroxide, steam/hydrogen peroxide mix, or peracetic acid (Davenport, 1989; Lysfjord *et al.*, 1995). One of the difficulties with chemical sterilants is the time necessary to carry out a typical cycle, which includes both a gas exposure phase and a subsequent evacuation or degas phase. Vapor-phase hydrogen peroxide mixed with steam has been shown to be an effective sterilizing agent with relatively short cycle times (Lysfjord *et al.*, 1995).

HEPA- or ULPA-filtered air within the barrier is used to maintain a positive pressure differential and protect the sterile, open product containers, both empty and filled, from particle contamination. The air velocities required in these barrier systems to achieve the desired number of air changes per hour is significantly less than for a clean room suite, because of the smaller air volume within the barrier. Air qualities of Class 10 or lower are claimed to be easily attainable because of the high air volume turnover rate.

The filling/stoppering line to be used inside a barrier system must be minimal in size and all parts must be easily accessible within "an arm's reach" from the sides of the barrier enclosure. The trend is to design the lines so that they are narrow and run linearly. The type of liquid filler that would potentially minimize the need for setup, adjustment, and trouble shooting is the time-pressure filler. This filling machine employs an accurate pressure control system in the formulated bulk holding vessel which feeds the filling needles and relies on precisely controlled on/off valves on the line running to each filling needle. The valves can be as simple as pinch clamps on the needle tubing, which would necessitate the least amount of preparation and setup for a filling order, because new tubing could be prepared and sterilized for each new order. Another design objective would

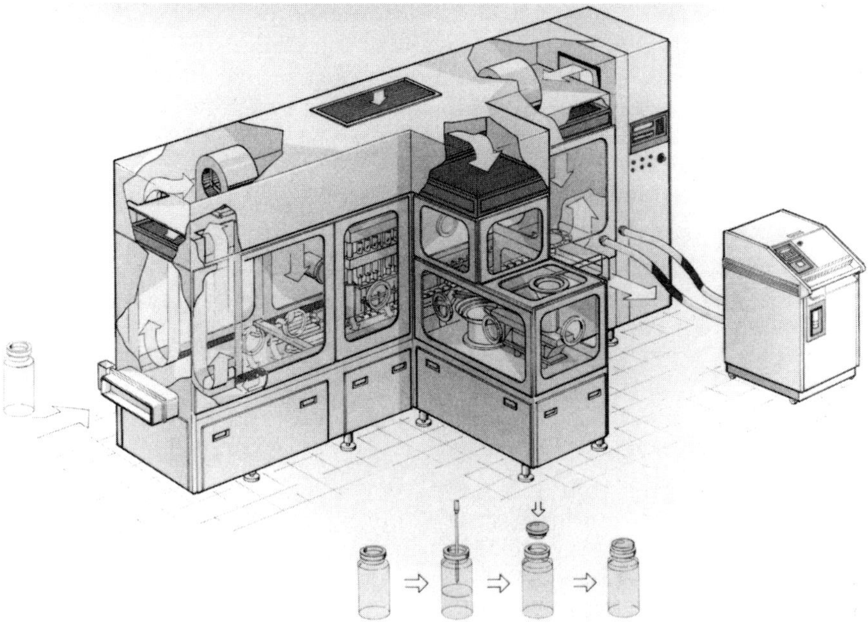

Figure 4. Example of barrier technology system for vial filling (Bosch MLF Machine). Reprinted with permission from TL Systems Corporation.

be to locate as much of the mechanical and electrical hardware for the filling/stoppering machine(s) as possible outside of the barrier system. In this way routine maintenance and repair could be done without having to breach the barrier system.

3. UNIQUE CHALLENGES IN THE ASEPTIC PROCESSING OF PROTEIN PHARMACEUTICALS

3.1. Maintenance of Physical and Chemical Stability during Processing

3.1.1. AVOIDING DENATURATION AND AGGREGATION

Denaturation is defined as the disruption of a proteins's native secondary and tertiary (or quartenary, if applicable) structure and it may occur to varying degrees, depending on the conditions which cause it to occur. It can be either a reversible or an irreversible event, depending on the extent of the unfolding and the subsequent interactions which may occur between the denatured protein molecules. Such interactions may lead to aggregation and possibly precipitation (Manning *et al.*, 1989).

Denaturation can occur as a result of many types of physical and chemical stresses, for example, interfacial exposures in conjunction with surface tension effects (of which foaming is a special case, because of the additional surface area that is generated), shear stress, freezing and thawing, pH extremes, high temperatures, and high salt concentrations. The following sections will discuss these stresses in greater detail as they affect the outcome of protein drug product manufacturing.

3.1.1.1. Foaming and Interfacial Denaturation

During the processing of protein-containing formulations, denaturation of proteins can occur at air/liquid, liquid/liquid, or solid/liquid interfaces (Lundstrom, 1983). Protein molecules come into contact with these interfaces through diffusion, convective flow, pumping, and agitation.

Assuring complete dissolution and a homogeneous distribution of the formulation components during compounding depends on proper mixing. Depending on the batch size to be manufactured and the tank geometry, the properly sized and designed mixing impeller is selected to achieve the optimal mixing dynamics while minimizing foaming of the solution. Proteins and peptides have varying degrees of surface activity and many formulations of these drugs also contain other surfactants to aid in preventing air/liquid interfacial denaturation. Foaming greatly increases the amount of air/liquid interfacial area in the solution and increases the potential for protein denaturation and insoluble aggregate formation. Even though foaming may be produced by high-shear processes, for example, a high-shear mixer entraining air, interfacial denaturation as a result of foaming is a distinct category of stress that can lead to denaturation in a manner which differs from shear-induced denaturation.

The surface tension of the liquid plus any expansion and contraction at the air/liquid interface is hypothesized to be sufficient to entangle the already conformationally modified proteins adsorbed to the interface, potentially leading to denaturation and aggregation (Kaplan and Fraser, 1953; MacRitchie, 1978). Such an occurrence could adversely affect the potency of the formulation and may also result in the formation of visible particulate matter. Proteins have regions which exhibit greater hydrophobic character than other parts of the molecule. With a protein in an aqueous solution, the hydrophobic regions will tend to be found in the interior of the molecule, an environment which excludes them from water. Thermodynamically, such an architecture minimizes the protein's free energy. However, this thermodynamic driving force also underlies the interfacial denaturation of a protein as it unfolds to expose or insert its hydrophobic domains into or onto a nonpolar region at an interface. Such locations can be found at the interfaces of the liquid and its surroundings, for example, the gas headspace in the container, at immiscible liquid-like silicone oil droplets on the surfaces of the container or suspended within the liquid, and containment walls, which are made of hydrophobic materials such as polystyrene, polyvinyl alcohol, polyvinyl chloride, or other polymers.

3.1.1.2. Foaming during Compounding and Filtration

Foaming generally occurs as a result of vortexing, which causes air entrainment and bubble formation. Vortexing can be minimized or eliminated by selecting the proper impeller for the application. Impeller diameter and pitch should be considered in relation to the height/width ratio (H/W) of the tank and solution height in the tank. The impeller diameter utilized is generally one-fourth to one-half the tank diameter (Tatterson, 1991, pp. 2–5).

The design of the impeller blades with either a fixed angle (axial flow turbine impeller), a variable, radial pitch angle (such as a square-pitch marine-type impeller), or a radial flow, flat-blade turbine impeller will affect the "pumping" capacity of the impeller and its tendency to cause vortexing (Oldshue, 1997). The variable, radial pitch impeller, termed a "marine impeller," will have greater "pumping" capacity, which will make it suitable only for tanks with relatively large H/W ratios and solution heights $>0.7H$. The fixed-pitch impeller will have less of a tendency to cause vortexing, because there is less "pumping" capacity near the rotor shaft. This phenomenon occurs because the lower angular velocity of the blades near the rotor shaft is not compensated for by an increasing blade pitch, as it is by the marine impeller. This makes the fixed-pitch impeller better suited to tanks with smaller H/W ratios and lower solution heights.

Other procedures that should be considered to optimize tank and impeller setup are properly setting the rotor shaft speed (for reproducible speed control, controllers should have built-in tachometers), positioning the mixing shaft at an angle to and offset from the longitudinal axis of the tank (see Fig. 5), and locating the mixing impeller as close to the tank bottom as practical (Tatterson, 1991, pp. 8, 221). Impellers of varying designs, examples of which are shown in Fig. 6, are currently available commercially.

Magnetically coupled stirrers such as the MagMixer™ MBI (Lightnin, Rochester, NY) or the NA-mixer® (Nov Aseptic AB, Nödage, Sweden) are alternative types of mixing devices for stainless steel tanks. A shaft, on which the specially designed mixing impeller rotates, must be permanently welded in place on the inside bottom of the tank. On the outside of the tank is mounted the magnetic drive unit which causes the magnetic impeller inside the tank to rotate on the shaft-bearing surface. The impellers are specially designed for use with suspensions or solutions and are sized appropriately for the tank geometry. The advantage of this design is that it generally requires much slower rotational speeds than conventional impellers to achieve optimal mixing. This also results in less foaming.

For smaller batch sizes, <20 L in size, a magnetic stir bar may be utilized in place of an impeller on a shaft. Stir bars must be selected with as much care as impellers to address the same issues as have already been discussed.

Foam can also be formed when two solutions are combined (with one or both containing a surface-active compound) by pouring one into the other directly, or pumping one solution into another through a dip tube that does not extend below the liquid level in the receiving vessel. Such transfers occur when thawed bulk protein, which is stored frozen, must be pooled to yield sufficient material for a drug product batch. The thawed active pharmaceutical ingredient (API) is poured or pumped into a compounding vessel and then formulation buffer is added to dilute the protein to the desired potency (Note: this operation may be done in reverse order; see Section 3.2.1.2 for a discussion of other potential stability issues.) During parenteral production there are other instances of protein-containing solutions being moved from one vessel to another, as when solutions are combined to produce a suspension or to carry out filtration, all of which can yield foam.

3.1.1.3. Foaming during Filling

Foam generated during the filling of a lyophilized formulation can result in the formation of a cake with an unacceptable appearance. This can generally be avoided by utilizing a "bottom-up" or "diving" filling needle movement so that the distance from the bottom of the filling needle to the top of the liquid in the vial is minimized throughout the fill. The flow through a typically sized filling needle used for up to a 10-ml fill

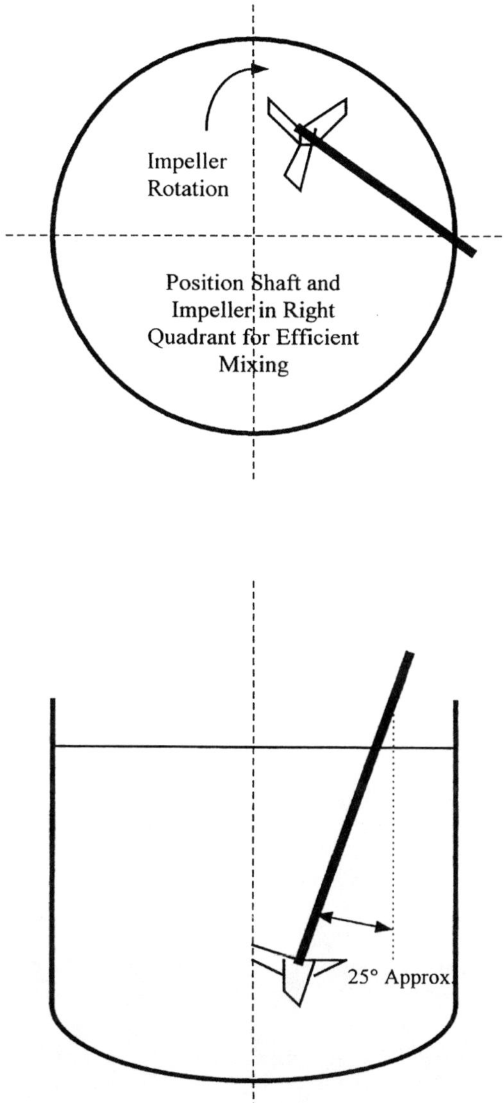

Figure 5. Demonstration of eccentric mixer positioning in a tank to achieve optimal mixing.

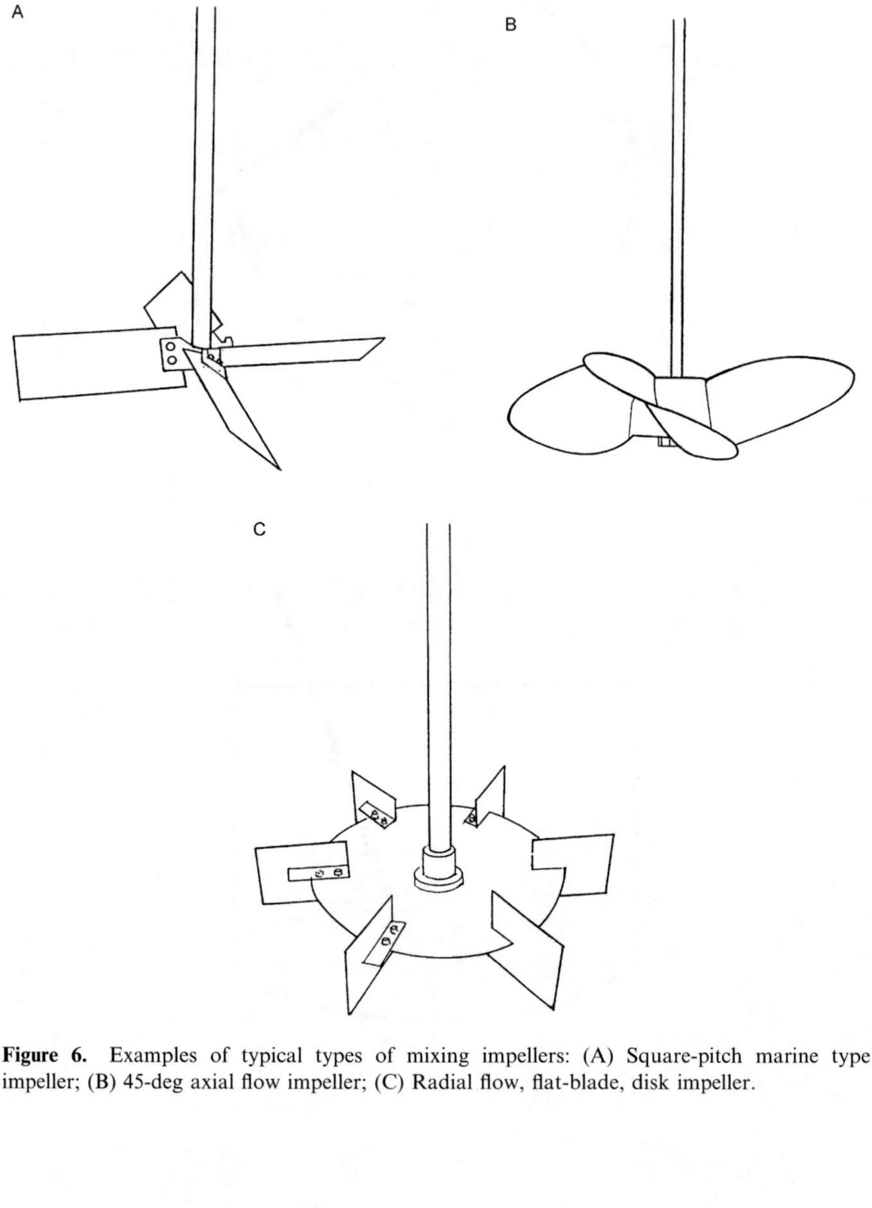

Figure 6. Examples of typical types of mixing impellers: (A) Square-pitch marine type impeller; (B) 45-deg axial flow impeller; (C) Radial flow, flat-blade, disk impeller.

(filling speed of 150–300 vials/minute on line having eight needles) is turbulent in nature [as shown by Reynolds Number calculations (Bird *et al.*, 1960, p. 42); data not shown] and will easily entrain air once it leaves the filling needle. Utilizing larger diameter filling needles and/or slowing the filling rate will also help reduce the formation of foam during the filling operation.

3.1.1.4. Foaming and Comminution of Suspensions

Foaming and difficulty in wetting suspension particles are two problems often encountered when a sterile powder is being combined with a vehicle during compounding (M.J. Akers *et al.*, 1987). The powder must be combined with a minimum volume of vehicle and entrapped air must be removed so that complete wetting can occur. After this has been achieved then the remainder of the vehicle can be added without causing excessive foam formation.

Suspension formulations must be agitated and recirculated during filling in order to maintain a homogeneous distribution of solids. A specialized agitation system must be used for sterile holding tanks which does not compromise the integrity of the tank and the sterility of its contents. Examples of systems that have been used for this application are VIBRO-Mixers and magnetically coupled mixers. As previously described, the stirring rate of a magnetically coupled mixer is generally slow enough so that foam is not generated. On the other hand, a VIBRO-Mixer (B. Braun Biotech International, Melsungen, Germany) can generate foam (see Fig. 7). A VIBRO-Mixer consists of a mixing plate mounted on a shaft that extends into the liquid. The plate is perforated by perfectly round holes, which form the open ends of truncated cones, or venturi openings that have been fabricated into the plate. To mix a solution or suspension the plate is moved vertically in a reciprocating action at high frequency, producing continuous jets of liquid from the venturi openings; the degree of agitation generated is controlled by varying the amplitude of the reciprocating movement. Plate location in the tank, orientation of the venturi openings (forcing flow up or down), amplitude of the plate vibration, and tank bottom design can affect both mixing efficiency and the tendency for foam formation.

A recirculation system employing a peristaltic pump, flexible tubing, and a stainless steel manifold may be used for delivering the suspension formulation to the filling pumps and maintaining its homogeneity. The tubing diameters, the pump, and the manifold dimensions must all be

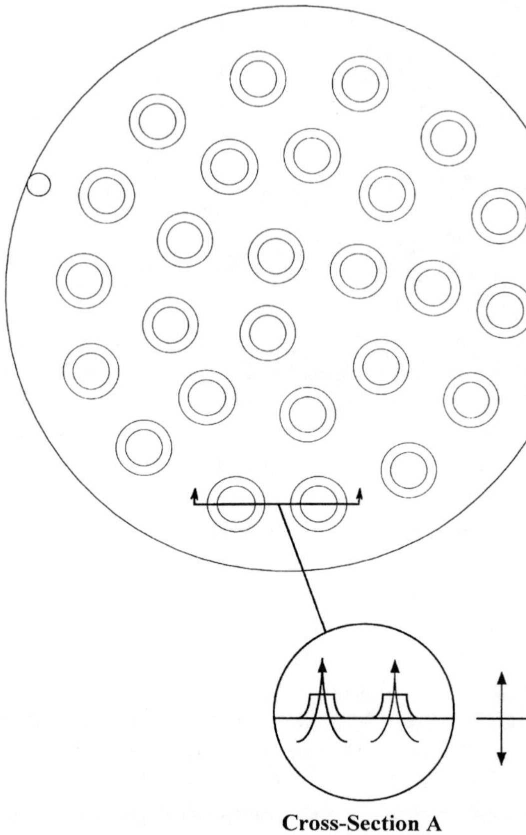

Figure 7. The VIBRO-Mixer stirrer, which consists of a flat plate perforated with hollow truncated cones which are shaped like venturi openings. When the plate is vibrated in a reciprocating manner in a noncompressible fluid, a continuous jet pumping action is created by liquid being accelerated out of the narrow end of the venturi cones.

properly sized so that cavitation and corresponding foaming does not occur at the recirculation rates necessary to maintain suspension homogeneity and adequately supply product to meet the anticipated filling rates.

Another consideration for suspension formulations is that both mixing and recirculation systems must be designed to minimize the comminution of the suspension particles. This is especially critical for crystalline suspensions, because the particles may be fragile, for example, needle-like crystals, and may not be able to withstand a great deal of abrasion or shear stress. Because particle size affects the dissolution rate of the suspension, a significant size reduction from fracture or abrasion could affect the pharmacokinetic profile of the dosage form.

3.1.2. SHEAR STRESS-INDUCED DENATURATION

Shear stress in fluids is the result of a force being applied parallel to a cross-sectional area of interest, in this case a protein molecule, which produces continuous deformation or flow as layers of the fluid move over each other at different velocities. Such a stress could potentially yield sufficient torque to unfold or denature the molecule. Shear stress may be imparted to a protein in solution during drug product manufacturing from actions such as the flow of the liquid through tubing, flow through a narrow-bore filling needle or pump valve orifice, flow through a sterilizing membrane filter, or the revolving of a mixing impeller or stirrer in a solution.

The author has noted the relative scarcity of literature describing investigations on the shear stress-induced denaturation of proteins, a subject which warrants additional work. However, there are three relevant investigations dealing with the effects of shear stress during flow through tubing and during sterile filtration that are worthy of mention.

Charm and Wong (1970) investigated the effects of shear stress on protein-containing solutions flowing through a tube. The flow velocities of the fluid in the tube are depicted in the schematic shown in Fig. 8. The mean shear rate, dv_z/dr, is defined as the rate of change of the fluid velocity along the longitudinal axis of the tube as a function of the radial position in the tube. Empirically they found that the loss of protein activity is proportional to the mean of the product of the mean shear rate and the time of exposure to the shear, $\langle (dv_z/dr) \times t \rangle$. This product is determined by

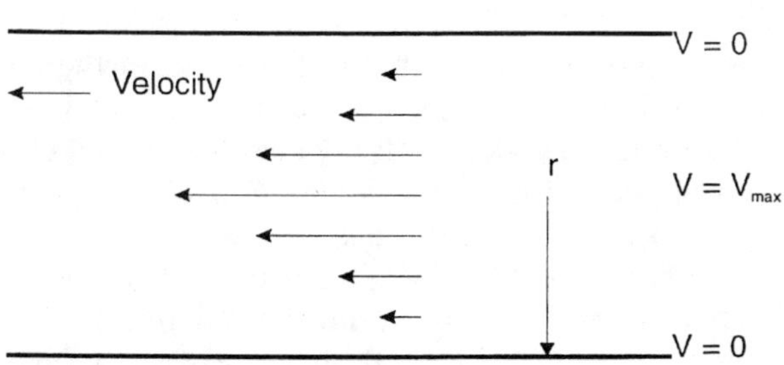

Figure 8. Flow of fluid through a tube.

the tubing length-to-radius ratio (L/r) times the number of passes (N) through the tube:

$$\text{loss of protein activity} \propto (dv_z/dr) \times t = (8/3)(L/r) \times N$$

It follows that at a constant shearing stress the protein activity would decrease with longer exposure times, and for a constant exposure time, greater amounts of shear stress would lead to decreasing protein activities. In addition, the loss of protein activity was found to be additive for intermittent exposures to a shear stress.

For the two proteins investigated, catalase and carboxypeptidase, mean shear rate–time products $>10^5$ were required to cause demonstrable protein activity losses of 5–10%. Higher shear rate–time products resulted in greater activity losses.

Pikal *et al.* (1991) applied this tubing flow approximation to the single passage of protein-containing solutions through a sterilizing-grade membrane filter (pore size = 0.22 μm). Using Charm and Wong's empirically derived formula,

$$\text{loss of protein activity} \propto (8/3)(L/r) \times 1$$

where the radius of a pore $= 0.11 \times 10^{-4}$ cm (0.22 μm filter) and the length of the pore = 0.013 cm, gives

$$(8/3)(0.013\,\text{cm}/0.11 \times 10^{-4}\,\text{cm}) = 3.15 \times 10^3$$

The shear rate–time product was approximately 10^3–10^4, suggesting that sterile filtration probably does not produce significant inactivation of proteins because the Charm and Wong data indicated a shear rate–time product of $>10^5$ was required to bring about protein activity loss. They experimentally demonstrated this conclusion by pumping human growth hormone (hGH) solutions through capillary tubes that would approximate the pores of sterilizing-grade filters and showed that no aggregates were produced. Truskey *et al.* (1987) came to a similar conclusion in their studies evaluating the effects of membrane filtration on the conformation of human immunoglobulin G (IgG), bovine insulin, and bovine alkaline phosphatase. The results suggested that any conformational changes which were found to occur were due to the interaction of the protein and the membrane and not to hydrodynamic shear forces.

Impellers or stirrers used during solution compounding or for suspension mixing can also potentially impart significant shear stress to a protein-containing solution. The fluid shear stress is a function of the product of the fluid shear rate and the viscosity of the solution (Oldshue,

1997). The maximum and the average shear rate $\langle (dv_x/dy) \rangle$ around the periphery of an impeller can be calculated if fluid velocity measurements are plotted as a function of the distance from the impeller blade. In practice, these fluid velocities are difficult to measure, so the following relationship gives a general rule of thumb to follow in selecting impeller sizes and speeds for proteins having differing shear sensitivities.

The power P_{app} applied to a system from a turning impeller is proportional to the pumping capacity Q times the velocity work term VW, which is related to the shear rates around the impeller:

$$P_{\text{app}} \propto Q \times VW$$

The power P_d drawn by impellers in low viscosity solutions is proportional to N^3D^5, where N is the impeller speed and D is the diameter of the impeller:

$$P_d \propto N^3D^5$$

The pumping capacity of an impeller Q is proportional to ND^3.

By combining the above considerations, one finds that the ratio $(Q/VW)_P$, flow to velocity work (or shear rate) at constant power, turns out to be proportional to $D^{8/3}$. This indicates that larger impellers rotating at slow speeds give greater pumping action and less shear than smaller impellers rotating at higher speeds.

In addition, the shear rate VW of the impeller is proportional to N^2D^2. The relationship of shear rate to impeller diameter result from the fact that the tip of the mixer, where tip speed is greatest, will produce the largest differential flow velocities and thus the largest shear rate. The tip speed of the impeller is defined as (Tatterson, 1991, pp. 2–5)

$$V_{\text{tip}} = \pi DN$$

where D is the impeller diameter and N is the rotational speed. Therefore, for different diameters of the same type of impeller operated at identical rotational speeds, the larger impeller will produce a greater shear rate than the smaller one, because of its higher tip speed.

In contrast to the availability of a shear-induced inactivation index empirically derived to estimate protein denaturation during flow through a tube (Charm and Wong, 1970), no empirically derived equation for calculating a shear-rate product for mixing was found in the literature. This would be a very beneficial area of research, which would assist in the appropriate selection of mixers and mixing processes for protein-containing solutions. Research on shear denaturation of a larger variety of proteins would allow more definitive conclusions regarding the importance of shear denaturation of proteins during normal processing.

3.2. Active Pharmaceutical Ingredient (API) Storage and Compounding Issues Affecting Protein Stability

3.2.1. PROTEIN AS API

The physical state of the bulk protein greatly influences the early steps in the drug product manufacturing scheme. The strategies for the storage of the purified proteins or peptides derived from bulk manufacturing processes vary from one manufacturer to another. The decision to store the material as a dried solid, either crystalline or amorphous, or as a frozen concentrate may depend on several factors. Some of these considerations are the relative stability of the protein in the dried versus the solution state, the protein's sensitivity to freezing and thawing, the cost of the facilities and equipment necessary to convert the protein into the solid state (e.g., bulk freeze-dryers, large-scale centrifuges, and drying ovens) and the economics of storing dried powder versus frozen solutions. For storage of bulk material as a frozen solid, the temperature of storage should be considered. Commonly used storage temperatures are −40°C and −80°C.

3.2.1.1. Compounding with API Powder

The compounding of a formulated bulk drug solution is the process of dissolving or diluting the proper quantities of the active drug substance, in this case either a protein or a peptide, and the excipients in the appropriate volume of water to achieve the desired concentration or potency. The details of this operation, for example, how the powder components are added, the order of addition of the components, tank and mixer geometry, and mixer design and speed, may all have an effect on the rate of dissolution of the components and the final stability of the active drug substance.

If possible, the dissolution of bulk drug powder should be done at a solution pH that would maximize its solubility and dissolution rate so that the operation could be completed expeditiously without prolonged mixing. Such a dissolution plan may include a pH excursion away from the intended final pH of the formulation into a pH range that would maximize the protein's solubility in order to quickly solubilize the bulk drug. At the completion of dissolution the pH would be adjusted to the final formulation pH. As an example, dissolving zinc insulin crystals during compounding is accomplished by reducing the solution pH to the range of 3.0–3.5 to accelerate dissolution and then raising the pH to neutrality, which is the final formulation pH (Brange, 1987). Such a plan must be fully investigated during the formulation and process development phases by defining and evaluating the pH–solubility and pH–degradation rate profiles of the protein. Time limits may need to be instituted for such pH excursions during compounding, because the kinetics of a particular degradation mechanism for the protein may be accelerated significantly at the extremes of pH. This is the case for zinc insulin, where the rate of one deamidation reaction is greatly increased in the lower end of the pH range which yields maximal solubility (Fisher and Porter, 1981). In addition, if the pI of the protein is traversed during the pH adjustment, precipitation may occur. This may only be a transient phenomenon; however, it should be simulated in the laboratory before such a pH adjustment is attempted at larger scale.

3.2.1.2. Compounding with Frozen API

If bulk protein is stored frozen, there are several compounding-related issues that should be investigated during process development in order to achieve maximal physical and chemical stability of the protein. These include the method of thawing, the temperature/time relationship for the thaw, and the order of addition of the bulk protein and dilution buffer. Thawing should be carried out in a reproducible manner, in a recirculating water bath, a vibratory shaker bath, an environmental chamber, or under some other controlled condition. Dilution of the thawed bulk protein can be done either by adding the protein to the buffer or vice versa.

Achieving the desired protein potency in the formulated drug product can be accomplished by increasing the protein concentration from infinite dilution (adding API protein to dilution buffer, which mimics the dissolution of dried bulk protein powder in solution) or diluting from the higher bulk protein concentration (adding dilution buffer to pooled API protein). The argument for using the latter procedure is based on the fact that the solution stability of many proteins is enhanced as protein concentration increases (Hanson and Rouan, 1992). In these cases, very dilute protein solutions, such as when API protein is added to dilution buffer, may actually be deleterious.

3.2.2. EXCIPIENTS

Excipients, for example, buffer salts, amino acids, polyols, and preservatives, should be added in an order which would result in a positive or at least a neutral influence on the active protein, but should not result in their own accelerated degradation. For example, dissolving a sugar such as sucrose in a very low pH solution may be detrimental, because it is much more susceptible to hydrolytic degradation under acidic conditions (Pazur, 1970). Therefore, if possible, such an excipient should be added after the compounding pH has moved closer to neutrality.

The exact form of the excipients that are specified during development of the drug formulation and production processes must be employed in its manufacture for clinical or commercial use. For example, different hydrates of buffer salts may have very different dissolution characteristics. The solubility in cold water of the heptahydrate of dibasic sodium phosphate is quite different from the dodecahydrate, 104 g/100 ml and 4.15 g/100 ml, respectively (Weast, 1971). If the incorrect hydrate is used, these differences may result in incomplete dissolution of the excipients during a set mixing period in compounding. This could result in a pH that is far from that intended for the protein or peptide at that step in the process and could affect either its physical or its chemical stability.

Another example of the importance of using the correct excipients relates to the fact that the solubilities of some proteins are often significantly affected by the ionic strength of the medium in which they are dissolved. Ionic strength is generally most dramatically affected by buffer salts, other salts added to the formulation, and the salt which is generated from the titration with acid or base from the starting pH to its final value. If the compounding instructions call for the addition of one buffer species, say, dibasic sodium phosphate, Na_2HPO_4, and the other species is added instead, say, monobasic sodium phosphate, the salt generated from the additional titrant needed may have an adverse impact on the protein's solubility. Such an error could lead to haze or visible precipitate in the formulation.

3.3. Low Concentration of Protein in Drug Product

Proteins of pharmacological utility, most of which are produced via recombinant DNA or hybridoma technology, are macromolecules which are generally very similar to those found in the human body. These polypeptides, such as insulin, hGH, interleukin-11, interferon γ 1-B, or antihemolytic factor VIII, are all found in low concentrations within the body. The signaling, regulatory, modulatory, or enzymatic activities through which they exert their effect do not require high physiological concentrations.

Administration of doses on the order of tens of micrograms to a few milligrams are common for these very potent molecules (Facts Comparison Staff, 1996). This would translate into drug product formulations that contain a few micrograms or milligrams of active substance per milliliter. For a protein such as insulin having a molecular weight of approximately 6000 Da, a formulation containing a 3-mg/ml concentration would correspond to 5×10^{-7} M. For a protein of molecular weight 60,000 Da in a formulation containing 3 mcg/ml, the molarity would be 5×10^{-11}.

3.3.1. CONSEQUENCES FOR DISPENSING AND COMPOUNDING

For these low-concentration formulations, relatively small amounts of active protein are added during compounding of small clinical batches. Therefore, the dispensing of the protein as active pharmaceutical ingredient must be done accurately on balances having the appropriate level of sensitivity. For example, if the batch record calls for the protein to be weighed to the nearest 0.1 g, then the balance used should have a readout that measures to the nearest 0.01 g.

Dried proteins as active pharmaceutical ingredients may be hygroscopic to varying degrees. This is important to consider during the storage and dispensing operations, because the specific activity of the API may change due to moisture adsorption for those that are hygroscopic. In order to avoid such an occurrence, storage of API should be in air-tight containers and dispensing done in controlled temperature and humidity areas.

During the compounding of a protein formulation the container in which the API is received from dispensing should be thoroughly rinsed after it is emptied into the mixing vessel. For these very potent macromolecules in small batch sizes, the mass of API to be added may not be very great and the amount of drug left as a residue in a dispensing container may bc significant. This will help assure that the correct potency is achieved in the formulated drug product. Proper mixing in the compounding tank is necessary before an in-process sample is taken for potency determination (see Sections 3.1.1.2 and 4.3).

One strategy that can be used to help assure that the desired protein potency is achieved during compounding is to add only 70–80% of the theoretically anticipated amount of excipient solution to the mixing tank, add the API, mix, and then perform an in-process potency assay. The results will dictate how much additional excipient solution must be added to reach the required potency. The success of this strategy depends on the accuracy and precision of the in-process potency assay.

3.3.2. ADSORPTION

Adsorption of small amounts of a protein, already at low concentrations in a formulation, to any of the fill/finish equipment surfaces or the container/closure surfaces could significantly reduce the potency in all or some of the filled vials. The compatibility of a protein and its formulation with the materials used in the fabrication of the equipment employed in the fill/finish process should be investigated during process development and scale-up activities. Adsorption to the components of the container/closure system should be investigated as a part of formulation development and the determination of an appropriate bench-scale manufacturing process.

3.3.2.1. Possible Adsorption Scenarios

Adsorption may be a reversible or an irreversible physical reaction which may lead to conformational and/or chemical modifications to protein molecules. Possible interactions between protein molecules and the surfaces of the equipment in which the manufacturing takes place are depicted below:

$$P + S \leftrightarrows P - S \qquad \text{Reversible (protein conformation unchanged)} \tag{1}$$

$$P + S \longrightarrow P - S \qquad \text{Irreversible (protein conformation unchanged, but elution from surface does not occur)} \tag{2}$$

$$\begin{array}{r} P + S \longrightarrow P^* - S \\ P^* - S \longrightarrow P^* + S \end{array} \qquad \text{Irreversible (protein conformation modified and elution of modified form possible)} \tag{3}$$

$$P + S \longrightarrow P^* - S \qquad \text{Irreversible (protein conformation modified, but elution of modified form does not occur)} \tag{4}$$

The reversible reaction shown in (1) may not result in any detectable reduction in protein concentration unless the rate constant for the dissociation or elution backreaction is significantly lower than the forward association reaction.

Reactions (2) and (4) are probably indistinguishable without a great deal of analytical difficulty (infrared spectroscopy in the reflectance mode may detect the difference); however, they yield a reduction in concentration and/or potency of the protein in the formulation. A finite number of binding sites on the equipment surfaces will eventually lead to their saturation, at which point the adsorption will cease. For a formulation containing a very low protein concentration, on the order of a few micrograms per milliliter, attaining full binding saturation may yield a significant reduction in concentration. This can be addressed by making the number of adsorptive binding sites as few as possible, which can be achieved by minimizing the equipment surface-to-formulation-volume ratio (S/V) for each of the process steps. For example, the sizes of the mixing containers in which the compounding is done should be only slightly larger than the batch size to be manufactured, and the length of any tubing that demonstrates some adsorptive properties should be minimized.

The irreversible reaction (3) is very undesirable, because it reduces the potency of the active compound and generates degradation products. It is possible that this reaction yields inactivated or denatured protein at a liquid/solid interface. Protein inactivation was reported by Edwards and Huber (1992) who studied the adsorption and irreversible inactivation of β-galactosidase at the wall–liquid interface of the containers in which it was stored. The rate of inactivation of dilute solutions of the enzyme was proportional to the wall–liquid surface area and the wall composition (glass or polypropylene). Denaturation of adsorbed proteins may lead to aggregation of the desorbed protein species P* into amorphous or gelatinous particles. Just such a reaction has been seen to occur with the adsorption and aggregation of insulin in implanted pumps and catheters, which led to their blockage (Lougheed *et al.*, 1980).

3.3.2.2. Adsorption to Process Equipment

Equipment adsorption studies can be done in a number of ways. Small quantities of the formulation can be exposed to scaled-down versions of the equipment to be employed in the clinical or commercial manufacturing operation, or small "coupons" of the materials of construction can be immersed in minimal volumes of the formulation. Scaled-down equipment or coupons should be made of exactly the same material and have the same surface finish as the full-scale equipment. For metals such as stainless steel, the identical alloy (e.g., 316, 316L, 304) with the same surface finish (e.g., #3 finish, electropolished) should be used. Small formulation volumes are used so that even minimal adsorption would be amplified, making it detectable analytically. The exposure time for these studies should be significantly longer than that anticipated for actual manufacturing, so that some safety factor could be built into the results. The temperatures of the exposure studies should cover the range commonly encountered during manufacturing and storage, which could vary from room temperature down to refrigeration temperatures.

Membrane filters, and their metal housings or plastic capsules, used for sterile filtration of the formulated bulk protein solutions should also be tested for adsorptive properties (Brophy and Lambert, 1994; Martin and Manteuffel, 1988; Pitt, 1987). The testing can be done in a dynamic mode where samples of the filter effluent can be collected as a function of time after the beginning of filtration. The samples should be analyzed for the concentrations of both the active protein as well as the excipients. Changes in either could adversely affect the formulation's immediate potency or its long-term stability. Figure 9 shows the results of one such experiment where two types of filtration media (Filter Media B and C) display much greater saturable binding activity than the third (Filter Medium A). The amount of protein bound can be calculated from these curves and its impact on the final protein concentration in the formulated bulk solution can be evaluated.

Measuring the flow decay during the sterile filtration of a formulation is another method for evaluating the potential for adsorptive interactions with filtration media. As seen in Fig. 10, two of the filters (Filter Media A and C) evaluated for use with the protein formulation in question exhibited significant flow decay, whereas the other filter (Filter Medium B) maintained its starting flow rate throughout the time interval of the test. The three filters were made of different polymeric materials and the ones that demonstrated the flow decay may have interacted with the protein in such a way that denaturation and aggregation occurred, leading to physical blockage of the filter. The filter yielding the least flow decay should be the one chosen for the production operations, provided that it is found to be fully compatible with the formulation.

3.3.2.3. Adsorption to Container/Closure Systems

Adsorption of protein to glass vials or stoppers may be addressed in a variety of ways. The glass treatments or alternatives to glass vials described below alleviate the adsorption problem by reducing or eliminating the

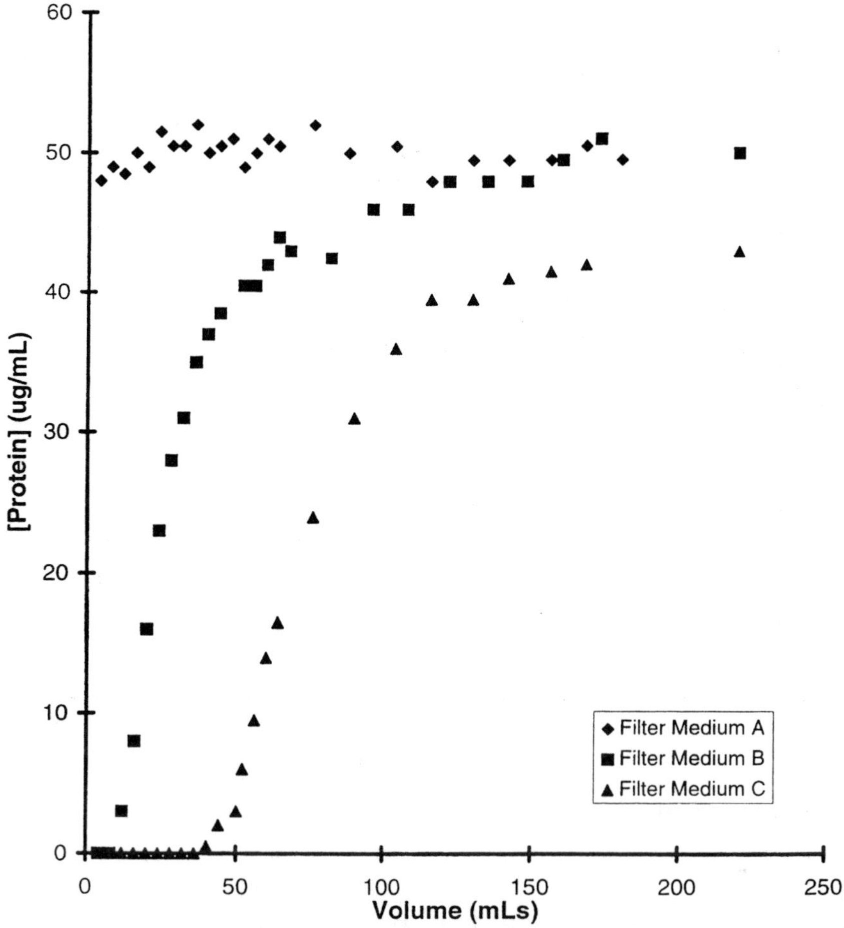

Figure 9. Dynamic binding of a protein to three types of filter media.

electrostatic binding sites on the glass surface. The solutions to such adsorption problems generally have complicating ramifications for the manufacturing process. If a protein adsorbs to a surface via a predominantly hydrophobic mechanism, which, for example, is the case for insulin (Sluzky *et al.*, 1991), many of the alternatives listed here would be counterproductive. These approaches, for the most part, provide a hydrophobic covering or alternative to glass surfaces.

The type of glass routinely used in the manufacture of vials and ampoules for parenteral products is Type I borosilicate glass. This glass is

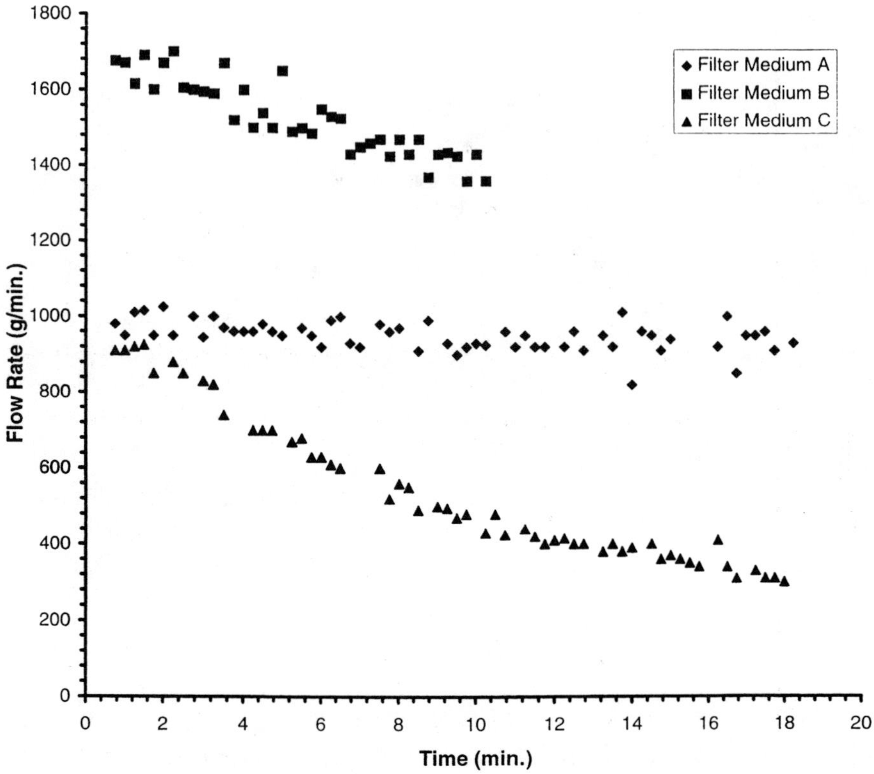

Figure 10. Filtration flow rates for three types of filter media (same media as Fig. 9).

potentially the least reactive and has the least amount of extractables as compared to soda-lime glass (Adams, 1977). Type I borosilicate glass can also be produced with different expansion coefficients as a result of different ingredients and proportions of ingredients being added to the glass formulation. However, regardless of the expansion coefficient of the glass, essentially the same types and numbers of electrostatic binding sites will be present on the surface of the glass, namely the $\equiv$ Si–O$^-$ Na$^+$ groups.

A "sulfur" treatment is offered by glassware suppliers, which is purported to react with and reduce the number of reactive groups on the glass surface. The treatment involves spraying a dilute solution of ammonium sulfate, $(NH_4)_2SO_4$, into the container before the newly formed container is placed into a lear (oven) to anneal the glass. The high temperature of the lear vaporizes the ammonium sulfate and converts it into ammonium hydrogen sulfate, which in turn reacts with many of the $\equiv$ Si–O$^-$ groups, yielding a less reactive glass surface (Roseman *et al.*, 1976).

Molded plastic vials fabricated from different polymeric resins are an alternative to glass vials. A polymer called "CZ resin" manufactured by Diakyo of Japan is used in the production of exceptionally clear vials. These vials have been found not to be adsorptive for one recombinant protein which adsorbs to glass with high affinity (unpublished data). One potential drawback in the use of CZ vials, however, is their inability to withstand dry heat oven depyrogenation temperatures. The vials must be depyrogenated during washing and then autoclaved and dried prior to filling. The CZ vials are also significantly lighter than glass vials, which may make their handling difficult on filling equipment designed for glass. In addition, the vials have been shown to be suitable for lyophilization, even though their heat transfer coefficients are less than those of glass (unpublished data). As with all prospective parenteral container candidates, overall compatibility between the protein formulation and the plastic vials must be demonstrated before they could be considered for use.

Siliconization of the glass vial interior is another method for minimizing protein adsorption. Siliconization is a commonly used treatment for syringe barrels to supply lubricity to facilitate plunger stopper movement for the expulsion of the syringe contents. In general, though, the amount of silicone oil required for lubricity is greatly in excess of that needed to prevent protein adsorption. For syringe barrels a 1–2% silicone emulsion dilution (the starting emulsion contains approximately 35% silicone oil plus emulsifying and stabilizing agents, so the oil concentration in the diluted emulsion is 0.35–0.60%, w/v) is sprayed into the barrels after washing. The syringe barrels are then heat-treated at 250–300°C for several hours to bring about "bonding" of the silicone oil to the glass surfaces. Such a high concentration of silicone oil results in significant levels of "free" or unbonded silicone oil. This is undesirable for use with protein-containing formulations, because of the potential for interfacial denaturation of the protein on silicone oil droplets (see Section 3.1.1.1). In fact, a silicone emulsion concentration 1/100th of that used for syringe barrels may be sufficient to prevent protein adsorption in glass vials and not leave excess silicone oil on which denaturation can take place. The exact emulsion concentration to be used must be determined during process development.

Rubber stoppers must generally be siliconized to allow them to be used in automated filling/stoppering equipment and to prevent sticking and clump formation during autoclaving. Their movement in vibratory bowls and along stainless steel tracks as they are moved to the stopper placement head requires that they have sufficient lubricity. Again, with protein formulations an excess of silicone oil on the stopper can bring about interfacial denaturation and visible particulate formation. Appearance problems linked to excess silicone oil can occur in both solution and lyophilized formulations. Hazy solutions resulting from siliconized stoppers have been linked to the presence of surfactants in the formulation which disperse the silicone oil from the stoppers into small droplets, leading to haze formation (unpublished data). Pavanetto *et al.* (1991) found that the particle levels in parenteral solutions were directly related to the amount of silicone oil applied to the stopper. Groves (1993) stated that the major source of particulate matter in drug products is silicone oil from stoppers, and therefore the amount of silicone oil placed onto stoppers must be minimized. For proteins that are exquisitely sensitive to the presence of silicone oil, Teflon-coated stoppers are available, which should have sufficient lubricity for good machinability and may not require siliconization to prevent sticking during autoclaving. Siliconized or Teflon-coated stoppers should prevent protein adsorption if the adsorption mechanism is electrostatic in nature.

3.4. Microbial Growth-promotive Nature of Protein Formulations

Many protein-containing formulations are manufactured in single-dose configurations where a preservative is not required. These formulations contain not only the active protein, but other excipients, for example, sugars, sugar alcohols, and amino acids, which may make them well-suited to support microbiological growth. This is a critical consideration during the compounding phase of manufacturing and during any hold periods that may occur prior to sterile filtration of the formulated bulk drug product. It is imperative that the bioburden on processing equipment and in raw materials to be used in compounding be as low as possible so that significant microbial growth does not occur during this processing step. Compounding of the formulated bulk may take several hours to complete and, in most cases, it occurs at room temperature. These conditions may allow for the growth of bacteria or fungi in the formulated bulk. This could result in the production of microbial metabolic by-products, such as enzymes, toxins, and endotoxins, which would adversely affect the quality of the drug product.

Effective cleaning procedures for process equipment may reproducibly yield low bioburden levels. However, the greatest assurance of low (zero) bioburdens is obtained by sterilizing the equipment after cleaning. Attention to the bioburden levels in raw materials is also critical. It may be wise to have raw material specifications call for bioburden testing such that every batch used in production is known to have acceptable bioburden levels.

Multidose products are required to include a preservative in their formulations. Growth of microbial contaminants in the formulated bulk during compounding would be prevented by the presence of the preservative; however, high bioburden levels on equipment or in raw materials should still be avoided.

3.5. Equipment Cleaning

For equipment used in the manufacture of multiple products, cleaning effectiveness is critical for preventing cross-contamination between batches of different products. Avoiding carryover contamination is no less critical if a single-protein product were manufactured using dedicated equipment. The consequences of inadequate cleaning could be that protein material, both native and nonnative, could slough off improperly cleaned surfaces into a succeeding batch of product and adversely affect its quality.

A strategy which avoids the necessity for cleaning reusable equipment during drug product manufacturing is to use as many disposable items as possible. This allows the manufacturer to avoid the cross-contamination or carryover issues for those items that can be discarded after every batch. Tubing and capsule-sterilizing filters are two of the most commonly used disposables.

3.5.1. DEVELOPING A CLEANING PROCEDURE

There are two major considerations when cleaning procedures are being developed for equipment used in the manufacturing of protein formulations. First, if at all possible, protein-containing solutions should not be allowed to dry on equipment surfaces. Second, cleaning solutions and exposure temperatures should be selected to maximize the protein's solubility and not cause denaturation or precipitation of the protein.

Air or evaporative drying of a solution or suspension formulation of a protein on a surface can result in denaturation and aggregation of the protein. In contrast to a protein formulation developed for lyophilization, solution and suspension formulations would not contain additives to protect the physical stability of proteins in the dried state. Evaporative drying can be compared to the changes that occur during the freezing of a solution. The removal of water results in increases in solute concentrations, which may include high salt concentrations and significant pH changes (Franks, 1990), both of which could result in protein denaturation. The denatured protein mixed with other dried solutes could be in the form of amorphous, crystalline, or gelatinous material. This solid-phase material may have very different solubility in water or other solutions than the native protein or the excipients originally had. Therefore, manufacturing equipment should be cleaned, or at least soaked, as quickly as possible after completion of the processing step during which it was used.

Cleaning, whether using a manual procedure or a CIP system, should follow a well-developed plan for solubilizing and rinsing away both native and nonnative protein molecules as well as excipient solutes. If they can be avoided, detergents should not be used in the process, because it is then necessary to demonstrate that all detergent residuals are also removed from the equipment after the cleaning is completed. In most cases the most difficult solutes to remove from the equipment will be protein. This is especially true if denatured protein has dried onto a surface.

The sequence of cleaning steps, including pHs and temperatures and water rinses, can be planned by examining the pH–solubility profile and the thermal denaturation temperature, T_m of the active protein (Middaugh and Volkin, 1992). If the native protein is soluble in water at the concentration found in the formulation, then the first step in a cleaning procedure should probably be a warm-water rinse to dilute and remove most of solutes on the equipment. The temperature of the rinse should be warm, but no higher than 5–10°C below the T_m of the protein. If the protein has a very low solubility in water, then a water rinse should not be used early in the procedure.

A rinse at the pH of maximum solubility should follow next. In the case of one currently marketed recombinant protein, maximum solubility is at the extremes of the pH range. In this case, a HCl solution in the pH 1–2 range was chosen for the initial rinse after the water rinse. An acidic rather than a basic medium was chosen for this step, because this was the compounding strategy used for initially solubilizing the bulk crystalline powder. A caustic rinse using a NaOH solution followed. Because it was felt to be easier to rinse away acidic rather than caustic residues, another HCl rinse followed. A demineralized water rinse and a water-for-injection rinse ended the cleaning procedure.

For smaller pieces of manufacturing equipment that can fit into a ultrasonic cleaning bath, many or all of the cleaning steps should be carried out there. The ultrasonic bath is effective in solubilizing solutes adhering to equipment surfaces, because of the cavitation generated by the ultrasonic waves in the washing solution.

3.5.2. CLEANING VALIDATION

Cleaning validation must be done on product contact equipment prior to manufacturing the first clinical lots, especially if the equipment is not dedicated to the product being manufactured. The FDA's *Guide to Inspections of Validation of Cleaning Processes* should be consulted prior to planning a cleaning validation exercise (U.S. FDA, 1993b). The validation testing must utilize a meaningful and reproducible sampling procedure, such as evaluating residues in the final rinse water, or swabbing a defined surface area with an appropriate, protein-solubilizing solvent. Development of these procedures should include demonstrating that both native and nonnative proteins can be detected, as well as determining the protein recovery percentage. The advantage of the rinse sampling technique is that a much larger surface area is being sampled than by the swab method. However, the dilution effects from the large volume of the rinse may make any residuals undetectable. Many validation protocols will include both rinse water and swab samples to take advantage of the benefits offered by both.

The strategy for setting acceptance criteria for cleaning validation has historically been based on one of the following considerations:

- *Toxicity or safety*, where the no-effect dose is determined and then is divided by a safety factor to set the limit for a specified surface area
- *Pharmacologically based*, where the lowest clinical dose is established and is then divided by a safety factor (1000 is the most commonly used safety factor)
- *Cleaning process capability-based*, where the lowest cleanable level is established after "heroic" cleaning of the manufacturing equipment

Another important consideration in setting residual specifications is the analytical detection capability. Assay detection limits should be considered in selecting an appropriate analytical method for the cleaning validation exercises; however, the detection limit should not be the sole basis on which a residual limit is established (Jenkins and Vanderwielen, 1994).

The analytical methods selected to quantitate the residual proteins in the samples must have the required sensitivity in order to meet the acceptance criteria. These methods can be either nonspecific [e.g. total organic carbon (TOC), protein-specific dyes or spectrophotometric methods] or product-specific in nature [e.g., enzyme-linked immunosorbent assay (ELISA), reversed-phase high-pressure liquid chromatography (RP-HPLC)]. The ELISA method utilizing monoclonal antibody detector probes is generally so specific that only native protein molecules are detected and not denatured ones. Use of the appropriate panel of polyclonal antibodies as detector probes can broaden the scope of detection to both native and nonnative protein molecules.

3.6. Adverse Effects from Disinfectants or Sterilants

Aseptic filling areas and lyophilizers (that are not steam-sterilizable) must be disinfected at specified intervals to minimize microbial contamination. In addition, it may be standard practice in some aseptic filling operations to place disinfectants on tubing and sterile fittings prior to making an aseptic connection. For protein formulations, very small amounts of these disinfectants or sterilants may cause chemical modifications to the protein.

Kirsch *et al.* (1993) investigated the potential of a variety of commonly used disinfectants to induce protein degradation. The studies consisted of spiking very low concentrations of alcohols, hydrogen peroxide, phenolics, aldehydes and some combination commercial disinfectants/sterilants into a commercial formulation of hGH. The protein was characterized before and after lyophilization to evaluate the relative effects of the chemical agents in solution and then after the stresses of freezing and drying. Alcohols (methanol and isopropyl alcohol) caused no significant changes in the protein. Hydrogen peroxide caused significant degradation to the monomeric hGH in the lyophilized form, but did not cause aggregate formation (as would be expected, it had no effect on hGH in solution). Formaldehyde and glutaraldehyde both caused massive covalent aggregation. Phenol and other phenolic derivatives caused varying amounts of both reversible and irreversible dimer formation in both the solution and freeze-dried forms. Similar levels of dimer were found with combination commercial products which contained several phenolic derivatives, plus an aldehyde in one case. These results indicate the relative susceptibilities of one particular protein pharmaceutical to residues of chemical disinfectants/sterilants that are commonly used in parenteral drug product manufacturing. Other proteins may exhibit different spectra of susceptibilities to all of the agents described above, with the exception of the aldehydes.

As described in the previous paragraph, aldehydes are very reactive compounds that may chemically modify and crosslink proteins both in solution as well as in a freeze-dried cake (Lubig *et al.*, 1981). Many European pharmaceutical companies have commonly used formaldehyde vapor as a means of sterilizing their aseptic filling areas. Following adequate aeration of the area, the residual levels may be reduced sufficiently so that no adverse effects can be detected in protein products that may be manufactured there. However, sterile holding tanks containing bulk formulated protein should not be stored in the aseptic areas during such formaldehyde vapor treatment. Most tanks will have air vent filters in place, which may either allow for the ingress of the vapor during the treatment or may adsorb formaldehyde vapor and then slowly release it into the headspace gas in the holding tank. Another potential concern is that some types of tubing used for solution transfer, such as silicone tubing, may absorb the vapor if left in the aseptic area during the gas treatment and then allow the aldehyde to diffuse into any solution that passes through it. Given the sensitivity of proteins to aldehydes, the use of the latter in areas or on equipment used for protein pharmaceutical manufacturing seems risky (Fraenkel-Conrat and Olcott, 1948a,b; Habeeb and Hiramoto, 1968; Saleh *et al.*, 1989).

3.7. Surfactants in Protein Formulations

Many protein formulations contain surfactants in order to stabilize the protein conformation and prevent interfacial denaturation during dosage form manufacturing, shipping, storage, and end-user manipulations (Levine *et al.*, 1991; Manning *et al.*, 1989). The mechanism of this protein-stabilizing characteristic has variously been described as being due to the exclusion of protein molecules from interfaces by the surfactant molecules and/or the binding of surfactant molecules to the protein, thereby enhancing the stability of the protein's native state (Wang and Hanson, 1988).

Surfactants are surface-active compounds, that is, they readily adsorb to the air–liquid interface in a solution. This phenomenon occurs because surfactant molecules are linear, having distinct hydrophobic and hydrophilic ends. They may align at an air/liquid interface with the hydrophobic end extending into the air and the hydrophilic end in the solution. The hydrophobic end is mainly a chain of methylene groups of varying lengths, and the hydrophilic end can be either ionic or nonionic in nature. Nonionic surfactants interact with protein molecules via hydrogen bonding, which is pH-independent, whereas the electrostatic interactions between ionic surfactants and proteins arise from oppositely charged moieties on the respective molecules. The ionization state of ionizable groups, especially on the protein molecules, is pH-dependent, whereas ionic groups on the surfactant molecules, for example, sulfate or sulfonate sodium salts are, within the scope of this discussion, pH-independent.

Surfactants reduce the surface tension of a solution, thereby increasing its "wetting potential." This characteristic, as well as their ability to interact with hydrophobic molecules to solubilize or disperse them in solution, greatly enhances a solution's "extracting potential" from container/closures or process equipment used in drug product manufacturing. The presence of surfactants in a protein formulation may require a more rigorous approach to the optimization of stopper cleaning and siliconization and the selection and cleaning of other polymeric or elastomeric equipment, such as tubing and gaskets. Surfactants may be a difficult excipient to remove from multiuse product contact equipment during cleaning operations, because of the tenacity with which they may adsorb to a variety of equipment surfaces.

As previously described, silicone is placed on virtually all stoppers to allow for machinability and prevent sticking during autoclave sterilization (see Sections 2.1.3.2 and 3.3). Silicone oil from stoppers (or from siliconized vials or syringes/cartridges) can be dispersed by a surfactant in a formulation. This dispersion, depending on the amount of silicone oil extracted, can generate a hazy or opalescent appearance due to light scattering. As a rule of thumb, the amount of silicone oil applied to stoppers used for protein formulations should be minimized for appearance reasons as well as to avoid the potential for interfacial denaturation of the protein.

Protein denaturation and aggregation is postulated to occur at the interface between two immiscible liquids (Sluzky *et al.*, 1992), such as an aqueous solution and dispersed silicone oil droplets. In a case where there is an excessive amount of silicone oil dispersed into a surfactant-containing solution, the surfactant may actually exacerbate the potential denaturation problem, because by dispersing the silicone into smaller droplets it is actually increasing the amount of available surface area on which denaturation can occur. The polymeric molecules of silicone oil, with which the hydrophobic ends of the surfactant molecules interact, may also reduce the effectiveness of the surfactant molecules in excluding protein molecules from a liquid/liquid interface as compared to an air/liquid interface. This may occur due to a reduction in the packing efficiency of surfactants at the silicone oil interface in comparison to that at an air interface.

4. DEVELOPING ASEPTIC PROCESSES FOR DRUG PRODUCT MANUFACTURING

In the overall scheme of drug product development, manufacturing process development should play an increasingly greater role as a drug product candidate moves from the production of informal stability lots to clinical lots and finally to commercial batches. It is likely that the drug product formulation, and perhaps even the dosage form, may change (e.g., lyophilized to liquid formulation) as clinical trials progress. Such a scenario would suggest that minimal process development effort be expended to understand and optimize manufacturing parameters during Phase I and II trials. Once the product is defined for Phase III pivotal trials, then an extensive effort will be required to lay out the production process in detail, scale it up, define the critical process parameters, and then determine their optimal target values and allowable ranges. All of these efforts should culminate in a validated commercial drug product manufacturing process.

4.1. Bench-Scale and Early Clinical Trial Manufacturing Process Development

Responsibility for the development of the bench-scale manufacturing process lies with the formulation development and process development scientists. Definition of the bench-scale process depends on the following considerations:

- Compatibilities of the formulation with the materials of construction of potential process equipment
- Selection and preparation of the container/closure components
- Type of sterilizing filter selected
- Any sensitivities of the protein observed during preformulation or early formulation development work that must be accommodated in the bench-scale process (e.g., extreme shear sensitivity requiring greater attention to mixing, pumping, and filling parameters)
- If applicable, a lyophilization cycle developed in a laboratory-scale freeze dryer

The bench-scale process devised with these items in mind should be suitable for manufacturing informal stability lots, tox material, and other preclinical supplies.

Most of the technical details of the bench-scale process, with the exception of equipment sizes and the lyophilization cycle parameters, may be transferred to a cGMP pilot or commercial production facility for the manufacturing of cGMP drug product for Phase I, Phase II, and early Phase III clinical trials. Equipment should be selected to match as closely as possible the clinical batch sizes to be manufactured. Before transferring the lyophilization cycle to a larger facility it would be prudent to run the bench-scale cycle in a freeze-dryer approaching in scale the one to be used for the clinical batch, and, if possible, in the actual freeze-dryer to be used for the clinical fill/finish operation. The primary objective would be to verify the sufficiency of the primary and secondary drying times which had been previously defined in a smaller freeze-dryer. Such a technical qualification could be done with vials filled with formulation buffer alone or as a "seeded buffer" load, where some vials containing active protein are seeded into a load of vials containing only formulation buffer. The "seeded" batch approach would conserve valuable API, but would allow for an evaluation of residual moisture level variability in a full-scale load as well as a limited evaluation of the lyophilization effects on protein molecular integrity.

4.2. Phase III Clinical and Commercial Manufacturing Process Development

Once the decision has been made to move a product and a dosage form to market, the role of a process development group and a pilot development facility increases significantly. At this point, a plan should be devised to optimize and scale up the bench-scale process to a commercial manufacturing operation. The technical details of a bench-scale process would consist of, at the very least, the container/closure system, a sterilizing filter type, and, if the product is to be lyophilized, an unoptimized cycle. The following general process development scheme might be a first step in assembling a more detailed plan:

Bench Scale

1. Optimize the process at bench scale
2. Determine the compatibilities of potential materials of construction with the formulation
3. Define the protein's and formulation's sensitivity to potential processing stresses
4. Determine critical process parameters (CPP) at bench scale using a design-of-experiments (DOE; Torbeck and Browning, 1996) approach where the effects of multiple variables are evaluated in a matrix*
5. Define the allowable CPP ranges at bench scale

*An article by Li *et al.* (1993) provides an example of the use of DOE to investigate a process development problem. The study reports on the use of an eight-run Plackett–Burman design matrix to determine the critical parameters affecting particle contamination from siliconized stoppers.

Pilot Scale

6. Test validity of bench-scale CPP at pilot scale
7. Evaluate all process steps for scale-up problems
8. Define allowable hold times and temperatures during and between process steps
9. Define and validate the unit operations for component cleaning and sterilization
10. Define the cleaning, preparation, and sterilization of product contact equipment

Commercial Scale

11. Perform technical qualification fills at commercial scale
    a. Evaluate all process steps for scale-up problems
    b. Verify CPP and ranges from pilot scale

4.3. Prevalidation Process Development Studies

The following prevalidation studies for each process step are examples of a more detailed version of the plan laid out above. These studies should have the objective not only of defining the CPP and their allowable ranges but also of building flexibility into the manufacturing process. This is not meant to be an exhaustive list but suggests representative topics for consideration during planning for a solution or lyophilized product.

1. Active pharmaceutical ingredient
    a. Establish stability profile
        i. For frozen protein API: effects of multiple freeze-thaws, thawing temperatures and times, hold times at different temperatures, other handling stresses
        ii. For dried powder API: effects of storage temperature and time, humidity levels (water vapor adsorption isotherm), storage method after dispensing
    b. Define compatibility with equipment and processing variable ranges
2. Formulation excipients
    a. Define optimal storage temperature and humidity ranges and expiry time or retest frequency
    b. Define optimal dispensing procedure and storage afterward
3. Rubber closures (assuming compatibility of primary closure with formulation has been established)
    a. Establish storage temperature range before processing
    b. Establish expiration dating prior to processing

c. Establish closure preparation procedure
   i. Define wash cycle (equipment, time, temperature, use of detergents, rinse times)
   ii. Determine minimum silicone levels
   iii. Define optimal siliconization process
   iv. Define time/temperature of autoclave sterilization cycle
   v. For lyophilization stoppers, define optimal drying time/temperature after autoclaving to minimize absorbed moisture in the closures
d. In conjunction with formulation scientist, identify secondary closure; carry out steps a–c

4. Compounding of formulation buffer (for diluting frozen protein API)
   a. Define order of addition of excipients
   b. Optimize mixing vessel (size and height/width ratio) and mixer configuration (type, impeller, mixing speed range)
   c. Define minimum and maximum volume
   d. Define required mixing time and temperature
   e. Define allowable hold time of completed buffer
5. Compounding of bulk drug product
   a. For frozen API + formulation buffer
      i. Define order and manner of addition of each solution
      ii. Determine effects of shear stress and agitation on protein
      iii. Optimize mixing vessel (jacketed tank required?) and mixer configuration
      iv. Define maximal hold times and temperature range with and without a filtration step to control bioburden.
   b. For dried powder API
      i. Define order and manner of addition of API and excipients
      ii. Determine effects of shear stress and agitation on protein
      iii. Optimize mixing vessel and mixer configuration
      iv. Define maximal hold times and temperature range with and without filtration step to control bioburden
6. Sterile filtration (assuming compatibility of primary filter with formulation has been established)
   a. Perform bacterial retention test
   b. Confirm bubble point
   c. Establish compatibility with scaled-up filtration apparatus
   d. Determine maximum flow rate
   e. Assess effects of filtration on protein
   f. Establish maximum time for filtration to occur and evaluate effects on formulation

   g. Define maximal hold times and temperatures after filtration
   h. In conjunction with formulation scientist, identify secondary filter; carry out steps a–g
7. Aseptic filling
   a. Determine effects of filling speed on protein; optimize pump and needle size
   b. Establish maximum filling rate
   c. Determine formulation compatibility with materials for construction of pumps
   d. Establish maximum filling time and temperature
   e. Define maximum hold times and temperature for interrupted fill
8. Lyophilization (if applicable)
   a. Optimize the bench-scale lyophilization cycle to make it more economical and robust
   b. Define the targets and allowable ranges for all cycle parameters
   c. Evaluate cycle parameters for scale problems
   d. Define minimum/maximum number of vials per load
9. Capping
   a. Evaluate seal integrity of capped vials from commercial process [several types of tests for seal integrity have been described by Hofmann (1993, pp. 139–141), including dye leak testing, microbial media fills, microbial broth challenges, and residual seal force testing]
10. Cleaning of product contact equipment
   a. Define the limit for residual protein levels after cleaning
   b. Design sampling method and analytical testing for protein residuals
   c. Define cleaning agents and order of use
   d. Evaluate cleaning procedure for excipient removal and modify procedure if required
   e. Validate the cleaning method, the sampling method, and the analytical method

The results from these prevalidation studies would form the basis and the rationale for much of the sampling and testing plan embodied in the Drug Product Process Validation Protocol.

5. CONCEPTUAL FRAMEWORK FOR DRUG PRODUCT PROCESS VALIDATION

The validation of drug product manufacturing is a FDA requirement for licensure of all products and is described as “establishing documented evidence which provides a high degree of assurance that a specific process will consistently produce a product meeting its predetermined specifications and quality attributes" (U.S. FDA, 1987, p. 2). Such an exercise also has an economic benefit for the pharmaceutical company, because it generally results in a better understood and controlled manufacturing process, which should translate into fewer product losses from poorly developed processes. The Validation Master Plan is the key planning document and is assembled to define the overall strategy to be followed in preparing for, writing, and then executing and evaluating the Product Validation Protocol, and also for carrying out or reviewing the adequacy of facility, utility, and equipment qualifications and process validations, for example, lyophilizer validation, filling line validation, and autoclave sterilization validation.

5.1. Validation Master Plan

The objective of the Validation Master Plan is to define a comprehensive approach for demonstrating and documenting that the overall drug product manufacturing process is reproducible and results in drug product that will consistently meet predetermined release and stability specifications. The strategy laid out in the Master Plan, which is designed to lead to the completion of a successful validation effort, encompasses facility, utility, and equipment qualifications, formulation and process development experimentation, specific unit operation qualifications and validations, the writing and execution of a Product Validation Protocol, and the planning and production of sufficient numbers of drug product batches at the appropriate scale to demonstrate stability and product consistency.

The Master Plan for a prospective validation effort should include at least the following items:

1. A brief description of the product dosage form and how it will be used
2. A brief process outline
3. Definition of responsibility for writing, reviewing, approving and monitoring the Master Plan
4. Definition of the procedure (reviewing and approving) for amending the Plan and Protocol
5. Definition of responsibility for reviewing plans and results of formulation and process development prevalidation experiments
6. Definition of responsibility for planning and coordinating the manufacture of stability, product consistency, and product validation lots

7. Definition or responsibility for writing, reviewing, approving, and executing the product validation protocol
8. Definition of responsibility for collecting the product validation data, evaluating them, and writing the Final Report
9. Definition of responsibility for reviewing and approving the Final Report
10. Definition of the components minimally included in the Validation Protocol
11. Definition of the components minimally included in the Final Report
12. Postapproval strategy for revalidation of the manufacturing process
13. Definition of responsibility for assuring the completion or adequacy of facilities, utilities and equipment qualifications, and process validations

5.2. Fill/Finish Plan for Generating Stability Lots and Product/Process Consistency Lots

Regulatory requirements call for the production and stability testing of three representative product batches at either pilot or commercial scale in order to establish expiry dating at product approval. Preliminary data from these three lots will be included in the regulatory filing for licensure and will be continuously updated through and after approval. It is generally accepted that 12 months of data should be included in the submission; therefore, manufacturing of the stability lots must commence approximately 14 months before filing so that the testing can be completed and the results included in the regulatory document.

Consistency lots are required for the regulatory submission to supply the FDA with evidence that a company has sufficient manufacturing experience with the product to assure that it can be produced reliably. These data, process as well as release testing, are included in the submission as a part of the representative batch history section. According to the latest guidelines, these product batches may be pilot scale, provided the process is representative of the commercial-scale process.

5.3. Product Validation Protocol

Product validation is designed to demonstrate through an extensive sampling and testing plan, over and above release sample testing, that the manufacturing process is "under control." In statistical process control (SPC) parlance, being under control means that the inherent variability of each unit operation is defined and that the process output consistently remains within this boundary of expected results.

In order to demonstrate this process control, the Product Validation Protocol should be executed for three lots of product, ideally at the maximum commercial batch size. If possible, multiple lots of components and raw materials should be utilized to further demonstrate the robustness of the manufacturing process. In addition, if more than one piece of major equipment (e.g., lyophilizers, filling lines) are intended for use in commercial manufacturing, then each of them should be included in the validation exercises.

The Product validation Protocol should minimally contain the following items:

1. Process flow diagram
2. Description of the equipment employed for the manufacturing process
3. Rationale for the design of the Validation Sampling and Testing Plan based on summaries of the prevalidation work done by formulation and process development groups (see Section 4.3)
4. Validation sampling plan
   a. Instructions for numbers and types of samples to be taken
   b. Types of analytical assays to be run on each sample
   c. Statistical analysis methods to evaluate the data
5. Acceptance criteria

Validation will be considered successful if the process yields drug product that meets release specifications, all validation sample testing meets acceptance criteria, and the lots are dispositioned as releasable for three consecutive lots.

5.4. Product Validation Final Report

After the successful execution of the Protocol on three consecutive production lots, the data are assembled, analyzed, and evaluated and a Final Report is written. This report is made available to the FDA for their review, but is generally not included in the regulatory submission. Depending on the FDA agency involved in reviewing the submission, it may be necessary to have the Final Report completed for the preapproval inspection, but in all cases it must be completed by the time of product approval and launch.

The Final Report should minimally include the following components:

1. Product Validation Protocol
2. Manufacturing batch records for the three lots
3. Release sampling testing plan and results
4. All deficiency reports and reports of investigation that would explain their disposition
5. Disposition of the lots
6. The raw analytical data for the validation tests
7. A statistical analysis of the validation data
8. An evaluation of the release and validation results for the three lots
9. Conclusions

ACKNOWLEDGMENTS

The author wishes to recognize the thought-provoking conversations that helped formulate the considerations and the development/validation approaches described in this chapter. Among the individuals involved were Michael Boenitz, Donald Sipe, Fen Fang Bond, Jennifer Juneau, Christopher Riel, Richard Northey, Dr. Thomas Spitznagel, Dr. Lawrence Bush, Dr. Kasra Kasraian, Dr. Nicholas Warne, and Dr. Godfrey Amphlett.

REFERENCES

Adams, P. B., 1977, Surface properties of glass containers for parenteral solutions, *Bull. Parenteral Drug Assoc.* **31**(5):213–226.

Akers, J. E. 1988, PDA response to FDA's guideline on sterile drug products produced by aseptic processing, *J. Parenter. Sci. Technol.* **42**(4):114–117.

Akers, J. E., and Agalloco, J. P., Carleton, F. J., and Korczynski, M. S. 1988, Overview of the current state of technology on aseptic parenteral manufacturing, *J. Parenteral Sci. Technol.* **42**(2):53–56.

Akers, M. J., Fites, A. L., and Robison, R. L., 1987, Formulation design and development of parenteral suspensions, *J. Parenteral Sci. Technol.* **41**(3):88–96.

Bird, R. B., Stewart, W. E., and Lightfoot, E. N., 1960, *Transport Phenomena*, Wiley, New York.

Brange, J., 1987, *Galenics of Insulin*, Springer-Verlag, Berlin.

Brophy, R. T., and Lambert, W. J. 1994, The adsorption of insulinotropin to polymeric sterilizing filters, *J. Parenter. Sci. Technol.* **48**(2):92–94.

Center for Drug Evaluation and Research, 1987, Guideline on sterile drug products produced by aseptic processing, *Fed. Reg.* (181–332):60360.

Charm, S., and Wong, B., 1970, Enzyme inactivation with shearing, *Biotechnol. Bioeng.* **12:**1103–1109.

Davenport, S. M., 1989, Design and use of a novel peracetic acid sterilizer for absolute barrier sterility testing chambers, *J. Parenteral Sci. Technol.* **43**(4)**:**158–166.

Eckhardt, B. M., Oeswein, J. Q., Yeung, D. A., Milby, T. D., and Bewley, T. A., 1994, A turbidimetric method to determine visual appearance of protein solutions, *J. Pharm. Sci. Technol.* **48**(2)**:**64–70.

Edwards, R. A, and Huber, R. E., 1992, Surface denaturation of proteins:the thermal inactivation of β-galactosidase (*Escherichia coli*) on wall–liquid surfaces, *Biochem. Cell. Biol.* **70:**63–69.

European Pharmacopoeia, 1997, 3rd ed., p. 15.

Facts Comparison Staff, 1996, *Drug Facts and Comparisons*, Facts and Comparisons, St. Louis, MO.

Fisher, B. V., and Porter, P. B., 1981, Stability of bovine insulin, *J. Pharm. Pharmacol.* **33:**203–206.

Fraenkel-Conrat, H., and Olcott, H. S., 1948a. The reaction of formaldehyde with proteins. VI. Cross-linking of amino groups with phenol, imidazole, or indole groups, *J. Biol. Chem.* **174:**827–843.

Fraenkel-Conrat, H., and Olcott, H. S., 1948b. The reaction of formaldehyde with proteins. V. Cross-linking between amino and primary amide or guanidyl groups, *J. Am. Chem. Soc.* **70:**2673–2684.

Franks, F., 1990, Freeze drying: From empiricism to predictability, *Cryo-Letters* **11:**93–110.

Fry, E. 1987, FDA update on aseptic processing guidelines, *J. Parenteral Sci. Technol.* **41**(2)**:**56–60.

Groves, M. J., 1993. *Particulate Matter: Sources and Resources for Healthcare Manufacturers*, Interpharm Press, Buffalo Grove, IL.

Haas, P. J., 1995, Engineering design considerations for barrier isolation systems, *Pharm. Technol.* **19**(2)**:**26–38.

Habeeb, A. F. S. A., and Hiramoto, R., 1968, Reactions of proteins with glutaraldehyde, *Arch. Biochem. Biophys.* **126:**16–26.

Hanson, M. A., and Rouan S. K. E., 1992, Formulation of protein pharmaceuticals, in: *Stability of Protein Pharmaceuticals, Part B, In Vivo Pathways of Degradation and Strategies for Protein Stabilization* (T. J. Ahern, and M. C. Manning, eds.), Plenum Press, New York, pp. 216.

Hofmann, K. L., 1993, Sterile products validation, in: *Pharmaceutical Process Validation*, 2nd ed. (I. R. Berry and R. A. Nash, eds.), Marcel Dekker, New York, pp. 89–167.

Jenkins, K. M., and Vanderwielen, A. J., 1994, Cleaning validation: An overall perspective, *Pharm. Technol.* **18**(4)**:** 60–73.

Kaplan, K. J., and Fraser, M. J., 1953, Formation of fibres from protein monolayers, *Nature* **171:**559–562.

Kirsch, L. E., Riggin, R. M., Gearhart, D. A., Lefeber, D. S., and Lytle, D. L., 1993, In-process protein degradation by exposure to trace amounts of sanitizing agents, *J. Parenteral Sci. Technol.* **47**(4)**:**155–160.

Knapp, J. Z., and Abramson, L. R., 1990, Automated particulate inspection systems: strategies and implications, *J. Parenter. Sci. Technol.* **44**(2):74–107.

Levine, H. L., Ransohoff, T. C., Kawahata, R. T., and McGregor, W. C., 1991, The use of surface tension measurements in the design of antibody-based product formulations, *J. Parenteral Sci. Technol.* **45**(3):160–165.

Li, L. C., Parasrampuria, J., and Tian, Y., 1993, Particulate contamination from siliconized rubber stoppers—A statistical evaluation, *J. Parenteral Sci. Technol.* **47**(5):270–273.

Lougheed, W. D., Woulfe-Flanagan, H., Clement, J. R., and Albisser, A. M., 1980, Insulin aggregation in artificial delivery systems, *Diabetologia* **19**:1–9.

Lubig, W. E., Kush, P., Roeper, K., and Zahn, H., 1981, Mechanisms of protein reactions with gluteraldehyde, *Monatsch. Chem.* **112**:1313–1323.

Lundstrom, I., 1983, Surface physics and biological phenomena, *Phys. Scripta* **T4**:5–13.

Lysfjord, J. P., Haas, P. J., Melgaard, H. L., and Pflug, I. J., 1995, The potential for use of steam at atmospheric pressure to decontaminate or sterilize parenteral filling lines incorporating barrier isolation technology, *J. Pharm. Sci. Technol.* **49**(5):212–219.

MacRitchie, F., 1978, Proteins at interfaces, *Adv. Protein Chem.* **32**:283–290.

Manning, M. C., Patel, K., and Borchardt, R. T., 1989, Stability of protein pharmaceuticals, *Pharm. Res.* **6**:903–918.

Martin, J. M., and Manteuffel, R. L., 1988, Protein recovery from eluents of microporous membranes, *BioPharm* **1**(10):20–25.

Middaugh, C. R., Volkin, D., 1992, Protein solubility, in: *Stability of Protein Pharmaceuticals, Part A, Chemical and Physical Pathways of Protein Degradation* (T. J. Ahern and M. C. Manning, eds.), Plenum Press, New York, pp. 120–123.

Munson, T. E., and Sorensen, R. L., 1990, Environmental monitoring: Regulatory issues, in: *Sterile Pharmaceutical Manufacturing, Applications for the 1990's, Vol. 2* (M. J. Groves, W. P. Olson, and M. H. Anisfeld, M. H. eds.), Interpharm Press, Buffalo Grove, IL, pp. 169–182.

Nakata, T., 1994, Destruction of challenged endotoxin in a dry heat oven, *J. Pharm. Sci. Technol.* **48**(2):59–63.

Oldshue, J. Y., 1997, Agitation, in: *Fermentation and Biochemical Engineering Handbook* (H. C. Vogel, and C. L. Todaro, eds.), 2nd ed. Noyes Publications, Park Ridge, NJ, pp. 181–206.

Parenteral Drug Association, 1978, *Validation of Steam Sterilization Cycles*, Technical Monograph No. 1, Parenteral Drug Association, Bethesda, MD.

Parenteral Drug Association, 1981, *Validation of Dry Heat Processes Used for Sterilization and Depyrogenation, Technical Monograph No. 3*, Parenteral Drug Association, Bethesda, MD.

Parenteral Drug Association, 1996, *Process Simulation Testing for Aseptically Filled Products, Technical Monograph No. 22*, Parenteral Drug Association, Bethesda, MA.

Pavanetto, F., Conti, B., Genta, I., and Modena, T., 1991, Particulate contamination from siliconized rubber stoppers, *Int. J. Pharm.* **74:**175–181.

Pazur, J. H., 1970, Oligosaccharides, in: *The Carbohydrates, Chemistry and Biochemistry, Vol. IA* (W. Pigman, and D. Norton, eds.), Academic Press, New York, pp. 96.

Pikal, M. J., Dellerman, K. M., Roy, M. L., and Riggin, R. M., 1991, The effects of formulation variables on the stability of freeze-dried human growth hormone, *Pharm. Res.* **8**(4)**:**427–436.

Pitt, A. M., 1987, The nonspecific protein binding of polymeric microporous membranes, *J. Parenteral Sci. Technol.* **41:**110–113.

Roseman, T. J., Brown, J. A., and Scothorn, W. W., 1976. Glass for parenteral products: A surface view using the scanning electron microscope, *J. Pharm. Sci.* **65**(1)**:**22–29.

Saleh, A. M., El-Khordagui, L. K., Robb, D. A., and Florence, A. T., 1989, Effect of some drugs and additives on the cross-linking of bovine serum albumin by gluteraldehyde, *Int. J. Pharm.* **57:**205–210.

Sluzky, V., Tamada, J. A., Klibanov, A. M., and Langer, R., 1991, Kinetics of insulin aggregation in aqueous solutions upon agitation in the presence of hydrophobic surfaces, *Proc. Natl. Acad. Sci. USA* **88:**9377–9381.

Sluzky, V., Klibanov, A. M., and Langer, R., 1992, Mechanism of insulin aggregation and stabilization in agitated aqueous solutions, *Biotechnol. Bioeng.* **40:**895–903.

Tatterson, G. B., 1991, *Fluid Mixing and Gas Dispersion in Agitated Tanks*, McGraw-Hill, New York.

Torbeck, L. D., and Branning, R. C., 1996, Designed experiments—A vital role in validation, *Pharm. Technol.* **20:**108–114.

Trappler, E. H., 1993, Validation of lyophilized products, in: *Pharmaceutical Process Validation*, 2nd ed. (I. R. Berry, and R. A. Nash, eds.), Marcel Dekker, New York, pp. 445–477.

Truskey, G. A., Gabler, R., DiLeo, A., and Manter, T., 1987, The effect of membrane filtration upon protein conformation, *J. Parenteral Sci. Technol.* **41**(6)**:**180–193.

Tsuji, K., and Harrison, S. J., 1978, Dry heat destruction of lipopolysaccharide: dry heat destruction kinetics, *Appl. Environ. Microbiol.* **36:**710–714.

U.S. Food and Drug Administration, 1987, *Guideline on General Principles of Process Validation*, U.S. FDA, Center for Drugs and Biologics and Center for Devices and Radiological Health, Rockville, MD.

U.S. Food and Drug Administration, 1991, Use of aseptic processing and terminal sterilization in the preparation of sterile pharmaceuticals for human and veterinary use, *Fed. Reg.* **56**(198)**:**51354.

U.S. Food and Drug Administration, 1993a, *Guide to Inspections of Lyophilization of Parenterals*, Reg 07/01/93, U.S. FDA, Division of Field Investigations, Rockville, MD.

U.S. Food and Drug Administration, 1993b, *Guide to Inspection of Validation of Cleaning Processes*, U.S. FDA, Division of Field Investigations, Office of Regional Operations, Office of Regulatory Affairs, Rockville, MD.

U.S. Food and Drug Administration, 1996. Title 21, Part 211: Current good manufacturing practices for finished pharmaceuticals, *U.S. Code of Ferderal Regulations*, U.S. Government Printing Office, Washington, DC.

USP, 2000a, *United States Pharmacopeia*, *USP 24/NF 19*, Rockville, MD, ⟨1116⟩ Microbiological Evaluation of Clean Rooms and Other Controlled Environments.

USP, 2000b, *United States Pharmacopeia*, *USP 24/NF 19*, Rockville, MD, ⟨1211⟩ Sterilization and Sterility Assurance of Compendial Articles.

USP, 2000c, United States Pharmacopeia, *USP 24/NF 19*, Rockville, MD, ⟨1211⟩ Sterilization and Sterility Assurance of Compendial Articles.

Wang, J. Y.-C., and Hanson, M. A., 1988, Parenteral formulations of proteins and peptides: Stability and stabilizers, *J. Parenteral Sci. Technol. Suppl.* **42:**S3–S26

Weast, R. C. (ed.), 1971, *CRC Handbook of Chemistry and Physics*, 52nd ed., Chemical Rubber Co., Cleveland, OH.

Yamamoto, O., 1992, Effect of radiation on protein stability, in: *Stability of Protein Pharmaceuticals, Part A, Chemical and Physical Pathways of Protein Degradation* (T. J. Ahern and M. C. Manning, eds.), Plenum Press, New York, pp. 361–421.

4

Fundamentals of Thermal Sterilization Processes

Richard T. Wood

1. INTRODUCTION

A cell is considered dead when it can no longer reproduce under conditions known to promote growth of that microorganism. A strict definition of a sterilization process is one in which all viable cells in a given population are inactivated by some physical or chemical means, and inactivation is irreversible. However, as currently defined in the pharmaceutical industry, the effectiveness of a sterilization process is determined by the statistical probability of a non-sterile unit (PNSU) in a lot or batch of product.

Peptides and proteins typically cannot withstand the high temperatures required to destroy microorganisms in terminal sterilization processes. However, thermal sterilization processes are applied in the sterilization of process equipment and primary packaging items for peptide and protein finished dosage forms. The fundamental principles presented in this chapter apply to the sterilization of glass vials (or other glass packaging), rubber closures, filters and filter assemblies, process tubing, mixing tanks and other manufacturing containers, filling machines, freeze-dryer chambers, piping, and all other hard surface items (e.g., needles, nozzles, forceps, metal containers) used in sterile process manufacturing. The following discussion is aimed at providing the formulation scientist with a working knowledge of

---

Richard T. Wood • Pfizer, Inc., Terre Haute, Indiana 47808.

Development and Manufacture of Protein Pharmaceuticals, edited by Nail and Akers. Kluwer Academic/Plenum Publishers, New York, 2002.

the fundamentals of sterilization biology as well as basic engineering aspects of thermal sterilization processes. This is important with respect not only to manufacture of finished pharmaceutical dosage forms, but also fermentation processes and downstream purification of proteins, where minimizing the bioburden and endotoxin levels relies on thermal sterilization processes.

2. EFFECTS OF HEAT ON LIVING CELLS

Probably the best-known and most thoroughly researched sterilizing agent is heat. As temperatures increase beyond the range where biological activities occur at an optimum rate, cellular metabolism begins to slow down, and growth and reproduction eventually cease. For each type of cell there is a temperature limit above which vital proteins, enzymes, and nucleic acids are permanently damaged and irreversible cell death occurs.

Microscopic, prokaryotic cells that do not have a true membrane-bounded nucleus (e.g., the bacteria and blue-green algae) exhibit the greatest tolerance to heat. Certain thermophilic, or heat-loving, bacteria can grow at temperatures above 80°C. However, the vegetative cells of most bacteria are rapidly destroyed at temperatures between 80°C and 100°C. Microscopic eukaryotic cells, which have a true membrane-bounded nucleus, such as fungi and protozoa, are rapidly killed at temperatures in the range of 60–80°C. Most viruses are heat-labile, although the hepatitis B virus (HBV), which is considered to have above average heat resistance, requires at least a 3-min exposure at 75°C for inactivation (Li, 1986).

There are other biological entities that show a much greater resistance to heat. For example, the etiologic agent of Creutzfeldt–Jakob disease (called a prion), which is neither a cellular microorganism nor a virus, requires autoclaving for 1 h at 132°C for complete inactivation (Russell, 1989). Among the bacteria, aerobic organisms of the genus *Bacillus* and their anaerobic counterparts in the genus *Clostridium* exhibit considerable heat resistance. These bacteria have the ability to produce endospores, or intercellular resting bodies, which upon maturation are resistant to the destructive effects of heat as well as drying and harsh chemicals. Inactivation of 90% of a bacterial endospore population within 1 min requires temperatures in the range of 100–170°C for dry heat or 80–129°C for moist heat.

The thermoresistance of bacterial endospores is related to a number of factors that are not fully understood. When sporulation occurs, these bacteria undergo a biochemical shutdown and enfold their genetic material.

A series of biophysical changes occurs which includes a substantial dehydration and reduction in the size of the cell protoplast and formation of a thick cortex and coat surrounding the reduced protoplast. A unique chemical substance called dipicolinic acid (DPA; pyridine-2,6-dicarboxylic acid) is also formed. Calcium dipicolinate complexes with DNA and possibly with cellular enzymes during sporulation to protect them from damage while the spores are in the dormant state (Gerhardt and Murrell, 1978). Endospores may remain dormant, but viable for many years, and can revert back to the vegetative state in a matter of minutes under conditions that induce germination.

3. FACTORS AFFECTING RESISTANCE OF MICROORGANISMS TO THERMAL INACTIVATION

The major factors affecting the thermal resistance of microorganisms are the inherent resistance of the microorganism itself, environmental factors during the growth and formation of cells or spores, and environmental influences during the sterilization process itself (Pflug and Holcomb, 1991).

3.1. The Physical/Chemical Environment

Environmental factors that affect thermoresistance during bacterial endospore formation include high growth temperature and the presence of divalent cations (e.g., Ca^{2+}, Fe^{2+}, Mg^{2+}, Mn^{2+}), both of which tend to increase heat resistance. Fatty acids have also been reported to increase resistance (Pflug and Holcomb, 1991). On the other hand, heat resistance may decrease at pH conditions outside the range of 6.0–8.0 or when sporulation occurs in the presence of high salt or phosphate concentrations.

The heat resistance of mature spore populations is also affected by environmental conditions such as solute concentration, water activity a_w, pH, and the presence of chemical substances that can physically damage the spores or inhibit their ability to germinate (Joslyn, 1983). Again, the greatest thermal resistance is associated with a range of pH centered around pH 7, although the width of this range is affected by the species and by the composition of the medium. High concentrations of carbohydrates are generally associated with increased resistance. Studies of the influence of salt concentration on resistance to thermal destruction have produced mixed results (Pflug and Holcomb, 1991).

Spores encapsulated in crystals or organic matter usually show dramatic increases in heat resistance compared to unencapsulated spores. Thus, sterilization of spores occluded in soil may require a 10-fold longer exposure to heat at a given temperature than spores cultured from the same soil (Joslyn, 1983). The practical significance of this is that objects contaminated with spores originating from soil (e.g., in dust and dirt transported through unfiltered air or by direct contact with humans or inanimate objects) may prove more difficult to sterilize than spores present on biological indicators.

3.2. Moisture Level

Water plays a major role in the heat destruction of bacterial spores. Only two sterilization conditions are recognized with respect to water content: moist and dry heat. Saturated moisture conditions at a relative humidity (RH) of 100% (or a_w of 1.0) represent moist heat sterilization; all other levels of RH are dry heat sterilization. It has been shown that maximum heat resistance of bacterial spores at temperatures from 90°C to 125°C occurs at RH levels between 20% and 50%, and that heat resistance decreases rapidly above or below this RH range (Angelloti *et al.*, 1968).

4. THERMAL INACTIVATION OF MICROORGANISMS

4.1. Inactivation by Moist Heat

Moist heat sterilization is a very efficient and effective method of inactivating microorganisms in products and on objects that are compatible with this process. Moist heat sterilization generally implies the use of steam under pressure, although direct immersion in superheated water is sometimes utilized. Water takes up large amounts of heat energy (ca. 540 cal/g) when it changes from a liquid to a vapor, and gives up this energy (known as the latent heat of vaporization) when it condenses back again into a liquid. When steam contacts a cooler object, it releases its stored heat energy and forms condensate until the object reaches the temperature of the steam. This is a highly efficient heat transfer mechanism and is not matched by any other process.

The effectiveness of moist heat sterilization is believed to be due to the thermal denaturation or coagulation of critical cellular proteins and enzymes. Moisture facilitates this process, as evidenced by the lower temperatures required for moist heat sterilization compared to those for dry heat (Young, 1993).

4.2. Inactivation by Dry Heat

4.2.1. INTRODUCTION

As mentioned above, dry heat sterilization is any thermal inactivation process that occurs at less than saturated moisture conditions. Therefore, dry heat sterilization covers a broad moisture range from 99.9% to less than 1% RH. The heating medium for dry heat sterilization is usually air at ambient RH conditions, moved by forced convection. As moisture is taken up by the heated air, processed materials (and spores) will gradually dehydrate, causing microbial inactivation rates to vary (Wood, 1993).

Although the kinetics of dry heat sterilization processes is similar to that of moist heat sterilization, the mechanism appears to be different. In this regard, dry heat inactivation of microorganisms is attributed to oxidation rather than protein denaturation, which would account for the relatively high temperatures required for dry heat sterilization.

Dry heat sterilization, although less efficient than steam sterilization, is widely used in the pharmaceutical industry for sterilization of glass vials and product-contact equipment that will withstand the higher temperatures. Product components are generally not sterilized by dry heat, although exceptions include glycerin, oils, petroleum jelly, and some solid drug components such as talc and sulfonamides. The primary reason for using dry heat sterilization instead of steam sterilization is that dry heat will inactivate bacterial endotoxins.

4.2.2. DEPYROGENATION BY DRY HEAT

Lipopolysaccharide (LPS) from the outer cell membranes of Gram-negative bacteria is highly toxic to humans. This toxin is known as endotoxin, because it is released only when the cells rupture or lyse. Endotoxins are also called pyrogens, because they can cause a febrile response upon injection at low concentrations. Any process aimed at removing or inactivating LPS is called a depyrogenation process.

Dry heat is one of the most effective depyrogenation methods. However, the temperatures required for depyrogenation are quite high relative to moist heat sterilization—in the range of 170–250°C or higher—and the resistance of a given endotoxin is affected to some extent by the species of bacteria from which it arises. Rates of thermal inactivation of endotoxins are highly temperature-dependent and may be a complex function of temperature. For example, Ludwig and Avis (1990) reported that depyrogenation at 170°C required approximately 30 min to inactivate 90% of purified endotoxin, whereas the same amount of endotoxin is inactivated in 9 sec at 250°C. Dry heat depyrogenation processes are usually validated by direct inoculation of articles with known amounts of endotoxin followed by demonstration of a minimum three-log cycle reduction of the endotoxin.

5. STERILITY ASSURANCE

In the manufacturing of injectable pharmaceuticals, sterility is addressed not as an absolute concept, where a material is considered either sterile or nonsterile, but as a probability function. Sterility assurance is expressed in terms of the probability of a nonsterile unit within a given batch of product by the following equation:

$$\mathrm{PNSU} = N_0 - D_\mathrm{R} \tag{1}$$

where N_0 is the presterilization microbial population level, also known as the bioburden, and D_R is the decimal reduction in bioburden that occurs during a sterilization process. This equation shows that sterility depends upon control of the presterilization microbial load (bioburden) as well as the extent of bioburden inactivation that occurs during the sterilization process. It follows that no sterilization process should be required if microbial contamination can be prevented. However, excluding all microorganisms from a product or environment is very difficult, if not impossible. Moreover, proving experimentally that microbial contamination has not occurred cannot be accomplished with an acceptable degree of certainty.

Good manufacturing practice (GMP) for pharmaceutical products requires that written procedures be followed to minimize bioburden levels in order to minimize the initial bioburden (U.S. Food and Drug Administration [FDA], 1995). These practices include starting with clean, sanitized equipment, controlling the quality of the manufacturing environment, using raw materials and water that have a low bioburden, and restricting the length of the holding times, particularly for products that are susceptible to microbial growth. Good personal hygiene and thorough training in aseptic technique of the people involved in manufacturing are also very important GMP considerations.

6. THERMAL INACTIVATION KINETICS

Microbial death is a first-order process that can be represented with reasonable accuracy as a straight-line plot on semilogarithmic graph paper. Although microbial inactivation curves may be concave upward or downward (or have almost any shape), in most cases, a portion of these curves will approximate a straight line (Pflug, 1986).

6.1. Decimal Reduction Value (*D* Value)

The *D* value is defined as the time in minutes required for a one-log cycle or 90% reduction of a microbial population under specified lethal conditions. The *D* value is of fundamental importance to sterilization process analysis and can be determined by one of two methods: the survivor curve method or the fraction-negative method.

6.1.1. SURVIVOR CURVE METHOD

A minimum of two inoculated test specimens are each exposed at a preset temperature (ca. 121°C) for different time intervals. A constant-temperature, rapid-heating device such as a biological indicator evaluator resistometer (BIER vessel) is used for these studies. Surviving organisms are extracted for growth and enumeration from each set of treated specimens and from untreated controls. Survivors are recovered by pour-plating or by capturing them on membrane filters that are incubated on the surface of a suitable agar medium. At the end of the incubation period, the number of survivors N_s is determined by counting colonies, and the average counts (converted to Log_{10} values) are plotted against the corresponding exposure times in minutes, u, to construct a survivor curve as shown in Fig. 1.

The *D* value can be estimated by intersecting the survivor curve with lines drawn parallel to the x and y axes for one logcycle as shown in Fig. 1, or by calculating it from survivor curve data using the following equation (Parenteral Drug Association [PDA], 1978):

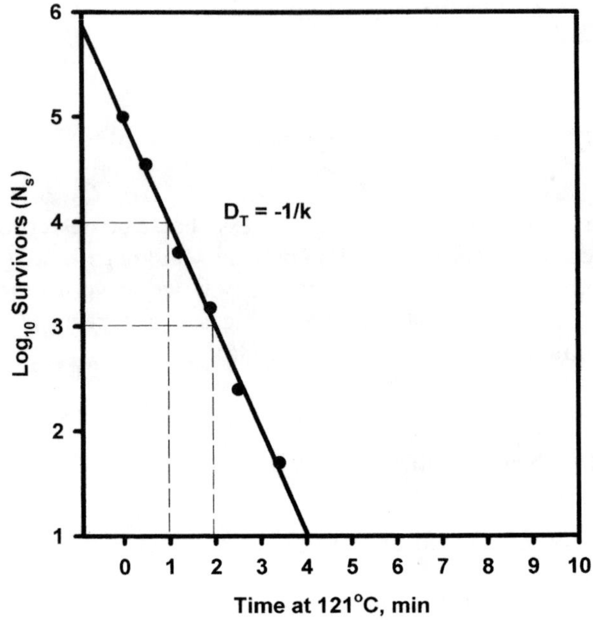

Figure 1. Hypothetical survivor curve for spores heated at 121°C.

$$D_{121^\circ C} = u/(\text{Log}_{10} N_0 - \text{Log}_{10} N_s) \tag{2}$$

For example, the $D_{121^\circ C}$ value at an exposure time of 5 min, where $\text{Log}_{10}N_0$ and $\text{Log}_{10}N_s$ are 5 and 2, respectively, is

$$D_{121^\circ C} = 5/(5 - 2) = 1.7\,\text{min}$$

It may also be seen in Fig. 1 that the D value is the negative reciprocal slope $(-1/k)$ of the survivor curve. The slope may be calculated from an unweighted least-squares linear regression analysis of the linear portion of the survivor curve,

$$k = \frac{\sum uy - (\sum u)(\sum y)/n}{\sum u^2 - (\sum u)^2/n} \tag{3}$$

where y is the average colony count at time u and n is the number of data pairs.

6.1.2. FRACTION-NEGATIVE METHOD

This is a most probable number method, and the D value can be estimated from the results of one exposure time. Individual samples in a group of specimens (minimum of 10) are tested for sterility following a timed exposure at a preset temperature. This approach permits the detection of survivors below the range that can be observed by the plate count technique. The fraction-negative method is based on the assumption that microbial inactivation is linear from N_0 to the inactivation endpoint. The D value (at temperature T) can be calculated using the following equation (Stumbo *et al.*, 1950):

$$D_T = u/\text{Log}_{10} N_0 - \text{Log}_{10} (2.303 \text{ Log}_{10} n/q) \tag{4}$$

where n is the total number of specimens processed and q is the number of sterile specimens following incubation.

Another fraction-negative approach that can be used to estimate the D value requires at least 10 specimens each exposed at a preset temperature at several different (equally spaced) time intervals (USP, 1995). Each specimen from each test group is incubated separately under appropriate conditions in a tube of sterile broth. At the end of the incubation period the number of tubes in each group showing no growth is recorded, and the data are arranged as shown in Table I.

The D value is calculated from the results that lie between the last group at the shortest exposure period showing no inactivation (0/10) and the first group at the longest exposure period showing all specimens inactivated (10/10). It may be seen from the table that these exposure times are 5 and 15 min, respectively. The heating time T required for complete kill (e.g., < 1 viable spore) is calculated using the equation

$$T = T_K - d/2 - \left[(d/10) \times \sum f\right] \tag{5}$$

Table I
Estimating D value by Fraction-Negative Analysis[a]

	Value at given exposure time @121°C							
	3 min	5 min	7 min	9 min	11 min	13 min	15 min	17 min
f/n	0/10	0/10	1/10	3/10	5/10	7/10	10/10	10/10
f	—	0	1	3	5	7	10	—

[a] f/n = number of tubes showing no growth (f)/total number of tubes (n) incubated.

where T_K is the exposure time at the lower limit of the fraction-negative range, d is the time difference in minutes between exposure periods, and f is the number of tubes showing no growth at each time point.

From the data in Table I, it may be seen that

$$T = 15 - 1 - [(2/10) \times 26] = 8.8 \text{ min}$$

The D value is then calculated using the following equation:

$$D = T/(\log_{10} N_0 + 0.2507) \tag{6}$$

Assuming $N_0 = 10^5$, one finds

$$D = 8.8/(5 + 0.2507) = 1.7 \text{ min}$$

6.2. Temperature Coefficient or *Z* Value

When the D value for a microbial population is determined at several different temperatures (under the same experimental conditions) a straight line is obtained if the log D (in minutes) is plotted against the corresponding temperature on a linear scale. The negative reciprocal slope of this line is called the temperature coefficient, or Z value, and it provides an indication of the change in microbial lethality with temperature. The Z value may also be calculated using the equation (Owens, 1993)

$$Z = T_2 - T_1/\text{Log}_{10}(D_2/D_1) \tag{7}$$

where D_1 is the D value at temperature T_1 and D_2 is the D value at temperatute T_2.

The Z value is defined as the number of degrees of temperature required for a one-log cycle or 90% change in the D value, and is the universal factor used in designing and monitoring pharmaceutical sterilization processes (Pflug, 1986). Experimental values of Z for moist heat processing of bacterial spores vary from approximately 8°C to 12°C, but a Z value of 10°C is usually assumed when calculating lethality. A Z value of 20°C is generally assumed for dry heat processing.

6.3. LETHALITY RATIOS

When the Z value has been determined for a microbial population (under controlled experimental conditions), it can be used to calculate

Table II
Lethality Ratios during Moist Heat Sterilization[a]

Process temperature (°C)	Lethality ratio (compared to T_R)
110	0.08
112	0.13
114	0.19
116	0.32
118	0.50
120	0.79
122	1.26

[a] T_R = 121°C, Z = 10°C.

lethality ratios (or lethal rates L) from temperature measurements recorded during a sterilization process, e.g.,

$$L = 10^{(T_i - T_R/Z)} \tag{8}$$

where T_i is temperature at time interval of measurement and T_R is the reference temperature.

For example, if T_R is 121°C and the Z-value is 10°C, then lethality varies with temperature, as illustrated in Table II.

6.4. Process Values (*F* Values)

Practical use can be made of microbial inactivation rate calculations to determine the F_T value, or process value, of a sterilization cycle. Thus, lethality rates during all phases of the sterilization cycle (e.g., heat up, cycle timing, and cool down periods) can be integrated using the following trapezoid rule approximation to estimate the equivalent time of a sterilization cycle at the reference temperature (Patashnik, 1953):

$$F_T = \Delta t\,(L_1/2 + L_2 + L_3 + \cdots + L_{n-1} + L_n/2) \tag{9}$$

Here Δt is the time interval of temperature measurement.

Temperature measurements must be taken at equal time intervals if Eq. (9) is used. Since lethality rates are very small at the beginning and end of the mathematical integration, the equation can be simplified to

$$F_T = \Delta t \Big/ \sum L \tag{10}$$

When the reference temperature is 121.1°C, the F_T value is referred to as F_0; for dry heat sterilization the reference temperature is 170°C, and the process value is referred to as F_H. Both F_0 and F_H are universally recognized values that can be used to compare different moist or dry heat sterilization cycles, respectively. By rearranging Eq. (2) and substituting F_0 for u, we have the following equation:

$$\text{Log}_{10} N_s = \text{Log}_{10} N_0 - F_0/D_T \tag{11}$$

This equation can be used to predict the biological outcome of a sterilization process using physical (temperature) measurements from thermocouples placed in a sterilizer load. In this case, D_T is the $D_{121.1°C}$ value for a real or hypothetical microbial population [e.g., a biological indicator (BI) or bioburden population] that is to be inactivated during the process. For example, let us assume that the F_0 is 8 min, and that a population of 10^6 spores having a $D_{121.1°C}$ value of 0.5 min is exposed to these conditions. From Eq. (11) it may be seen that the surviving population endpoint (or PNSU) is 10^{-10}:

$$\text{Log}_{10} N_s = 6 - 8/0.5 = -10$$

The total spore log reduction (SLR) that occurs during this sterilization process is 16. The SLR is equal to N_0 minus N_s, or F_0 divided by the $D_{121.1°C}$ value:

$$\begin{aligned} \text{SLR} &= \text{Log}_{10} N_0 - \text{Log}_{10} N_s \\ &= 6 - (-10) = 16 \end{aligned} \tag{12}$$

or

$$\begin{aligned} \text{SLR} &= F_0/D_{121.1°C} \\ &= 8/0.5 = 16 \end{aligned} \tag{13}$$

6.5. Biological Indicators

Valuable information about a sterilization process can be obtained from thermocouples placed in items exposed to heat. Although predictions concerning the death of micoorganisms can be made from thermocouple heat penetration data, the complete or partial inactivation of BIs is the best indication of whether a sterilization process is effective.

BIs are available in different forms. Since sterilization processes in the context of manufacture of protein pharmaceutical products refer primarily

Table III
Thermal Sterilization Process Biological Indicators

Sterilization method	Microorganisation (as spores)	Approximate D value (min)
Moist heat at 121°C	*Bacillus stearothermophilus*	1.5–3.0
	Bacillus subtilis ATCC 5230	0.3–0.7
	Bacillus coagulans	0.4–0.8
	Clostridium sporogenes	0.4–0.8
Dry heat at 150°C	*Bacillus subtilis* var. *niger*	> 1.0

to sterilization of equipment, bacterial spores are usually deposited on a carrier material such as paper, fiberglass, or stainless steel and placed on or in the articles being sterilized, with particular emphasis on areas that may present problems with respect to heat penetration.

Characterization of BIs involves enumeration of spores as well as their D and Z values under actual use conditions. Some commonly used BIs in moist and dry heat processing are shown in Table III.

7. STERILIZATION CYCLE DEVELOPMENT

7.1. Establishing an Endpoint

The goal of any sterilization process is to completely destroy all microorganisms. However, considering the probabilistic nature of microbial death and the difficulty in detecting low-level microbial contamination, it is not possible to prove that all cells have been destroyed during a sterilization process. The question then arises as to what sterilization endpoint should be used as a target. The FDA has solved this problem for the industry by defining sterility as reduction of a microbial population to 10^0 plus six additional log cycles (e.g., a PNSU $\leqslant$ 1 in 1 million units, or 10^{-6}) (U.S. FDA, 1987).

Putting aside engineering considerations for the moment, the physical and biological factors in a sterilization process that must be rationalized to achieve a 10^{-6} PNSU are fairly simple, as shown in Table IV.

When N_0 and D_T are known, the process value required for a 10^{-6} sterilization endpoint can be determined by rearranging Eq. (11) and solving for F_T:

$$F_T = (\text{Log}_{10} N_0 - \text{Log}_{10} N_s) \times D_T \tag{14}$$

Table IV
Biological and Physical Factors in Sterilization Process Development

Biological factors	Physical factor
Presterilization bioburden, N_0	Sterilization process, F_T
Bioburden heat resistance, D_T	
Sterilization endpoint, N_s (10^{-6})	

For example, a product containing 100 heat-resistant spores per container having a $D_{121.1°C}$ value of 1 min will require a moist heat sterilization process value F_0 of 8 min to reduce the bioburden to 10^0 plus six additional log cycles:

$$F_0 = [2 - (-6)] \times 1 = 8\,\text{min}$$

An F_0 of 8 min would be the minimum target sterilization endpoint for this process. An upper F_0 limit would also be established based on the performance characteristics of the sterilizer.

7.2. Bioburden-based Sterilization Processes

A bioburden-based sterilization process usually applies to terminal sterilization of final product, which does not apply directly to protein pharmaceuticals. However, the approach to a bioburden-based process is discussed here for the sake of completeness. In a bioburden-based process, the endpoint has been established based on the number and heat resistance of the bioburden typical of the material being sterilized. It should be remembered, however, that the goal of manufacturing, particularly of bulk proteins to be formulated into an injectable product, should be to keep microorganisms out of the product, particularly in view of (1) the growth-promotive nature of most protein formulations and the concern over bacterial endotoxin contamination, and (2) the fact that sterilization of final product must rely on membrane filtration, which has a lower degree of sterility assurance than terminal sterilization.

When there is minimal product bioburden to deal with, a hypothetical spore population can be assumed for the purpose of sterilization cycle development. For example, 100 heat-resistant spores per item having a $D_{121°C}$ value of 0.5 min would represent a conservative bioburden assumption. Process development would then be based on referencing a test organism (BI) of known heat resistance to the destruction of this hypothetical bioburden. Test cycles are run and the average physical F_0 values obtained from thermocouple probes placed in product containers are compared to the average SLR values from similar containers inoculated with BI organisms (for which the spore count and D value are known). The level of reduction of the bioburden is then estimated from the level of BI inactivation using the equation

$$\mathrm{SLR_A} = \mathrm{SLR_B} \times D_\mathrm{B}/D_\mathrm{A} \tag{15}$$

where $\mathrm{SLR_A}$ is the spore log reduction of the bioburden organisms, $\mathrm{SLR_B}$ is the spore log reduction of the bioindicator organisms, D_A is the D value (either known or assumed) for the bioburden organisms, and D_B is the D value of the bioindicator organisms.

For example, if D_A and D_B are 0.5 and 2.0 min, respectively, at 121.1°C and a four-log cycle reduction of the BI organisms occurs at an F_0 of 8 min during the sterilization process, then the corresponding SLR of the bioburden organisms will be approximately 16-log cycles:

$$\begin{aligned}\mathrm{SLR_A} &= 4 \times (2.0/0.5)\\ &= 16\,\log\ \text{cycles}\end{aligned}$$

PNSU is then estimated by rearranging Eq. (12) and solving for $\mathrm{Log}_{10}\, N_\mathrm{s}$:

$$\begin{aligned}\mathrm{Log}_{10}\, N_\mathrm{s} &= \mathrm{Log}_{10}\, N_0 - \mathrm{SLR}\\ &= 2 - 16 = -14\end{aligned}$$

This approach is further illustrated in Fig. 2, where a portion of the BI organisms has survived at the end of the sterilization cycle. This does not mean that the process was ineffective. On the contrary, the BIs served their intended purpose as "biological integrators." They confirmed what the thermocouples indicated, that an F_0 of 8 min was delivered to the container contents.

Some practitioners mistakenly believe that a 10^6 concentration of the most heat-resistant spores available must be inoculated into product for process validation studies, and that a 6D or 12D inactivation level is necessary to guarantee a valid sterilization cycle. At times BI D values are enhanced in product, and the F_0 needed for complete inactivation becomes a very large number. There is no reason to pursue this approach if presterilization bioburden is controlled, monitored, and characterized as described below.

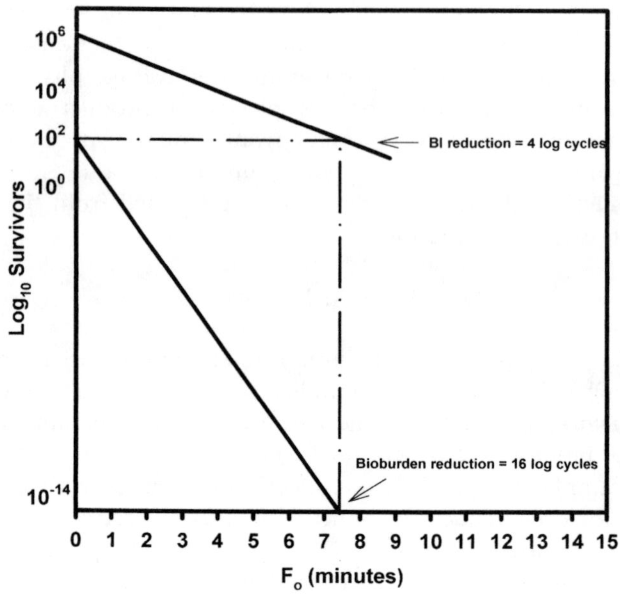

Figure 2. Bioburden lethality referenced to a biological indicator.

7.3. Bioburden Monitoring

It is very unlikely that a pharmaceutical product produced aseptically will contain heat-resistant spores prior to terminal sterilization if GMP practices are diligently followed. Even for products produced nonaseptically, the only microorganisms present prior to sterilization should be vegetative cells (in relatively low numbers) that have no significant heat resistance. Nevertheless, the bioburden of each batch of product produced nonaseptically should be monitored prior to terminal sterilization to ensure that microbiological quality is consistently maintained. This monitoring involves:

- Culling filled product containers from the end of filling operations
- Testing these samples for total aerobic microbial counts (TAMC)
- Testing for total spore counts (TSC) by heat-shocking sample aliquots at 100°C for 10 min, and culturing the heated samples both aerobically and anaerobically

A specification should be established for bioburden, and the TAMC data should be trended as a further check on the maintenance of low bioburden from batch to batch. The inability to recover any organisms from heat-shocked samples, which is the expected result, will indicate that there are no heat-resistant spores present to challenge the sterilization process.

In the unlikely event that bioburden organisms do survive the heat shock treatment, they will need to be identified and tested for D value. If the product isolates show a D value greater than the BIs used during process validation, then the sterilization process is inadequate, and steps should be taken to eliminate the heat-resistant bioburden. If these attempts are unsuccessful, the process must be revalidated using either the product isolate as challenge organism or a BI organism having a greater heat resistance.

7.4. Overkill Sterilization Processes

The approach usually taken in sterilization of process equipment is overkill sterilization, where bioburden is not considered when calculating process endpoints. However, this does not mean that bioburden is totally disregarded. On the contrary, it is still very important that good hygienic manufacturing practices be followed to control bioburden levels.

An overkill sterilization cycle has been defined as one that provides at least a 12-log cycle reduction of a heat-resistant organism (usually *Bacillus stearothermophilus*) having a D value of at least 1.0 min. Validation sometimes entails determining the process time required to inactivate BIs containing10^6 *B. stearothermophilus* spores, and then the process time is doubled.

Dry heat sterilization is often combined with depyrogenation processes, and depyrogenation conditions represent extreme overkill for bacterial spores, usually producing double-digit F_H values. Therefore, depyrogenation cycles involving a three-log cycle reduction of endotoxin do not have to be challenged with BIs (Young, 1993).

8. STERILIZATION ENGINEERING

8.1. Steam Sterilization Equipment

Steam sterilizers have progressed over the years from manually operated gravity displacement units to fully automated equipment with thermal load sensors and time–temperature integration controls, etc. High-vacuum sterilizers have a vacuum pump or steam ejector connected to the drain line to forcibily evacuate air from the chamber, and steam-pulsing units use two or more evacuations followed each time by a steam pulse to minimize air entrainment. Sterilizers for terminal sterilization of large- and small-volume parenteral solutions utilize either circulating water (water overspray) or steam–air–mixture (SAM) cycles. Controlled cooling and air over pressurization are used in these sterilizers, particularly during the processing of flexible plastic containers, to prevent container bursting.

8.2. Dry Heat Sterilization/Depyrogenation Equipment

Dry heat processing is used to sterilize and depyrogenate equipment and product components in pharmaceutical manufacturing. Static-batch-type ovens and continuous-belt, hot air tunnels that use forced convection through high-efficiency particulate air (HEPA) filters to trap airborne particulate matter are prevalent in the industry. This equipment has progressed from early sterilizer designs that used infrared radiation and gravity convection as their principal sources of heat. Modern dry heat sterilizers are fully automated and much more efficient than earlier equipment.

8.3. Process Validation

The validation philosophy developed by the pharmaceutical industry has had a tremendous impact on thermal sterilization processing. For a sterilizer to operate in a validated state of control, microbiological and engineering principles must be thoroughly understood; the equipment must be designed, installed, and operated properly; and documented evidence must be obtained through scientific testing to prove that the equipment will perform consistently and reliably. A documented record from early process design through maintenance of validation (representing a validation life cycle) must be maintained.

A thorough knowledge of the electrical, mechanical, and physical operating functions of a sterilizer is essential before attempting to validate it. This information is important when it comes to writing validation protocols and standard operating procedures. It is also useful in designing instrument calibration and maintenance programs and in formulating a revalidation policy to ensure that state of control is maintained at all times.

An example of the type of information needed to document the validation of a steam sterilization process includes the following (U.S. FDA, 1994):

1. An engineering description of the sterilizer
2. A description of the sterilization cycle(s); for example, saturated steam, water immersion, water spray, etc.
3. A description of cycle parameters and performance specifications; for example, temperature, pressure, time, minimum and maximum F_0, etc.
4. Loads defined; for example, loading patterns, product composition, container sizes, fill volumes, minimum and maximum loads, etc.
5. Methods and controls for monitoring the cycles; for example, thermocouples, pilot bottles, BIs, including number and location of each and acceptance/rejection criteria
6. Allowable minimum/maximum operating ranges for critical parameters
7. Automatic and manual control steps described
8. System checklists for possible recording of manual intervention
9. Lethality ranges for product(s)
10. Statistical evaluation of process variability
11. Vessel recertification program description; for example, description of the program for routine and unscheduled requalification, including frequency
12. Calibration program description

Biological and physical factors that must be considered when establishing a sterilization process endpoint were discussed in a previous section. It is also necessary to understand how the operating features and characteristics of a sterilizer influence these factors. In this regard, thermal qualification of a steam sterilizer cycle includes the studies dicussed below.

8.3.1. STERILIZER EMPTY CHAMBER HEAT DISTRIBUTION

Thermocouples are distributed geometrically throughout the chamber so that both horizontal and vertical zones are covered. The number of thermal monitors used and their location in the chamber should be described and included in a diagram. An additional thermocouple is placed in the exhaust drain in close proximity to the resident temperature sensor/controller. The sterilizer is run through several (usually three) maximum duration cycles. A relatively uniform temperature distribution should be observed during steady-state sterilization conditions (e.g., ±0.5°C between points). Uniformity, reproducibility, and conformance to the specifications of the sterilization cycle should be demonstrated.

8.3.2. HEAT PENETRATION OF EACH LOAD CONFIGURATION

The slowest-to-heat location in the total sterilizer load is determined by distributing thermocouples evenly throughout the load in product containers as well as in the chamber. These studies are conducted with the minimum and maximum loading configuration defined for the cycle. The effects of loading on thermal input are determined. The information that is obtained should include high and low temperatures (range), the average temperature during the dwell period, minimum and maximum F_0 values, and cycle dwell time. Container "cool spot" mapping is conducted for large-volume parenterals and products that are viscous liquids. Multiple thermocouples are distributed in product containers to obtain these data (Pflug, 1986).

Similar types of studies are also conducted with dry heat sterilizers, although heat distribution during steady-state conditions would be expected to vary more; for example, by as much as ±15°C in a forced-convection oven or continuous-belt tunnel. In addition, different types of materials and their loading patterns will have more of an effect on heat penetration under dry heat conditions than would be observed with moist heat.

9. PARAMETRIC RELEASE OF TERMINALLY STERILIZED DRUG PRODUCTS

In 1985, the FDA approved supplemental new drug applications for certain large-volume parenteral drug products, which substituted parametric release for routine lot-by lot end-product sterility testing. Parametric release was defined as "a sterility release procedure based upon effective control, monitoring, and documentation of a validated sterilization process cycle in lieu of release based upon end-product sterility testing (21 CFR 211.167)" (U.S. FDA, 1987).

Since this policy applies only to parenteral drug products which are terminally sterilized by moist heat, it does not apply to protein pharmaceuticals. Parametric release, in lieu of end-product sterility testing, is acceptable when all of the following general requirements are met and documented:

1. The sterilization process cycle has been validated to achieve microbial bioburden reduction to 10^0 with a minimum safety factor of an additional six logarithm reduction.
2. Integrity for each container/closure system has been validated to prevent in-process and postprocess contamination over the product's intended shelf-life.

3. Bioburden testing (covering total aerobic and total spore counts) is conducted on each batch of presterilized drug product.
4. Chemical or biological indicators are included in each truck, tray, or pallet of each sterilizer load.
5. If more than one critical parameter is not met, the batch is considered nonsterile, despite BI sterility.

In regard to point 4, chemical indicators are of two types, throughput indicators and multiparameter process indicators. Throughput indicators merely show by a color change whether an item has been exposed to minimum sterilization conditions. Multiparameter process indicators might integrate more than one process parameter, and behave similar to a BI. However, the FDA Guideline does not permit them to be used to measure cycle lethality.

10. SUMMARY

Information has been provided in this chapter on the fundamentals of thermal sterilization processing. The derivation of various equations presented in the text was not discussed; the reader is referred to the references cited for this information. Emphasis was placed on the need for good manufacturing practice and for control of presterilization bioburden. These considerations will have the greatest impact on product quality in the long run. An attempt was also made to dispel the notion that sterilization processes always require the most rigorous biological challenge for validation. Finally, parametric or process control release of terminally sterilized products was discussed. Parametric release is a concept that will find greater application in areas other than moist heat sterilization as the industry continues to derive full benefit from validated processes.

REFERENCES

Angellotti, R., *et al.*, 1968, Influence of spore moisture content on the dry heat resistance of *Bacillus subtilis* var. *niger, Appl. Microbiol.* **16:**735–745.

Gerhardt, P., and Murrell, W. G., 1978, Basis and mechanism of spore resistance: A brief review, in: *Spores VII* (G. Chambliss and J. C. Vary, eds.), American Society for Microbiology, Washington, DC, pp. 18–20.

Joslyn, L., 1983, Sterilization by heat, in: *Disinfection, Sterilization and Preservation,* 3rd ed. (S. S. Block ed.), Lea & Febiger: Philadelphia, pp. 3–46.

Li, Zhi-gui, 1986, The current status and prospect of research work of disinfection and sterilization in China, in: *Sterilization of Medical Products* (E. R. L.

Gaughran, R. F. Morrissey, and Wang You-sen, eds.), Polyscience, Montreal, Vol. IV, pp. 32–39.

Ludwig, J. D., and Avis, K. E., 1990, Dry heat inactivation of endotoxin on the surface of glass, *J. Parenteral. Sci Technol.* **44(1):**4–12.

Owens, J. E, 1993, Sterilization of LVPs and SVPs, in: *Sterilization Technology. A Practical Guide for Manufacturers and Users of Health Care Products* (R. F. Morrissey, and G. B. Phillips eds.), Van Nostrand Reinhold, New York, pp. 254–285.

Parenteral Drug Association, 1978, *Validation of Steam Sterilization Cycles*, PDA Technical Monograph No. 1, Parenteral Drug Association, Philadelphia.

Patashnik, M., 1953, A simplified procedure for thermal process evaluation, *Food Technol.* **7:**16–21.

Pflug, I. J., 1986, Wet-heat sterilization, including both the design of the process and equipment used to sterilize product, in: *Sterilization of Medical Products* (E. R. L. Gaughran, R. F. Morrissey, and Wang You-sen, eds.), Polyscience, Montreal, pp. 40–89.

Pflug, I. J., and Holcomb, R. G., 1991, Principles of thermal distruction of microorganisms, in: *Disinfection, Sterilization, and Preservation* (S. S. Block, ed.), Lea & Febiger, Philadelphia, pp. 751–810.

Russell, A. D., 1989, Fundamental aspects of microbial resistance to chemical and physical agents, in: *Sterilization of Medical Products* (R. F. Morrissey, and Y. I., Prokopenko, eds.), Polyscience, Morin Heights, Canada, pp. 22–42.

Stumbo, C. R., Murphy, J. R., and Cochran, J., 1950, Nature of thermal death time curves for P.A. 3679 and *Clostridium botulinum, Food Technol.* **4:**321–326.

U.S. Food and Drug Administration, 1995, *Code of Federal Regulations*, Title 21, Part 211: Current good manufacturing practice for finished pharmaceuticals, Food and Drug Administration, Department of Health and Human Services, Washington, DC.

U.S. Food and Drug Administration, 1987, *Parametric Release—Terminally Heat Sterilized Drug Products. Compliance Policy Guide.* 7132a.13, U.S. FDA, Office of Enforcement, Division of Compliance Policy, Rockville, MD.

U.S. Food and Drug Administration, 1994, *Guidance for Industry for the Submission Documentation for Sterilization Process Validation in Applications for Human and Veterinary Drug Products*, U.S. FDA, Center for Drug Evaluation and Research (CDER), Center for Veterinary Medicine (CVM), Rockville, MD.

USP, 1995, *United States Pharmacopeia, XXIII Official Monograph, in Biological indicator for steam sterilization, paper strip.* Mack Publishing, Easton, PA, pp. 173–175.

Wood, R. T., 1993, Sterilization with dry heat, in: *Sterilization Technology. A Practical Guide for Manufacturers and Users of Health Care Products* (R. F. Morrissey, G. B. Phillips, eds.), Van Nostrand Reinhold, New York, pp. 81–119.

Young, J. H., 1993, Sterilization with steam under pressure, in: *Sterilization Technology. A Practical Guide for Manufacturers and Users of Health Care Products* (R. F. Morrissey, and G. B. Phillips, eds.), Van Nostrand Reinhold, New York, pp. 120–151.

5

Membrane Filtration

Daniel J. Brose, Michael Dosmar, and Maik W. Jornitz

1. INTRODUCTION

The filtration of protein solutions, especially in the manufacture of biologically derived proteins, is a large and increasingly important step in the production of protein pharmaceuticals. The focus of this chapter is on the use of membrane filters for purification and sterilization of protein pharmaceuticals. However, we also review other important uses of membranes and nonmembrane filters in the production of protein pharmaceuticals. The reader is directed to several useful reference books that describe in more detail both the theory and the practice of filtration in the biopharmaceutical industry (Cheryan, 1986; Ho and Sirkar, 1992; Johnston, 1992a; Meltzer, 1987; Meltzer and Jornitz, 1998). Additional discussion of interactions between filters and protein formulations can be found in Chapters 2 and 3. The appendix to this chapter contains definitions of terms for those readers unfamiliar with terminology of membrane filtration.

Daniel J. Brose • Chemica Technologies, Inc., Bend, Oregon 97701. *Michael Dosmar* • Separations Division, Sartorius Corp., Edgewood, New York 11717. *Maik W. Jornitz* • Product Management, Sartorius AG, D-37075 Goettingen, Germany.

Development and Manufacture of Protein Pharmaceuticals, edited by Nail and Akers. Kluwer Academic/Plenum Publishers, New York, 2002.

1.1. Applications of Membrane Filters in the Production of Protein Pharmaceuticals

Protein biopharmaceuticals are typically produced by fermentation, cell culture, or extraction from tissue. Regardless of the origin of the biopharmaceutical protein, many steps are involved in purification of the protein before it is ready for pharmaceutical sale and use. The physical processes used to purify biopharmaceutical proteins may include filtration, adsorption, centrifugation, and chromatography. These processing methods are typically used to take a protein through the following sequence of steps in the purification scheme:

- Clarification: removal of cell mass from supernatant liquid
- Fractionation: isolation of desired protein from other solutes
- Concentration: concentration of desired protein in solution
- Diafiltration: removal or exhange of small solutes by sequential dilution and concentration steps
- Sterilization: removal of bacterial cells

Microfiltration (MF) and ultrafiltration (UF) membrane processes can be used to perform the above processing steps. Microporous filtration is sometimes used for clarification; UF is used for fractionation, concentration, and diafiltration; and MF is used for sterilization. In the sterilization of protein pharmaceuticals, MF is the method of choice, because proteins are denatured by thermal sterilization processes.

Membrane filtration processes can be divided into two distinct modes of operation. The first is a *single-pass dead-end* or *static mode* and the second is a *multiple-pass crossflow* or *tangential-flow* mode. Crossflow filtration is distinguished from dead-end filtration in that the flow path of the fluid being filtered is parallel to the membrane surface rather than perpendicular to the membrane surface, as shown in Fig. 1. In crossflow filtration, the flow parallel to the membrane surface generates shear, which limits the thickness of the filter cake or gel layer. In dead-end filtration, the filter-cake thickness increases with time, resulting in the eventual cessation of flow.

1.1.1. DEAD-END FILTRATION

Dead-end filtration is used primarily in microfiltration applications where the goal of the filtration is to remove particles greater than 0.1 μm, where the feed solution is relatively “clean,” and where the filtrate (or solute

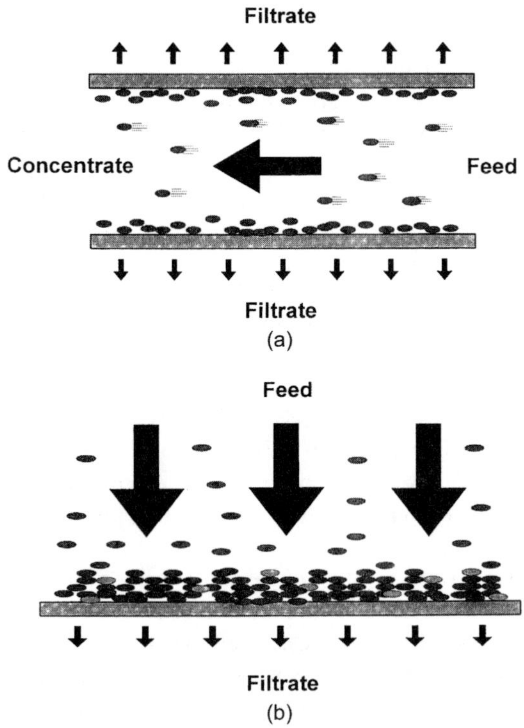

Figure 1. Comparison of (A) crossflow filtration and (B) dead-end filtration.

in the filtrate) is the product of interest. Crossflow microfiltration, on the other hand, is mainly used to separate or fractionate solutes or particles and where the feed solution has relatively high solids content. The product of interest may be the retained fraction or the filtrate. Sterilization by filtration is almost exclusively performed using the dead-end mode of operation.

In dead-end filtration, it is typical to sequentially remove particulates using a series of filters. The first stage of filtration is designed to remove the large particles, and progressively finer particles are removed as the stream progresses through the filter train. It is usually best to start a dead-end filtration operation at a low feed pressure. Although membrane filter cartridges can withstand relatively high pressure differentials, the overall volumetric throughput of a filter is maximized by operating at lower feed pressures. Serum products are usually sterile-filtered starting at inlet pressures of 5 psi or less and gradually allowing the pressure to build. When the inlet pressure reaches 15 psi, the filter is usually exhausted and little more solution will pass through the filter, even at elevated pressures.

1.1.2. CROSSFLOW FILTRATION

The traditional applications for crossflow filtration can be grouped into three basic types of operations: concentration, fractionation, and diafiltration. Concentration is the process of decreasing the solvent content of a solution to increase the concentration of solute. Typical concentration applications include the separation of cells from cell culture broth, the dewatering of proteins and peptides, and the harvesting of viruses.

Fractionation may be described as a separation process where a fraction of the solute in the process fluid is separated from the bulk. Examples of membrane fractionation are depyrogenation (endotoxin removal), lysate clarification, and molecular weight separation of polydispersed polymers such as proteins and carbohydrates. Depyrogenation of low molecular weight solutions such as buffers and pharmaceutical preparations can be accomplished by retaining the contaminating Limulus amebocyte lysate (LAL) reactive material in the retentate and allowing the pyrogen-free product to pass through the membrane (Brown and Fuller, 1993). Depyrogenation is typically accomplished using 20 K molecular weight cutoff (MWCO) or smaller pore size UF membranes.

Clarification of cell lysates is the fractionation or separation of cellular debris from the products of interest. Cell lysates can be very fouling in nature, and the success of the operation is very much affected by the selection of the appropriate membrane, pore size, and operating parameters.

Diafiltration is a process that combines dialysis and filtration. In most diafiltration operations the goal is to exchange one solute for another. In bioprocessing this usually means the exchange of one salt-containing buffer for another, generally prior to a column chromatography step. Diafiltrations may be run as either continuous or discontinuous operations. Continuous diafiltration is a constant-flow wash, where the addition of diafiltrate buffer is matched to the rate of permeate flow. In discontinuous diafiltration the batch is concentrated and then diluted with buffer in a stepwise operation. Aliquots of buffer are added after each successive concentration until the desired dilution is reached.

The selection of the appropriate membrane for a particular application must consider the goals of the process. The calculations of protein yield by a diafiltration process and by a concentration processes are different. Both calculations make use of the rejection coefficient R of the membrane for the protein product, which is defined as the percentage of protein in the feed solution that is retained by the membrane. Typically, membrane manufacturers determine rejection coefficients for a few standard proteins (and other macromolecules) by measuring the amount of protein that permeates the membrane at standardized conditions. The rejection coefficient is not an intrinsic property of a membrane; rather, it depends on the concentration polarization, the chemical nature and the structure of the protein, the solvent used, and the flow and pressure conditions. It is best for a membrane user to independently determine rejection coefficients for the specific protein of interest, with the appropriate solvent, and using actual pressures and flow rates.

The yield for a diafiltration application Y_{dia} is given as

$$Y_{\text{dia}} = \exp\left[\left(\frac{V_d}{V_0}\right)(R-1)\right] \tag{1}$$

The yield for a concentration process Y_{conc} is given as

$$Y_{\text{conc}} = \left(\frac{V_0}{V_f}\right)^{R-1} \tag{2}$$

where R is the rejection coefficient, V_{d} is the volume of the liquid added as diluent, V_0 is the initial feed solution volume, and V_{f} is the final feed solution volume. So long as the retention is less than 100% there will always be some yield loss of the protein product. Yield losses increase as the volume of diafiltration buffer (diluent) increases and as the concentration factor (V_0/V_{f}) increases.

The protein rejection that is measured for membranes from different membrane manufacturers can be quite different. Figure 2 shows dextran rejection curves for 30 K MWCO ultrafilters of different materials from two different suppliers. This figure shows that three different membranes with the same reported MWCO of 30 K actually have fairly different performance when tested for rejection of dextran ranging in size from 2000 to one million Da.

The importance of membrane rejection can be demonstrated through the following example. When using a UF membrane to separate two different-sized proteins from each other, typically one would choose a membrane that has a difference in rejection of at least 20% between the two proteins. For example, if one protein has a rejection coefficient of 0.99 and the second protein has a rejection coefficient of 0.8, one could use diafiltration followed by concentration to effect a 10-fold enrichment of the first protein relative to the second. [These and similar results can be calculated from Eqs. (1) and (2).]

An example of how membrane filtration is used repeatedly in the production of a protein pharmaceutical is shown in Table I, which shows the processing steps involved in the production of a protein pharmaceutical using mammalian cell culture. The intent of this example is not to include all possible process stages, but to illustrate the variety of membrane-based

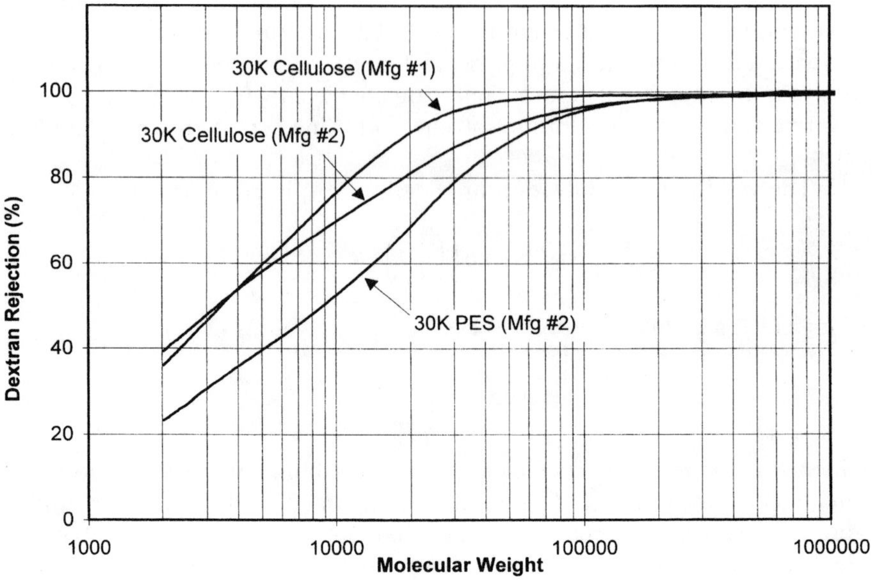

Figure 2. Comparison of 30 K MWCO membranes.

filtrations that are required in the production of a typical biopharmaceutical protein. Each filtration step in Table I takes a toll on the overall yield of protein. Indeed, it is critical that the filters have very low protein binding so that the minimum amount of protein is lost in each step. In addition, care must be taken to ensure that the process parameters are optimal so that the minimum amount of protein is lost in each step. Table II lists some of the process considerations that affect yield of protein in each of the filtration steps.

1.2. Requirements of Membrane Filters for the Production of Protein Pharmaceuticals

Biological protein-containing solutions are very complex and, as a result, the demands on filters used in their production can be stringent. Typical requirements of membrane filters used in the production of protein pharmaceuticals are (1) low adsorption of protein to the filter membrane, (2) high total liquid throughput, (3) minimum fouling of the filter, and (4) fulfillment of all applicable regulatory requirements.

Table I
Filtration Steps Used in Production of a Protein Pharmaceutical Using Mammalian Cell Culture

Process stage	Purpose of the filtration	Type of filter used
Medium preparation	Media and media components are often filtered in order to clarify, remove insoluble components, and sterilize the medium or medium components	Depth prefilters Membrane prefilters Sterilizing membrane filters
Separation of cell mass from the supernatant	If the protein of interest is excreted by the cells, filtration is used to separate cells from the protein containing supernatant	Depth and membrane cartridge filters Crossflow microfilters
Concentration of the supernatant	Dilute protein streams are concentrated prior to loading them onto chromatographic resins for protein purification; this step may also serve to remove unwanted low molecular weight proteins and peptides; serum albumin, which is used to promote cell growth, may also be removed	Crossflow ultrafilters
Column loading	Filtration is used to protect the expensive chromatographic columns from fouling by particulates	Depth prefilter capsules and cartridges
Buffer preparation	Buffers used in chromatographic columns are depyrogenerated by filtration	Crossflow ultrafilters (<20,000 MWCO)
Desalting	Eluted proteins in buffer are desalted to remove buffer salts or to change the buffer strength	Diafiltration with crossflow ultrafilters
Final packaging	A solution of the protein, along with formulation excipients, is sterile-filtered into a previously sterilized holding tank and filled into vials which are either sold as a ready-to-administer sterile solution or as a freeze-dried solid	Sterilizing membrane cartridge filters (dead-end)

Low protein adsorption is one of the most important requirements in the filtration of protein-containing solutions, especially when the protein concentration in solution is low (Datar *et al.*, 1992; Martin and Manteuffel, 1988). Protein adsorption is problematic for two reasons: (1) It adds expense to the process due to reduced yields and (2) it may be an uncontrolled and uncharacterized step in the production of a pharmaceutical product. Therefore, it is desirable to use membrane filters that have low and well-characterized protein adsorption. In some specific biotechnology and blood-processing applications, protein adsorption onto the filtration membrane

Table II
Factors Affecting Yield of a Protein Biopharmaceutical during Filtration

Process type	Filter type	Factors affecting yield
Cell harvesting	Crossflow microfiltration	Membrane polymer Filter preparation Filter surface area pK_a of the protein Temperature Feed pressure Recirculation rate
Clarification and prefiltration	Depth filter	Filter material charge Filter surface area Filter construction
Concentration	Crossflow ultrafiltration	Membrane polymer Filter preparation Filter surface area pK_a of the protein Temperature Feed pressure Recirculation rate
Diafiltration	Crossflow ultrafiltration	Membrane polymer Filter preparation Filter surface area pK_a of the protein Temperature Feed pressure Recirculation rate Stability of protein in the buffer system
Sterilization	Cartridge filter	Membrane polymer Filter preparation Filter construction Filter surface area pK_a of the protein Temperature Feed pressure

can cost thousands of dollars per batch due to the extremely high value of the product. Besides reducing the yield, protein adsorption also promotes fouling and premature blockage of the membrane filter. Blockage or clogging of the membrane is to be avoided, because it results in reduced process efficiency and added cost. Additionally, if the filter must be replaced during the processing of a batch of solution, expensive down-time and product losses may result. Specific flow rates are required to meet the processing schedules, and in some applications this processing time is crucial. In some applications, microorganism growth or enzymatic degradation can occur if the processing takes too long, and this in turn may cause formation of pyrogens or other undesirable byproducts that may effect yield. Cooling of the solution can reduce enzymatic activity and the growth of microorganisms, but cooling results in higher liquid viscosity and therefore reduced flow rates through the membrane filter.

Every step in the processing of protein-containing solutions needs careful evaluation in order to optimize all process parameters as well as to ensure that the process meets good manufacturing practice (GMP) requirements. Critical process parameters must be identified and controlled. By knowing the important parameters, such as protein adsorption and liquid throughput requirements, the prefiltration and the final membrane filtration steps can be designed to work together effectively.

All membrane filters and filter materials that are used in the production of a protein pharmaceutical must meet regulatory requirements according to the current edition of the *United States Pharmacopeia* (USP). This document contains specifications related to common tests, such as extractables testing, pyrogen testing, chemical compatibility, and testing for extraneous particulate matter (Reif, 1998). U.S. Food and Drug Administration (FDA) authorities usually need to see proper documentation of product bacteria-challenge tests to evaluate the retentive ability of the individual sterilizing-grade filter. In product bacteria-challenge testing, the test organism (*Brevundimonas diminuta*, previously *Pseudomonas diminuta*) is inoculated in the product, and the sterilizing-grade filter is challenged with 10^7 organisms per cm^2 of membrane area, according to ASTM F 838-83. Not every individual product needs to undergo such product bacteria-challenge testing; for example, when the product belongs to a similar group of products that differ only in the concentration of a specific ingredient, it is not always required to test each individual product. Such a situation was described by the FDA in the July 1995 *PDA Special Forum* (Bethesda, MD). If the product is supplied in forms with different concentrations of the active ingredient, and if the activity of the product is linearly related to the concentration, then only the lowest and highest concentrated product need to undergo a challenge test. If it happens that the product has any bacteriostatic or bactericidal effects on the organisms, then the challenge fluid parameters or challenge process needs to be modified to eliminate this effect and achieve a proper challenge level. Such test procedures are described in a report of the Parenteral Drug Association (PDA) (1998).

If a changeover is made to a sterilizing-grade filter manufactured by a different filter vendor than the one established in the existing validated process, then the filter user needs to requalify the filter (assuming that the new filter uses the same filter membrane material as the previous filter). If the filter material is changed, then the new filter needs to be revalidated. The requalification step usually includes steam-sterilization resistance tests, adsorption, flow rate, and total-throughput tests. Revalidation would additionally include the product bacteria-challenge testing, evaluation of product integrity test limits, and extractable and compatibility tests. All of these tests can be supported by the individual filter vendor.

2. OVERVIEW OF FILTRATION MEMBRANES

2.1. History of Filter and Membrane Development

Filtration of protein-containing solutions has its origins in the middle of the 19th century with the Pasteur–Chamberland porcelain filter. Using these bacterial-retentive porcelain filters, Charles Chamberland demonstrated the existence of filterable disease-causing agents, that is, viruses. In 1892 Ivanowski, using the filter technique of Chamberland, showed the causative agent of tobacco-mosaic disease to be a virus (TMV), and in 1898 Loeffler and Frosh showed foot-and-mouth disease also to be caused by a filterable virus.

Until the end of the 19th century, sterilization of heat-labile products was accomplished only through the addition of phenol or chloroform to the protein solution. Large-scale production of heat-labile products began in the 19th century with the manufacture of tetanus serum, but it was not until 1923 that attempts were made to sterilize these products by the use of filters. Until their prohibition, asbestos-pad filters were used for large-scale filtrations to sterilize heat-labile natural products.

Membrane microfilters (MF membranes) were first developed over 100 years ago. However, it was not until the work of Zigmondi and the establishment of Goettinger-Membranefilter Gesellschaft in 1927 that work progressed toward the commercialization of modern MF membranes. The first commercially available MF membranes were mixed esters of cellulose, which were produced in 1929 by the Sartorius Werke in Germany. From the first commercialization of membrane filters until the late 1960s, the choice of sterilizing-grade membrane filters was limited to flat-sheet configurations, usually in the form of disks and made from either cellulose nitrate, cellulose acetate, or regenerated cellulose. It was not until the banning of asbestos, coupled with the development of pleatable membrane materials in the late 1960s, that the choices of MF membrane materials and membrane formats multiplied. It should be noted that the acceptance of new membrane materials is fairly limited in the FDA-regulated arena. The most recently accepted new membrane material is polyvinylidene difluoride (PVDF), which was first introduced by Millipore in the 1970s.

Modern UF membranes have their origins in the 1950s, when reverse osmosis (RO) membranes were developed for water desalination (Michaels, 1986). The Loeb and Sourirajan process of membrane formation by phase inversion created the first asymmetric membrane, which is the basis of membranes used today for both RO and UF. The thickness of the rejection layer of the asymmetric membrane is 1000-fold less than that of the symmetric membranes used in MF (Cheryan, 1986). The development of asymmetric membranes by Loeb and Sourirajan made the present use of UF membranes in the biopharmaceutical industry practical (Loeb, 1981).

2.2. Types of Filtration Membranes

Membrane filters are made in a wide variety of pore sizes. Figure 3 shows the effective pore size for membranes used in reverse osmosis, nanofiltration (NF), ultrafiltration, and microfiltration. MF membranes typically differ from UF membranes in the morphology of the membrane cross section. The symmetry of sterilizing-grade microfilters usually ranges from uniform to slightly asymmetric. Ultrafilters, on the other hand, are highly asymmetric, with the rejecting layer consisting of a tight skin (0.5–10 μm thick) supported by a thick, spongy structure of a much larger pore size. Liquid generally flows through the membrane in the direction from the tight skin side toward the

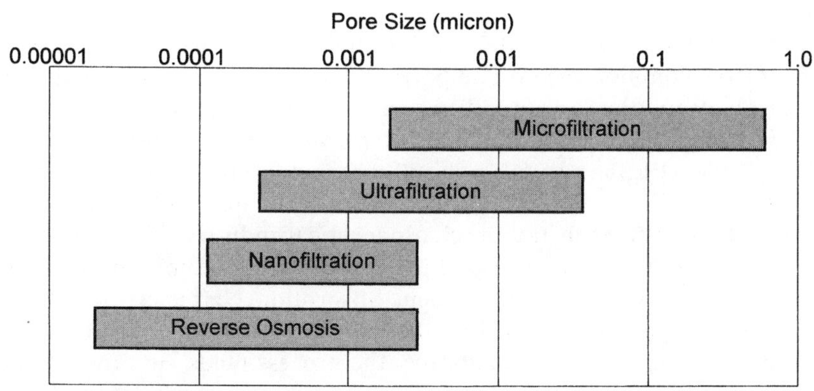

Figure 3. Typical pore sizes for membranes used in reverse osmosis, nanofiltration, ultrafiltration, and microfiltration.

open side of the UF membrane; flow in the reverse direction can result in the delamination or separation of the skin from the rest of the membrane.

Most commercial UF and NF membranes and many MF membranes are made by the phase-inversion process, where a polymer is dissolved in an appropriate solvent along with appropriate pore-forming chemical agents. The polymer solution is cast into a film, either on a backing or free-standing, and then the film is immersed in a nonsolvent solution that causes precipitation of the polymer. Other techniques for membrane formation include stretching the polymeric film while it is still in a flexible state and then annealing the membrane to "lock in" and strengthen the pores in the stretched membrane. Microporous polypropylene and Teflon® (polytetrafluoroethylene) membranes are made by this technique. Finally, track-etched membranes are made from polymers such as polycarbonate and polyester, wherein fission fragments from radioactive decay are bombarded onto the polymeric surface. This bombardment results in "sensitized tracks" where chemical bonds in the polymeric backbone are broken. Subsequently, the irradiated film is placed in an etching bath (such as an alkaline solution) in which the damaged polymer in the tracks is preferentially dissolved from the film, thereby forming cylindrical pores. The residence time in the irradiator determines pore density, and the residence time in the etching bath determines pore size. This etching results in cylindrical pores with very narrow pore-size distribution, albeit with low overall porosity. (Kesting, 1985, pp. 237–310; Strathman, 1986.) Each method of membrane formation has advantages and disadvantages relative to extractables, membrane integrity and reproducibility, and economics of production. In all cases, the membrane manufacturer must demonstrate that the membrane passes the pertinent USP tests before the manufacturer can sell it for use in pharmaceutical processing.

The materials used in MF and UF membranes today vary widely, and some of the common membrane polymers, in order from most hydrophilic to most hydrophobic, are: modified cellulose, regenerated cellulose, modified PVDF, cellulose acetate (CA), modified polysulfone, cellulose nitrate, polyethersulfone (PES), natural polysulfone, nylon, polypropylene, natural PVDF, and polytetrafluoroethylene. Membranes made of these materials are successfully used in the biopharmaceutical industry. These membrane polymers also differ in their useful pH ranges, temperature ranges, pressure ranges, surface charges, and protein adsorption characteristics. No one membrane polymer is suitable for all applications, and care must be taken to consider the nature of the membrane, the process fluids, and the operating conditions when selecting the membrane to use in a particular application. Characteristics of some of the more common membrane materials are shown in Table III.

Table III
Characteristics of Membranes Made from Various Membrane Polymers

Membrane polymer	Typical MWCO/pore size range	Useful pH range
Modified regenerated cellulose	5000 MWCO–0.45 μm	1–14
Polytetrafluoroethylene (PTFE)	0.02–40 μm	1–14
Polyethersulfone (PES)	1000 MWCO–0.65 μm	1–13
Polypropylene (PP)	0.1–8 μm	1–13
Regenerated cellulose (RC)	1000 MWCO–0.45 μm	1–10
Polysulfone (PS)	5000 MWCO–0.45 μm	1–10
Polycarbonate (PC)	0.015–12 μm	2–11
Nylon	0.04–0.65 μm	2–10
Polyvinylidene difluoride (PVDF)	6000 MWCO–0.65 μm	2–10
Cellulose acetate (CA)	1000 MWCO–5 μm	2–8

The effect of the membrane polymer on a protein-containing solution may be profound. Amphiphilic molecules in an aqueous environment tend to sequester their hydrophobic groups, exposing only their hydrophilic moieties. Native proteins, which are composed of both hydrophilic and hydrophobic amino acids, take on tertiary and quaternary structures as part of their functionality. Proteins, therefore, can expose hydrophobic regions, which can interact with hydrophobic membrane matrices. Hydrophobic interactions are favored when the ionic strength is high or when the temperature is elevated, and they are least favored in an ion-free environment.

Truskey *et al.* (1987) measured protein adsorption, circular dichroism, and biological activity of insulin, immunoglobulin G (IgG), and alkaline phosphatase before and after passing the proteins through a variety of MF membranes. (Shifts in circular dichroism and decreases in the activity of the enzymes are the result of conformational changes of the protein structure.) This study showed clearly that membranes with the greatest degree of hydrophobicity had the greatest effect on protein adsorption and protein deformation (Truskey *et al.*, 1987). Once hydrophobic interactions take place there is an increased tendency toward further binding to these exposed hydrophobic surfaces, which serve as the sites for subsequent binding. This leads to an increased rate of membrane clogging and fouling.

The porosity, pore size, and pore-size distribution of a membrane can also influence filtration performance. As the pore size is increased, the resistance to flow is decreased, but the tendency toward fouling is increased. The result of the increased flow through larger pores is an increase in convective flow of solute to the pores, which in turn causes pore plugging. Devereaux and Hoare (1986) showed that the decline in flux through a 0.2-μm membrane was greater than through a 50,000-MWCO membrane when filtering the same feed stream. It has also been shown that although the initial retention values for various proteins on an MF membrane may be low, over time their retention will increase. These consequences may be minimized through careful control of the operating conditions (Wolber *et al.*, 1988).

Numerous studies have analyzed the adsorption of proteins to MF membranes (Sarry and Sucker, 1992, 1993; Hu *et al.*, 1993; Kim *et al.*, 1994; Palecek and Zydney, 1994; Pitt, 1987; Truskey, *et al.*, 1987.) For example, Pitt (1987) measured the binding of bovine serum albumin (BSA) and sheep IgG to several commercial MF membranes. Protein binding in both "static" and "dynamic" experiments was measured, where the amount of bound protein was determined by analysis of membrane washes. Of the membranes tested, nylon had a very high BSA adsorption and PVDF and CA had BSA adsorption less than 5 $\mu g/cm^2$. Similar trends in protein adsorption were observed when the same tests were performed with IgG.

Related studies have examined the adsorption of proteins and preservatives in pleated capsules and cartridges, rather than on small pieces of flat-sheet membrane (Brose and Henricksen, 1994; Brose and Waibel, 1996). Modules containing MF membrane made of PVDF, nylon, PES, and CA were tested in flow-through experiments to quantitatively determine protein and preservative adsorption as a function of solute concentration in the feed solution. In all cases, the solute (protein or preservative) adsorption increased with increased solute concentration. As in the work reported by Pitt, nylon membranes demonstrated the highest adsorption for both BSA and IgG. The PVDF and CA modules demonstrated similarly low protein and preservative adsorption. It was also determined that the polypropylene materials used in construction of pleated-filter modules showed no significant protein or preservative adsorption.

3. MEMBRANE FILTRATION THEORY

3.1. Mechanisms of Membrane Filtration

In liquid filtration using MF membranes, two main retention mechanisms can be operative, either individually or in combination: adsorptive sequestration and sieve retention. Figure 4 shows schematically how these two retention mechanisms operate.

3.1.1. ADSORPTIVE SEQUESTRATION

Adsorptive sequestration relies on adsorption to the membrane-filter surface for retention and capture of macromolecules, colloids, cells, or

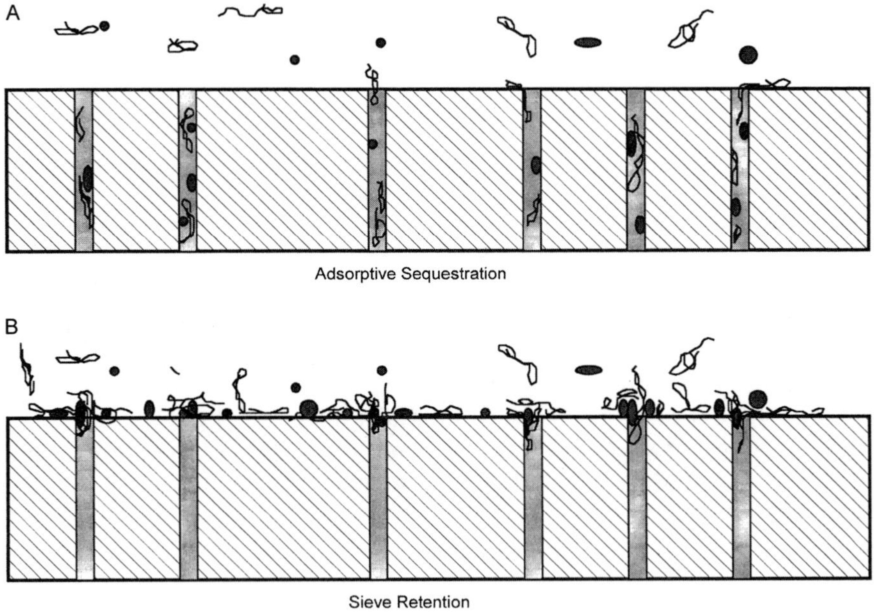

Figure 4. Schematic representation of (A) adsorptive sequestration mechanisms and (B) Sieve retention mechanisms.

particles that enter the pores. As a result, pore volume and pore size decrease in direct proportion to the volume of solution filtered, and eventually pores plug due to accumulation of deposited particles inside the membrane pores (this mechanism is also known as the "standard-blocking model"). The effective use of adsorptive sequestration for filtration purposes depends strongly on the correct choice of membrane material and solution properties. Some sieving membrane materials also have high adsorption potential (e.g., polyamide, polysulfone, and cellulose nitrate), and therefore both sieve and adsorptive retention play a role. Nevertheless, adsorptive retention can be influenced by the properties of the solution being filtered.

The most common mechanism by which adsorptive sequestration takes place is through hydrophobic interaction. This adsorption mechanism is favored when the ionic strength of the medium is high. An additional factor is the solution's surface tension or the presence of surface-tension-reducing agents in the liquid solution. These hydrophobic interactions are dramatically reduced once a surface-tension-reducing agent is introduced to the solution. Finally, sufficient contact time for the filtered liquid within the membrane matrix is of great importance, and therefore high filtrate flow rates represent a worst-case condition in terms of the retention efficiency of the filter.

3.1.2. SIEVE RETENTION

In a sieve-retention MF membrane, it is through physical exclusion that the largest pore in a membrane retains the smallest particle from the liquid. Usually a 100% efficient separation is not possible by this mechanism, and if it were possible, it would only be in the case where the pores in the membrane were so small that solutes would also be retained by the membrane. Because 100% efficiency is not possible, in sterilization applications the membrane is defined to be bacterial-retentive if the membrane has a log-reduction value (LRV) of 7 (LRV is defined as the base-10 logarithm of the ratio of the number of bacteria in the feed to that in the filtrate).

The sieve-retention mechanism is the more reliable of the two retention mechanisms due to the fact that the influence of the solution properties is not as significant as with the adsorptive-sequestration mechanism. In any case, product validation is always necessary in order to develop experimental evidence that the sterilizing-grade filter is functioning effectively under the specific process conditions and with the specific product being filtered. For this reason, bacterial-challenge tests are performed with the actual liquid product to be filtered. If the product is bacteriostatic or bactericidal, which can be determined by a viability test, a placebo solution that does not contain the inhibiting ingredient must be used.

3.1.3. COMPARISON OF RETENTION MECHANISMS

Table IV compares the advantages and disadvantages of the sieve and adsorptive mechanisms for membrane-filter retention. As the table indicates, there are numerous problems that can arise when using membranes that function according to the adsorptive-retention mechanism. Therefore, most users of membranes for critical applications, such as sterilization, do not rely on pure adsorptive capture, but instead validate membranes that function by the sieve-retention mechanism. Nevertheless, membranes are still used that rely on adsorptive capture, such as the zeta-potential filters, which are claimed to be able to remove pyrogens from solution. This is probably possible to a certain degree, but it is a highly controversial subject when one considers the difficulties that can arise upon saturation of the filter's adsorptive capacity. Upon saturation of the filter's capacity, shedding of material from the membrane may occur. Filters relying on adsorptive capture cannot be easily validated, because the filter's capacity

Table IV
Advantages and Disadvantages of Sieve-Retention and Adsorptive-Sequestration Mechanisms in Membrane Filtration

Retention mechanism	Advantages	Disadvantages
Sieve retention	Reliable at worst-case product properties Reliable separation even at high flows and pressure conditions Blockage, i.e., exhaustion, can be determined No unspecific adsorption, minimal loss of desired product, and little adsorptive fouling	Retentive only at the specific pore-size rating
Adsorptive sequestration	It is possible to retain particles smaller than the filter's indicated pore size Separation of colloidal substances is possible In some cases pyrogens can be removed	Highly influenced by product-specific properties Separated particles can be shed by varying process conditions Saturation of the active sites cannot be determined, no warning Unspecific adsorption will result in product losses and fouling Lower reliability in terms of absolute separation

is highly dependent on a number of process variables and environmental conditions.

When adsorptive filters are used in concert with a final-sieving filter, the validation burden is eased, because the adsorptive filter is not used as the final filter device. The function of such adsorptive prefilters is to remove colloids, lipids, and other contaminants that will promote the fouling of the final filter. Glass-fiber or cellulosic mixed-ester depth filters are very effective for protecting final-membrane filters. On the other hand, such filters can also adsorb valuable components from the product, such as proteins. In this instance, low-adsorptive prefilters need to be used, such as polypropylene, which functions according to the sieve-retention mechanism. The retention efficiency of such prefilters is critically dependent on the diameter of the polypropylene fiber used in the fleece and on the compression of the fibers in the fleece. The finer the polypropylene fiber and the higher the fleece compression, the greater is the sieve-retention capability.

3.1.4. MEMBRANE FILTER RATINGS

Microfiltration membrane filters are commonly labeled with a "nominal" pore size or retention rating. These labeling methods are often primarily for administrative reasons, that is, to aid in defining the capabilities of a specific filter. These labels should not generally be interpreted as absolute in their definition of pore size or retention capabilities. For example, the complete pore-size distribution of a "0.2-μm-rated" membrane filter will often include pores that range in size up to 0.45 μm. Nevertheless, the filter is still labeled 0.2 μm, and the reason for this classification is that the filter has been qualified via the bacteria challenge test (ASTM F 838-83) with *Brevundimonas diminuta*, and by this test the filter has been shown to produce a sterile filtrate when challenged with 10^7 organisms per cm^2 of membrane area. Membrane filters that pass this bacteria challenge test are usually labeled 0.2-μm sterilizing-grade filters.

Filters rated as 0.45 μm are commonly classified via a challenge test with *Serratia marscecens*, and 0.1-μm filters are classified via a challenge test with *Acheloplasma laidlawii*. Prefilters that use a depth-filter-like fleece are classified based on tests with challenge particles. Usually, a challenge test with latex beads of various sizes is used and the retention capabilities of the filter are determined by using a particle counter. One needs to be aware that there is no common standard in terms of labeling prefilter types. This causes confusion within the industry, and therefore many filter vendors have begun to switch to a more defined retention classification, which, for example, uses the retention of 99% of a specific latex bead size as the measure of filter retention characteristics.

3.2. Theoretical Description of Membrane Filtration Operations

Membrane filtration operations that function according to either the adsorptive-sequestration or the sieve-retention mechanism have been theoretically described. Liquid flow through porous media and filtration membranes is typically modeled using Darcy's law, which states that flow through a porous medium is directly proportional to the pressure driving force and cross-sectional area and is inversely proportional to liquid viscosity and the resistance to flow of the porous medium (Belter *et al.*, 1988; pp. 22–30; Johnston, 1992b; Green and Duwez, 1951; Vertenstein and Ronis, 1984). This relationship in mathematical form is

$$Q = \frac{\Delta P \cdot A}{\mu R} \tag{3}$$

where Q is the liquid flow rate, ΔP is the transmembrane pressure, A is the membrane-filter surface area, μ is the viscosity, and R is the resistance to fluid flow. Assuming that the "effective" membrane surface area and the resistance of the membrane to fluid flow change as the membrane pores plug, then initially $Q_0 = \Delta P \cdot A_0/\mu R$ and at any later time $Q_r = \Delta P \cdot A_t/\mu R_r$ (Hermia, 1982; Visvanathan and Ben Aim, 1989). The relationship between A_t, R_t, and the volume of solution filtered can be established if an assumption is made regarding the mechanism of pore blockage and membrane fouling, that is, sieve filtration or adsorptive sequestration.

Several assumptions are needed in developing the mathematical model for either of these two filtration mechanisms. For the adsorptive-sequestration mechanism, it is assumed that macromolecules, colloids, cells, or other particles enter the membrane pores and can adhere to the internal pore surface. As a result, pore volume and pore size decrease in direct proportion to the volume of solution filtered, and eventually pores plug due to accumulation of deposited particles. For the sieve-filtration mechanism, it is assumed that pores are not blocked by adsorption of particles on or in the membrane, but instead that a cake forms above the surface of the membrane and the thickness of this cake increases proportionally to the volume of solution filtered. The resistance of this cake to fluid flow is proportional to its thickness, and eventually the cake becomes thick enough that flow ceases on a practical time scale.

Use of these assumptions in combination with Darcy's law results in a unique equation that relates the cumulative volume of filtered solution V to filtration time t, initial pressure-normalized liquid flux before any filtration has occurred J_0, membrane area A_0, and operating pressure P. The derivation of performance equations corresponding to these assumptions is described in the literature (Hermia, 1982; Visvanathan and Ben Aim, 1989).

For filtration based on the adsorptive-sequestration model, the mathematical relationship that predicts filtrate volume as a function of time is (Brose *et al.*, 1994)

$$V(t) = \frac{J_0 A_0 P t}{1 + J_0 K_{\mathrm{Ads}} P t} \tag{4}$$

where K_{Ads} is an *adsorptive* filter-plugging constant, which is dependent on the characteristics of the solution being filtered, with units of $\mathrm{cm}^2/\mathrm{ml}$; V has units of ml; J_0 has units of $\mathrm{ml/min\text{-}cm^2\text{-}psig}$; A_0 has units of cm^2; P has units of psig; and t has units of minutes. A similar equation can be derived for filtration based on the sieve-retention model:

$$V(t) = \frac{A_0}{K_{\mathrm{Sieve}}} \sqrt{1 + 2 J_0 K_{\mathrm{Sieve}} P t} - 1 \tag{5}$$

where K_{Sieve} is the *sieve* filter-plugging constant, which is dependent on the characteristics of the solution being filtered, with units of cm^2/ml; and all other variables are as described above.

Equations (4) and (5) both can be rearranged with the value t/V as the independent variable, resulting in forms of the equations that allow easy plotting of data and interpretation of the appropriate theoretical model. Rearrangement of Eq. (4) for the adsorptive-sequestration model results in

$$\frac{t}{V(t)} = \frac{1}{J_0 A_0 P} + \frac{K_{\text{Ads}}}{A_0} t \tag{6}$$

Rearrangement of Eq. (5) for the sieve-retention model results in

$$\frac{t}{V(t)} = \frac{1}{J_0 A_0 P} + \frac{K_{\text{Sieve}}}{2 J_0 A_0^2 P} V \tag{7}$$

Experimental data should be plotted with t/V versus t and with t/V versus V. If the plot of t/V versus t results in a straight line, then the adsorptive-sequestration model is operative and the slope is K_{Ads}/A_0 and the y-intercept is $1/J_0\ A_0P$. If the plot of t/V versus V results in a straight line, then the sieve-retention model is operative and the slope is $K_{\text{Sieve}}/2J_0\ A_0{}^2P$ and the y-intercept is $1/J_0\ A_0P$. In some cases, neither correlation will give a straight line, and it can be assumed that both mechanisms are operative to some extent. A conservative approach to scale-up would be to use whichever model underpredicts flux, resulting in an overdesign of the system.

Once a decision is made as to which mechanism to use for scale-up, values of J_0 (an indication of the flux before any filtration has occurred) and K_{Ads} or K_{Sieve} (indications of the "filter-plugging" rate of the system being filtered) can be determined. Knowing these membrane-specific and solution-specific parameters allows prediction of scale-up performance. In Section 4.6, experimental data are correlated using these models, and the resulting information is used to predict scale-up performance of a sterilizing-grade filter.

3.3. Effect of Operating Conditions

Operating conditions can have a profound effect on the performance of a membrane-filtration process. In this section, we consider the effect of operating conditions, such as feed-stream composition, temperature, viscosity, pH/ionic conditions, pressure, and flow rate, on filter performance.

3.3.1. FEED-STREAM COMPOSITION

As discussed previously, the nature of the feed stream can have dramatic effects on the retentive nature of a filtration membrane, especially when the filter is functioning according to the adsorptive-sequestration mechanism. In either mode of filtration, the retention of large molecular weight components can increase the retention of smaller molecular weight components. Blatt *et al.* (1971) demonstrated that human serum albumin (67,000 Da) retention on a 100,000-MWCO membrane was nearly zero. However when γ-globulin (160,000 Da) was added to the feed stream, the albumin retention rose. Albumin retention showed a linearly increasing correlation to increasing concentrations of the γ-globulin (Blatt *et al.*, 1971). Porter (1981) similarly showed that the retention of oval bumin, chymotrypsin, and cytochrome c is increased when a 1% solution of albumin is added to the feed mixture. Similar observations have been made regarding the microfiltration of complex solutions. Most importantly, the actual retention capability of a specific membrane with a specific feed mixture should always be measured experimentally.

3.3.2. TEMPERATURE AND VISCOSITY

The effect of temperature on filtrate flux can usually be explained by the changes in solution viscosity. Temperature has an inverse effect on viscosity; that is, as temperature is increased, viscosity decreases. Flux is generally inversely proportional to the viscosity of the feed solution, and as a result flux is directly proportional to temperature.

Biopolymer films that form during filtration of proteinaceous solutions are gelatinous in nature. The temperature of the solution affects the plasticity of these gels. When the solution is kept cold and viscosity is high, they behave like rigid particles; when they are warmed, they are free to deform from the stresses of the filtration process. Polarizing gel layers are compressible, and the pressure of filtration can cause the gel layer to collapse and completely clog the membrane. The temperature of the solution influences the compressibility of the gel layer. When the solution is kept cold (e.g., 4°C) the gel layer is less compressible than when it is warm. As a result, filtration of biopolymer-containing solutions proceeds far better when the solution is kept cold. The increased throughputs when running cold far outweigh the reduction in flux as a result of the increased viscosity of the solution.

Temperature also influences the solubility of proteins. Cryoglobulins are a class of serum proteins that precipitate when the temperature of the solution is lowered to a critical limit. Freezing and thawing of protein solutions may also result in the generation of protein precipitates. Adding heat to the solution can cause the disruption of native-structure hydrogen bonds, causing unfolding of some portions of the protein, which in turn results in the formation of aggregates that may precipitate.

If the temperature is raised, protease activity may be activated or accelerated, which will result in protein degradation and precipitation of the proteolytic fragments. Finally, at higher temperatures the adsorption of protein to the membrane is enhanced, resulting in an increase in membrane fouling.

3.3.3. pH AND IONIC CONDITIONS

Protein solubility generally increases as the solution pH moves away from the isoelectric point and at extremes in solution ionic strength. As the isoelectric point is approached or as the ionic strength changes dramatically, the protein may precipitate from solution and foul the membrane. pH values far from the isoelectric point result in a charged protein, which can cause increased protein–membrane or protein–protein interaction. Under these conditions, proteins may start to agglomerate or bind to the membrane, which will adversely affect both the product and the process. Changes in protein conformation will, in general, affect adsorption to membrane surfaces. McDonogh (1990) showed that membrane fouling is maximized at the isoelectric point and that this phenomenon is concentration-dependent.

Buffer exchanges and diafiltrations are common in the processing of protein solutions. When performing these operations, it is important to maintain a constant ionic composition of the solution in order to avoid precipitation of the protein. This is especially important when proteins are being concentrated by UF, because if the salts concentrate along with the proteins, the proteins may begin to salt out. Similarly, if the salt level falls too far, precipitation may also take place. Calcium and phosphate are directly implicated in the formation of insoluble calcium salts that serve as bridging agents for protein deposits (Marshall *et al.*, 1993).

Removal of certain ionic species (e.g., by diafiltration) may also result in the formation of precipitates. This can happen for multimeric proteins when the quaternary protein structure is lost as the result of the removal of cations such as calcium or magnesium, which can serve to bridge the protein subunits. When serum proteins are dialyzed against distilled water, flocculation occurs due to precipitation of a class of proteins called euglobulins, consisting primarily of IgM.

3.3.4. PRESSURE AND FLOW RATE

The driving force for liquid flow through a membrane filter is the transmembrane pressure, which is the average pressure difference between the feed and filtrate streams. Usually, the higher the differential pressure, the higher is the flow through the filter system (depending on the fluid properties). Commonly the transmembrane pressure increases over the course of a constant-flow filtration due to blockage of the filter membrane. The choice of the applied transmembrane pressure at the start of filtration is crucial in protein solution filtration, and therefore filterability tests should be performed to find the most efficient filter setup and pressure conditions. Generally it is not practical to experimentally develop complete flow versus pressure profiles, but nonetheless it is necessary to experimentally determine some operating conditions that are reproducible.

Typically, one of three operating procedures is used in operation of membrane filters: (1) constant-flow mode, (2) constant-pressure mode, and (3) intermittent-flow mode. All modes have their advantages and disadvantages. These modes of operation are described more fully in Section 4.

3.3.5. FILTRATION TIME

All of the above-described parameters combine to form an important mix of process conditions, which needs to be optimal to achieve the highest yield and consistency within the process. Considering that the duration of a sterile-filtration operation should not exceed 8 h (U.S. FDA, 1991) and in some cases should be shorter if the bioburden of the raw material is high, one needs to choose an appropriate filter size and system to efficiently and rapidly complete the sterile filtration. It is good practice to design the filter system such that it is used at about 80% capacity when the batch is completed within the specified filtration time.

In some cases, the duration of the filtration can be much greater than 8 h. For example, some "blow–fill–seal" filling processes can run for up to 5 days. When filtering over these long time periods, the starting bioburden needs to be low. As a result, the filtration system usually contains two sterilizing filters in series to prevent any possible bacterial "growthrough" in the sterilizing filter. Bacterial growthrough is often discussed and debated, but has never been scientifically established. Nonetheless, growthrough is a common worry of filter users and regulatory agencies. For this reason, the validation process needs to be comprehensive, especially in terms of the product bacteria-challenge test conditions used during the lab-scale evaluations.

3.4. Filter Fouling

Fouling of a membrane filter is defined as the irreversible degradation of the filtrate flow, resulting in a significant reduction of flux even after rinsing of the filter. The fouling process can be attributed to membrane compaction and to the deposition of macromolecules onto or in the membrane pore structure. Membrane compaction is the result of high transmembrane pressure, resulting in a decrease in membrane permeability. The deposition of feed-stream components is the result of the adsorption of solutes such as proteins to the membrane, causing the formation of a monolayer either on the membrane surface or in the pores. These initial deposits are important and will determine the subsequent buildup of fouling layers or cakes (McGregor, 1990; Romero and Davis, 1988). The deposition of material onto the membrane leads to changes in the behavior or retention characteristics of the membrane. Over time there may be a shift to a lower apparent pore size, and the membrane selectivity will therefore decrease as the retention increases.

Fouling of filters is a serious concern when the membrane is to be used repeatedly or for a long period of time. Typically, this is not the case for sterile filtration, where cartridges and capsules are used only once before being discarded. In crossflow filtration, on the other hand, the filters and membranes are used repeatedly over many days. In both crossflow and dead-end applications, protein adsorption is a common cause of fouling (Levy and Shehan, 1991; Marshall *et al.*, 1993). The nature of protein in the feed stream affects the membrane permeability. Spherical depositions, for example, those laid down by γ-globulin, hinder membrane permeability less than the deposition of protein sheets, for example, β-keratin deposits. Because feed streams are usually complex mixtures of many proteins, the result is the formation of protein cakes made up of lattice-like structures in which larger components form an infrastructure and the smaller components fill the interstitial spaces (Marshall *et al.*, 1993). The nature of this cake is influenced by hydrodynamic conditions, and the cake compacts over time. Because of the consequences of fouling, it is useful to minimize the effect of fouling though the careful selection of membrane polymers and operating conditions.

4. THE PRACTICE OF STERILE FILTRATION

With the historical and theoretical framework established thus far, it is now possible to discuss the practical aspects of implementing a sterile microfiltration process for production of protein pharmaceuticals.

4.1. Commercially Available Filtration Membranes

As described in Section 2, a variety of polymeric membrane materials are in common use today for both UF and MF. Each membrane-filter product is unique in terms of the polymer used and casting process used in its manufacture. For sterilizing-grade final membrane filters (that is, 0.2-μm MF membranes), the most important production-related quality issue is the possible presence of extractables or endotoxins introduced during the manufacturing process. Suppliers of membranes to the biopharmaceutical market generally employ standards and controls that avoid this type of contamination problem.

Equally important to the choice of a membrane is the potential interaction of the membrane material with the product(s) being filtered. Membrane-filter materials have specific characteristics that may lead to physiochemical interactions with the solution or solutes. Table V lists the different commercially available membrane-filter materials currently accepted in the biopharmaceutical industry, along with advantages and disadvantages associated with each. All filter materials need to withstand a minimum temperature of 121°C due to steam-sterilization requirements. Most of the described membrane materials withstand steam temperatures above 134°C, which is also a common sterilization temperature.

4.2. Filter-Module Types

Filtration membranes used for sterile filtration are almost always dead-end processes that employ flat-sheet membranes, either in "disk" format or "pleated" format. Crossflow membranes, on the other hand, are commonly available in flat-sheet formats (cassettes, plate-and-frame, and spiral modules), hollow-fiber, and tubular formats. Because of the importance of crossflow filtration in biopharmaceutical processing, we present a short description of these module types, followed by a description of sterile-filtration module types.

4.2.1. CROSSFLOW-FILTRATION MODULES

4.2.1.1. Flat-Sheet Membrane Modules

Flat-sheet membranes used for crossflow filtration can be configured in cassette and plate-and-frame modules, spiral-wound modules, and rotating-surface modules. Cassette modules are the most common type of membrane device used for crossflow filtration in the pharmaceutical industry, and these modules consist of flat-sheet membranes mounted into a frame work as shown in Fig. 5. In this assembly, each flow path is made up of two membranes that

Table V
Commercially Available Membrane Materials Used in the Biopharmaceutical Industry

Membrane material	Advantage	Disadvantage
Cellulose acetate	Very low nonspecific adsorption (low fouling) High flow rates and total throughputs	Limited pH compatibility Not dry-autoclavable
Cellulose nitrate (nitrocellulose)	Good flow rate and total throughputs	High nonspecific adsorption Limited pH compatibility Not dry-autoclavable
Modified regenerated cellulose (Hydrosart)	Very low nonspecific adsorption (nonfouling) Moderate flow rates and total throughputs especially with difficult-to-filter solutions Broad pH compatibility	Ultrafilters not dry-autoclavable
Nylon 66	Good solvent compatibility Good mechanical strength Broad pH compatibility Dry-autoclavable	High nonspecific protein adsorption Low hot-water resistance Moderate flow rate and total throughput
Polycarbonate	Good chemical compatibility	Moderate flow rates Low total throughputs Difficult to produce
Polyethersulfone	High flow rates and total throughputs Broad pH compatibility	Moderate to low nonspecific adsorption, depending on surface modifications Limited solvent compatibility
Polypropylene	Excellent chemical resistance High mechanical resistance	Hydrophobic material High nonspecific adsorption due to hydrophobic interactions
Polysulfone	High flow rates and total throughputs Broad pH compatibility	Moderate to high nonspecific adsorption Limited solvent compatibility
Polytetrafluoro ethylene	Excellent chemical resistance High mechanical resistance	Hydrophobic material High nonspecific adsorption due to hydrophobic interactions High-cost filter material

Table V (*continued*)

Membrane material	Advantage	Disadvantage
Polyvinylidene-difluoride	Low nonspecific adsorption Dry autoclavable Good solvent compatibility	Moderate flow rate and total throughput Hydrophobic base, made hydrophilic by chemical surface treatment; may lose hydrophilic modification due to chemical attack High-cost filter material
Regenerated cellulose	Very low nonspecific adsorption (nonfouling) Very high flow rates and total throughputs	Limited pH compatibility Not dry-autoclavable

are facing each other. The upstream flow path must be sealed from the downstream permeate side of the membrane. Stacks of pairs of membranes are layered one on top of the other, and the permeate side of each membrane is supported by a porous spacer. Flow paths are usually open and may be in parallel, in series, or a combination of both. Preassembled cassettes come in a variety of sizes ranging from 0.01 to 50 ft^2. Systems utilizing these designs may have surface areas in excess of 1000 ft^2.

Cassette modules are generally easy to clean, have relatively high packing densities, support a wide variety of membrane materials, are amenable to incorporating turbulence-promoting spacers, and provide good crossflow performance. Drawbacks to this design are that fabrication of modules can be labor-intensive, and thus expensive, and packing densities are not as great as hollow-fiber modules. However, the advantages to this design often outweigh the disadvantages, and this design is widely used in the biopharmaceutical industry.

Spiral-wound modules utilize pairs of flat-sheet membranes bounded on the up- and downstream sides by screens similar to those in cassette systems. The membrane sandwich is sealed at three edges so that the feed is isolated from the permeate. The fourth side of the membrane sandwich is attached to a perforated permeate-collection tube. The membrane pairs are then rolled around the perforated collection tube, thereby creating the spiral. Feed enters at one end of the spiral, flows tangentially along the axis of the cartridge, and discharges at the other end. Permeate flows at a right angle to the feed flow toward the center of the spiral and is collected in the core of the spiral. Spiral-wound modules are available in a variety of surface areas, starting at 1 ft^2 and going up to 50 ft^2. Surface area can be increased

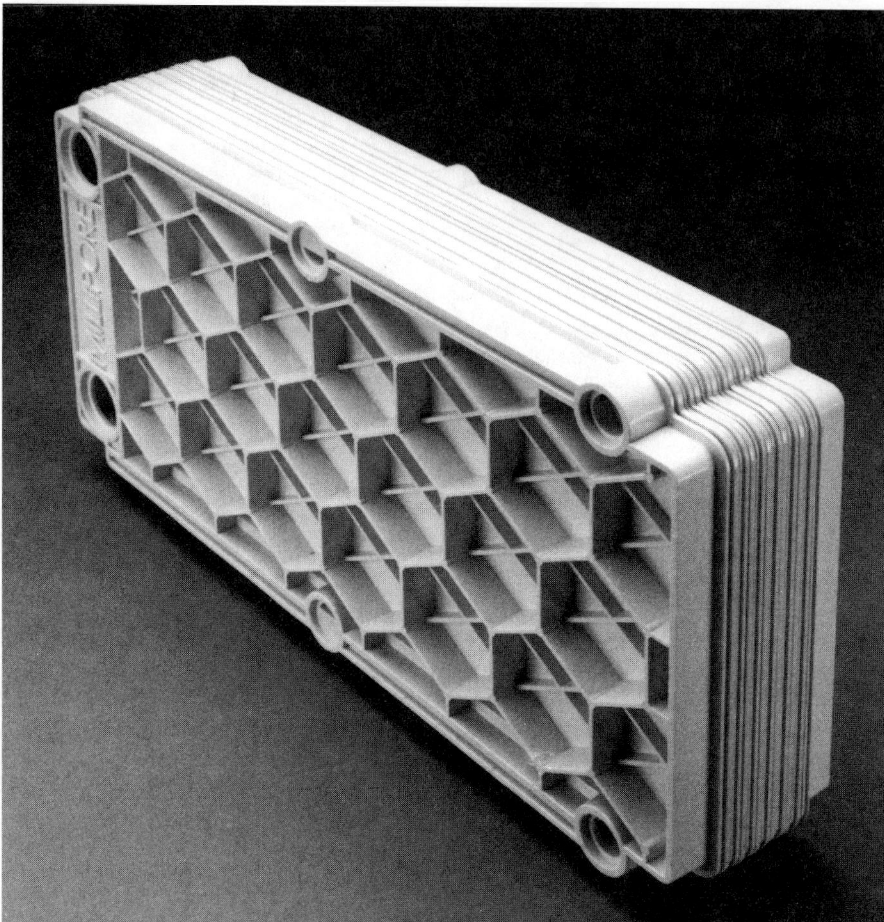

Figure 5. Cassette module for crossflow filtration. Courtesy of Millipore Corporation.

either by increasing the number of membrane pairs, increasing the diameter of the spiral wrap, or increasing the length of the flow path. Spiral cartridges are placed in snug fitting housings. The cartridge is sealed to the housing wall via an O-ring or brine seal. Feed flow on the outside of the cartridge is either blocked by an O-ring or restricted by a screen so as to provide similar resistance to flow as is seen in the feed flow channels between the membranes.

Spiral-wound systems feature relatively high membrane-packing densities and low cost. This design is extremely successful for particulate-free process streams, but is difficult to clean when particulates clog the flow paths or collect at the O-ring seals. Other drawbacks to this design include difficulty to sterilize and long flow paths, making high flow with low pressure drop impossible.

4.2.1.2. Hollow-Fiber and Tubular Membrane Modules

The fundamentals of hollow-fiber and tubular crossflow systems are very similar, with the essential difference being that tubes are considerably larger in diameter than are hollow fibers. Hollow fibers can be manufactured in various sizes, ranging in internal diameter from 0.5 to 3 mm; tubes have internal diameters of 3 mm up to several centimeters. Hollow fibers and tubes usually have an asymmetric pore structure, as observed when looking at a cross section of the fiber or tube. The smaller pores can be on either the inside or the outside of the fiber, with the pores increasing in size throughout the wall structure. The porous wall functions to support and strengthen the small-pore selective layer.

Hollow fibers and tubes are grouped and sealed into tubular shells to form modules as shown in Fig. 6. In the case where the selective layer is on the inside (lumen) of the fiber, the feed solution enters the lumen of the fiber at one end and flows down the length of the fiber, and retentate exits at the other end. Permeate is collected on the outside (shell side) of the fiber.

A solution flowing through a hollow fiber will experience a greater pressure drop than solution flowing through a larger-diameter tubular membrane. However, hollow-fiber modules also have much greater membrane surface area per volume than do tubular modules. Hollow-fiber modules are also cheaper than tubular modules. Tubular modules have low packing densities, but, because of their large lumen, are able to handle very high particle loads.

4.2.2. DEAD-END FILTRATION MODULES

Membrane filters used in sterile-filtration applications, whether prefilters or final filters, are generally operated in the dead-end mode.

4.2.2.1. Pad-Filter Cartridges

This design incorporates flat-sheet "pad" filters into a module as shown in Fig. 7, which shows a construct of multiple pairs of filter disks, where each filter-disk pair is fitted over a drainage/support layer. The filter pairs are sealed at the outer edge, and the downstream side of the filter is connected to

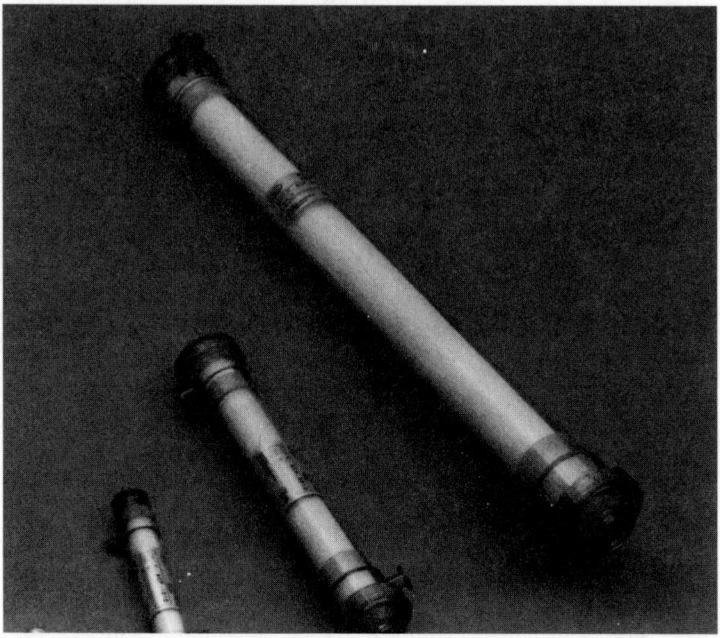

Figure 6. Hollow-fiber module for crossflow filtration. Courtesy of Pall Corporation.

common core. Pad-filter cartridges typically are used with filters having nominal retention ratings from 0.2 to 50 μm. They are distinguished by having relatively high flow rates and high dirt holding capacities. These filters are typically used for removal of precipitates and cells from solution. Manufacturers of these types of cartridges are Seitz, Cuno, Millipore, Cellulo, and Sartorius.

4.2.2.2. Pleated Flat-Sheet Cartridges

In production-scale filtration, the pleated geometry of flat-sheet membranes (Fig. 8) is preferred. Pleated cartridges consist of filter layers that are pleated to form the "accordion-like" structure shown in Fig. 8 and placed in a cylindrical polypropylene "cage," which is made up of a core, an outer support, and the two end caps. The filter layers typically consist of a prefilter, one or more membrane filters, and a supporting/draining mesh layer, each measuring about 2.5 m long by 25 cm wide. After the filter layer is pleated, the two ends are sealed together to form a cylinder that fits between the core and the outer support of the cage. The top and bottom are sealed to the end caps, thereby forming a complete pleated cartridge.

Figure 7. Pad filters and module housing. Courtesy of Cellulo Co.

In operation, the direction of feed flow in a pleated cartridge is radially from the outside inward toward the center of the core (see Fig. 9). Filtrate then flows out of the cartridge and out the outlet port on the cartridge housing. Also shown in Fig. 9 is a cartridge design that incorporates a heterogeneous double membrane, which is sandwiched between the support filter and the prefilter. The heterogeneous double membrane consists of two membrane filters of different pore sizes that are placed one directly on top of the other. This results in a gradual prefiltration as solution flows first through a coarse prefilter, then through a membrane prefilter with pore size on the order of 0.45 μm, and finally through the 0.2-μm final filter. The particulate burden that would normally be placed on the final filter is correspondingly reduced, giving better dependability and substantially prolonged service life.

Pleating of flat-sheet membranes results in a very large filtration area, with from 0.6 to 0.9 m^2 of membrane area in a cartridge measuring 25 cm high by 7 cm in diameter. Larger areas are obtained by doubling or tripling the height of such a so-called 1-high (10 in) cartridge, to give 2-high (20 in),

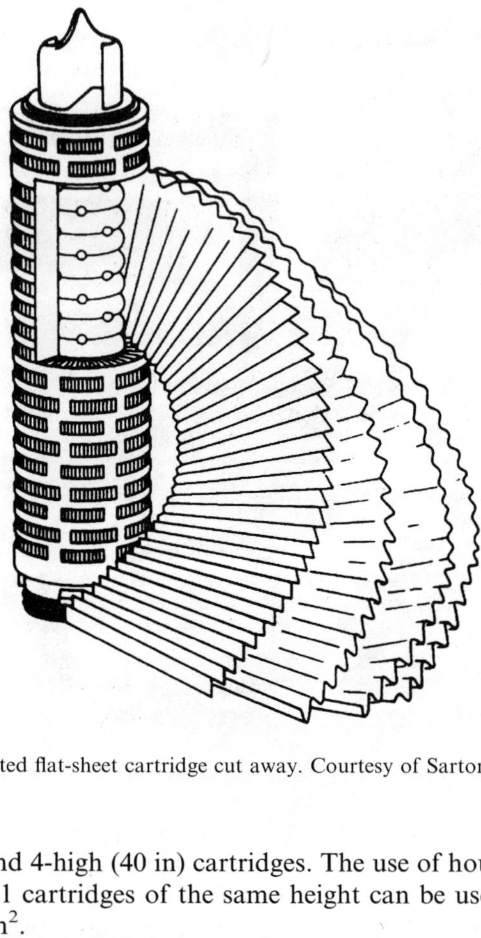

Figure 8. Pleated flat-sheet cartridge cut away. Courtesy of Sartorius Corporation.

3-high (30 in), and 4-high (40 in) cartridges. The use of housings in which 3, 4, 5, 7 , 12, or 21 cartridges of the same height can be used gives filtration areas up to 75 m^2.

Pleated modules for biopharmaceutical applications are available from Millipore, Pall, and Sartorius. Cartridges are available with a wide variety of membrane-filter types and combinations in the pleated element. For example, prefilters are available that function on the principle of adsorptive sequestration. These cartridges, which are often used as protection for sterilizing-grade membrane filters and chromatography columns, can be obtained using filters made from polypropylene, glass fiber, cellulosic fibers, and many other materials. These cartridges are designed to provide graded filtration so as to maximize throughputs and dirt holding capacity.

Also commonly available are "membrane prefilters" and "final filters." Membrane prefilters are designed to protect membrane final filters. They are constructed with combinations of filters that employ both adsorptive

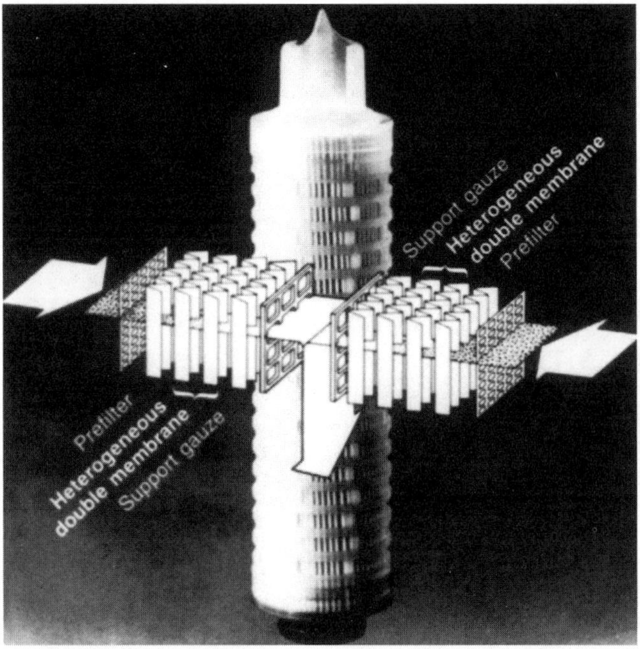

Figure 9. Direction of flow in a pleated cartridge. Courtesy of Sartorius Corporation.

sequestration and sieve retention. Sterilizing-grade final filters are similar to membrane prefilters, with the exception of the designation of the final membrane and the quality control requirements. The membranes used for sterilizing applications are rated as 0.2 μm absolute or tighter; for example, 0.1, 0.07, or 0.04 μm.

4.2.2.3. Cartridge Housings

Pleated membrane elements can be obtained in two common formats: disposable capsule formats, and elements for insertion into stainless steel housings. Capsule-type modules (Fig. 10) are disposable devices that consist of a pleated element encapsulated in a disposable housing made of a thermoplastic polymeric material. Because the capsules are disposable, they are often supplied sterile and are "ready to use" straight from the manufacturer. Capsules are typically available in sizes ranging from 0.015 to 2.1 m^2. Capsules have historically been used mainly in laboratory applications, however, they are quickly becoming more accepted for use in process applications (Soelkner and Rupp, 1998).

Figure 10. Capsule-style modules for sterile filtration. Courtesy of Sartorius Corporation.

Stainless steel housings for cartridge elements are generally used when the cost associated with cleaning validation is lower than the cost of using a disposable capsule filter. Stainless steel housings are designed for single cartridges or for multiple cartridge, and housings can be purchased for the standard cartridge lengths of 5, 10, 20, 30, and 40 in. Figure 11 shows some of the more common designs of stainless steel cartridge housings. Universal designs of filter housings are available and used by the major manufacturers, ensuring that any vendor's cartridge can be used in the housing. Housing manufacturers are responsible for ensuring that the designs are in compliance with FDA, USP, and ASME standards (Tarry, 1998).

4.3. Flow and Pressure Operating Modes

As described previously in this chapter, one of the following operating procedures is generally used in the operation of dead-end membrane filters:

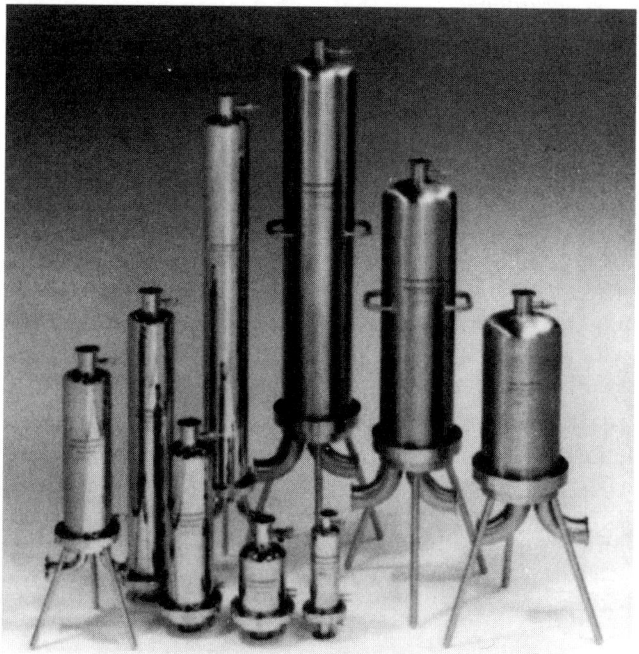

Figure 11. Stainless steel housings for use with pleated membrane elements. Courtesy of Sartorius Corporation.

(1) constant-flow mode, (2) constant-pressure mode, and (3) intermittent-flow mode. A description of each of these modes follows.

4.3.1. CONSTANT-FLOW MODE

The constant-flow mode is used when the filtrate is supplied to a process (e.g., a filling line) that requires a constant volumetric flow rate of feed liquid, and where there is no sterile holding tank between the filter and the filling line. This fixed volumetric flow needs to be constantly supplied to the filling line, and a change in flow would be detrimental. To achieve constant flow, there can be no fouling of the filter, or if fouling does occur, the differential pressure needs to be increased as the flow decreases. Differential pressure can be increased by increasing pump speed or, if the feed is supplied from a pressurized tank, by increasing the head pressure in the tank. It is generally necessary that the filtration system be sized so that a complete production batch can be filtered without shutting down the line and replacing the filters.

4.3.2. CONSTANT-PRESSURE MODE

This mode of filter operation is often used when fixed-speed pumps (centrifugal) are used to supply the medium to the filter. The constant-pressure mode is appropriate to use if the filtrate is fed to a sterile holding tank, from which liquid is drawn for the filling operation. Alternatively, if no holding tank is used, the constant-pressure mode of operation is appropriate only if the solution to be filtered is relatively clean or if the prefiltration prior to the final filter is highly effective at removing contaminants. In the constant-pressure mode, the flow rate will decrease over time due to gradual blockage of the filter. This decreased flow can still be sufficient for supplying a vial-filling operation; however, this mode of filtration is not often seen in applications of protein-solution filtration when no holding tank is used.

4.3.3. INTERMITTENT-FLOW MODE

Intermittent-flow operation is usually seen only in combination with the use of a holding tank directly in front of a filling line. The fluid is supplied to the filter using either a pump, which intermittently switches on and off, or a pressurized vessel with a valve that intermittently opens and closes. In either of these designs, it is difficult to avoid a pressure pulse that can promote premature blocking of the filter, especially in serum-filtration applications. The inlet pressure in intermittent-flow-mode filtration is usually constant, but pressure surges can occur whenever the pump or the valve is switched on or off. Flow will decrease over time, and the appropriate filter combination needs to be determined in order to assure that the entire batch can be filtered in the required time period (e.g., 8 h).

Although existing fill lines may dictate which mode of operation is used, experience shows that the best filtration mode for protein solutions is the constant-flow mode. The advantage to this mode of operation is that one can choose the initial starting pressure. It is best to start with a low pressure differential, because this will maximize total throughput and yield. Constant flow through the filter decreases the chance of filter fouling.

4.4. Prevention of Premature Filter Fouling

The main result of premature filter blockage and fouling is the loss in production yield coupled with an increase in filtration cost. Several different processing steps and optimization procedures can reduce filter fouling significantly. Typically, this is accomplished through effective prefiltration, filter pretreatment, and proper selection of membrane-filter materials.

4.4.1. PREFILTRATION

Prefiltration is the most common and best method for preventing fouling of a sterilizing-grade membrane filter. Different prefilter materials and constructions are commercially available, many having different and complementary functions and properties. It is important to choose a prefilter that best fits the application. There is no one prefilter or sterilizing-grade filter that will work optimally in every application. The prefilter(s) and sterilizing-grade filter train should be determined by filterability trials. Filter manufacturers offer their services and expertise for performing filter studies with the individual liquid-protein product.

Depth filters made of glass fiber, borosilicate, and cellulosic mixed esters are often used in serum and plasma applications. These filters typically adsorb some protein from solution, but they very effectively remove other contaminants (lipids and insoluble matter) from solution. For applications where the protein concentration is low and low protein adsorptivity is required, other prefilter types should be used, such as polypropylene depth filters, CA membrane filters, or hydrophilic PVDF membrane filters. These filters feature low protein adsorption and they reduce the contaminant mainly by sieve retention. Depending on the application, two or more of the low-protein-adsorbing prefilter types may be required ahead of the sterilizing-grade final filter. Filter trains are generally connected in series, and the medium is filtered sequentially until finally being sterilized by the terminal filter. The optimal filter-train design needs to be determined by filterability studies.

4.4.2. MEMBRANE PRETREATMENT

Prefilters and membrane filters can also undergo a pretreatment step to avoid or reduce filter fouling. This approach is commonly found when raw material is being filtered rather than in the final sterile filtration of a pharmaceutical product. Pretreatment can consist in either flushing the filters with an appropriate buffer solution or protein solution, or cooling of the filter. The buffer must be appropriate for the product being filtered. If the wrong buffer solution is used, it can lead to modification or precipitation of the protein or other active ingredients during the filtration process. Each specific application has its own optimum buffer solution.

Prerinsing of the filters with the appropriate buffer will generally increase both throughput and yield. Preflushing with a buffer or protein solution can minimize nonspecific protein adsorption and the formation of precipitates. For example, it has been shown that a preflush with human serum albumin reduced the losses in yield of Factor VIII, a blood clotting factor, because nonspecific adsorption was reduced. Filtration of protein solutions at reduced temperatures (e.g., 4°C) results in improved throughputs. Though the viscosity of a protein solution increases with decreasing temperatures, fouling also decreases. Filtration systems in serum- or plasma-processing plants are often located in cold rooms at temperature conditions that reduce fouling, reduce premature blockage of the filter, and limit microbial growth.

Appropriate pretreatment of membrane filters depends on the specific application. Some applications, such as the final-filling step, do not allow a pretreatment of the filter system due to dilution of the product by the pretreatment buffer or solution. In these instances, the filter-system setup and filter material need to be chosen accordingly through filterability trials followed by proper scale-up.

4.4.3. SELECTION OF MEMBRANE-FILTER MATERIAL

As described previously, there are a variety of prefilter materials that are suited for specific applications. The adsorptive properties of depth-filter materials can be used to remove lipids in applications with high protein concentrations. However, these filter types can be detrimental when it comes to products that require low-adsorptive filter types to reduce losses in yield. The same is true for membrane-filter types. Some membrane materials have a higher nonspecific adsorption than others. For example, nylon is known to be more adsorptive than polysulfone (PS) and PES, which in turn are more adsorptive than hydrophilic PVDF and CA. Higher nonspecific adsorption means not only losses in yield, but also a higher tendency toward fouling. (Adsorptive filters have a tendency to plug sooner than nonadsorptive filters.)

Besides the adsorptive properties of the filter membrane, the construction of the filter element itself may influence the fouling of the filter. For instance, if the pleating geometry is too tight, the filter may plug due to insufficient access to the entire filter surface area. Prefilter fleeces and membranes used in the filter construction must be designed to optimize the distribution of the liquid over the entire membrane surface, protect the final membrane, and have low adsorption of protein.

4.5. Cleaning of Crossflow Filters

Dead-end cartridges and capsules used for sterilization are typically not cleaned and reused; however, crossflow membranes are generally designed to be reused. Thus it is generally necessary to clean the crossflow membrane when fouling occurs or between filtration batches, so that cross-contamination between batches does not occur. Methods used to clean the membranes must not damage or alter the membrane, and the cleaning agents must be completely removed from the membrane before reuse.

The goal of the cleaning procedure is to remove product from the membrane and to restore the membrane flux as close to the original flux as possible. The selection of a cleaning protocol must take into account the nature of the product and the agent responsible for fouling the membrane. It is critical to examine the interaction of membrane foulant and cleaning agent, because it is possible to irreversibly fix fouling proteins to the membrane matrix or to render a membrane unsuitable for the application for which it is intended by leaving a residual charge on its surface. Also critical to the process is the ability to test for the presence of any residual cleaning or sanitizing agents because of their potential deleterious affects on subsequent runs (Kirsch *et al.*, 1993).

Cleaning may be considered from the perspective of time, temperature, and cleaning chemicals. As a rule, the longer a cleaning agent is allowed to work, the more effective it is. This is especially true when the cleaning agents contain proteases. The higher the temperature, the more reactive is the cleaning agent. The limitation of raising the temperature must be balanced between the reactive nature of the cleaning agent and the stability of the fouling proteins. Too high a temperature can lead to protein denaturation. When proteins become denatured and precipitate on or within the membrane matrix, cleaning becomes considerably more difficult. The proper selection of cleaning chemicals and their proper use will result in the desired effect (Table VI). Just as one must optimize the operating parameters for production, the same can be said for the cleaning process.

The cleaning procedure should begin immediately after the end of the filtration run. Allowing protein-containing solutions to remain in a drained membrane system can lead to protein denaturation and adsorption at the

Table VI
Biological Foulants and Recommended Membrane Cleaning Agents

Biological foulant	Membrane cleaning agent
Antifoamant	Cold water and P3 Ultrasil 53 or Calgonite CMR or P3 Ultrasil 11 or P3 Ultrasil 91
	Sodium hydroxide
Biopolymers and proteins	Isotonic saline and urea and citrate tetraborate or sodium hydroxide
	P3 Ultrasil 11
	P3 Ultrasil 91
Blood	Citrate solution and isotonic saline
Cell debris and denature proteins	Isotonic saline solution
	P3 Ultrasil 11 or Calgonite CMR
	Sodium hydroxide
	P3 Ultrasil 91
Lipids	P3 Ultrasil 53 or Calgonite CMR
	P3 Ultrasil 11
	P3 Ultrasil 91
Polysaccharides	Citrate tetraborate hypochlorite solution
	P3 Ultrasil 53 or Calgonite CMR
	Sodium hydroxide
	Hypochlorite

protein–membrane–air interfaces. It is generally advisable to rinse out as much of the product as possible at the end of the run, prior to introducing the cleaning agents. For protein-containing solutions it is generally recommended to rinse the membranes with an isotonic saline solution.

4.6. Sizing, Design, and Scale-up of Sterile Filtration Systems

As previously emphasized, it is important to perform filterability experiments with the actual solutions to be filtered and with the actual filters anticipated for use in the final process. These experiments can be performed at a reduced scale, provided that the reduced-area filter geometries match the production-scale process. Once data are collected and plotted as described in Section 3.2, the operative model for filtration can be determined, that is, adsorptive sequestration, or sieve retention. When the appropriate model is determined, the initial flux J_0 and the filter-plugging constant K_{Sieve} or K_{Ads} are determined and used for prediction of volumetric throughput at various scales of operation. This allows users of filtration devices to perform laboratory tests of their solution at small scale and, based on the results of these tests, predict the performance of larger scale devices for filtration of the same solution.

The model used to correlate filtration results contains two parameters (initial flux and filter-plugging constant) that must be determined experimentally. Once these parameters are known, it is possible to use the appropriate performance equation (Eq. 1 or Eq. 2) to predict the performance (flow rate and throughput) of membrane devices of other sizes. The following step-by-step procedure is used in such a scale-up procedure (Brose *et al.*, 1994):

1. With the process solution, perform at least three separate filtration tests using new disk filters or filtration capsules for each test; perform each test at a constant but different feed pressure, for example, 5, 10, and 15 psig feed pressure.
2. In each constant-pressure filtration test, measure filtrate volume V as a function of time t until the filtrate flow rate drops to approximately 10% of its initial value, and then plot t/V versus t and t/V versus V for each constant-pressure test. Using the plot with the best straight lines, determine the slopes and y-intercepts of the lines for each constant-pressure test.
3. Divide the reciprocals of the y-intercepts determined in step 2 by the membrane area and the operating pressure, and average the results for an average value of J_0.
4. If the plot of t/V versus t results in straight lines, multiply the slopes determined in step 2 by the membrane area, and average the three results, resulting in an average value for K_{Ads}. Alternatively, if the plot of t/V versus V results in straight lines, multiply the slopes determined in step 2 by $2J_0A_0^2P$, and average the three results for an average value of K_{Sieve}.
5. Using the values of J_0 and either K_{Ads} or K_{Sieve} determined in steps 3 and 4, use the appropriate performance equation (Eq. 1 or Eq. 2) to determine cumulative filtrate volume V for any size filter as a function of filter-membrane area A_0, operating time t, and operating pressure P.

Example 1. Bovine Serum. Using the above procedure, the following scale-up prediction was reported for sterile filtration of bovine serum (Brose *et al.*, 1994). Sterilizing-grade filtration capsules with identical cellulose acetate membrane (0.2-μm-rated pore size) were tested in three different sizes: 500, 2000, and 4000 cm^2. Raw (unfiltered) bovine serum was fed directly to the capsules, without prefiltration, at feed pressures of approximately 5, 10, or 15 psig. The filtrate weight was measured versus time until the filtrate flow rate was less than 10% of the initial flow rate. Each capsule was tested by this procedure using the same batch of feed solution, thereby assuring that differences between batches would not influence the results.

The time-course data that were obtained in filterability experiments with 500-cm^2 capsules were found to correlate best when t/V was plotted versus t, as shown in Fig. 12. From these data, the initial flux J_0 was 0.27 ($\pm$0.05) ml/min-cm^2-psig and the filter-plugging constant K_{Ads} was 0.62 ($\pm$0.02) cm^2/ml.

Bovine serum was chosen for these experiments in order to plug the filters rapidly. In actual production, prefiltration of the serum with membrane filters of larger pore size prior to the final sterile filtration would result in significantly greater throughputs.

Using these constants in Eq. (1) scale-up predictions were made as a function of operating time. Figure 13 compares actual data with the predicted curves for filtration of raw bovine serum with larger capsules (2000 and 4000 cm^2). The predicted curves fit the experimental data very well.

Example 2. Ascites Fluid. In another set of experiments, an ascites fluid was tested in small-scale filterability studies to predict the appropriate size for a

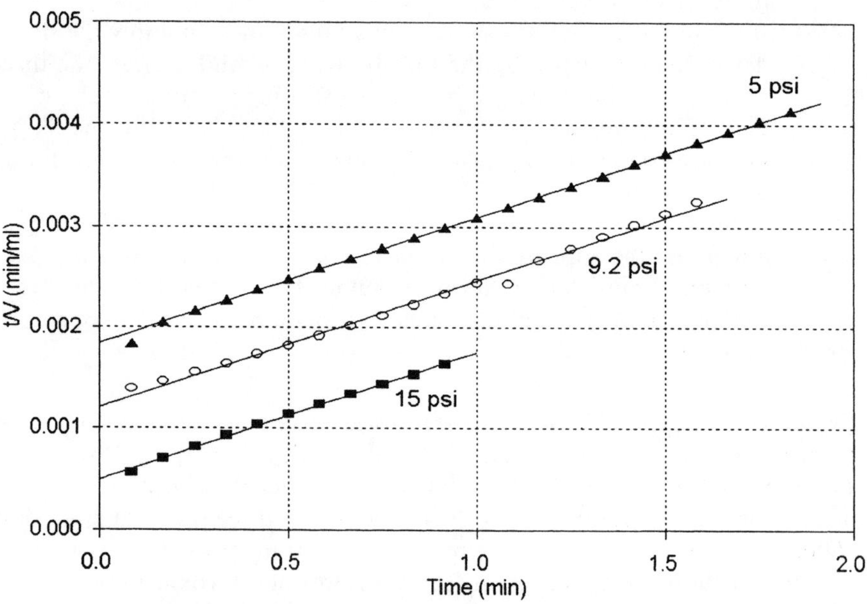

Figure 12. Linearized data from scaleup tests with bovine serum (500-cm^2 capsules).

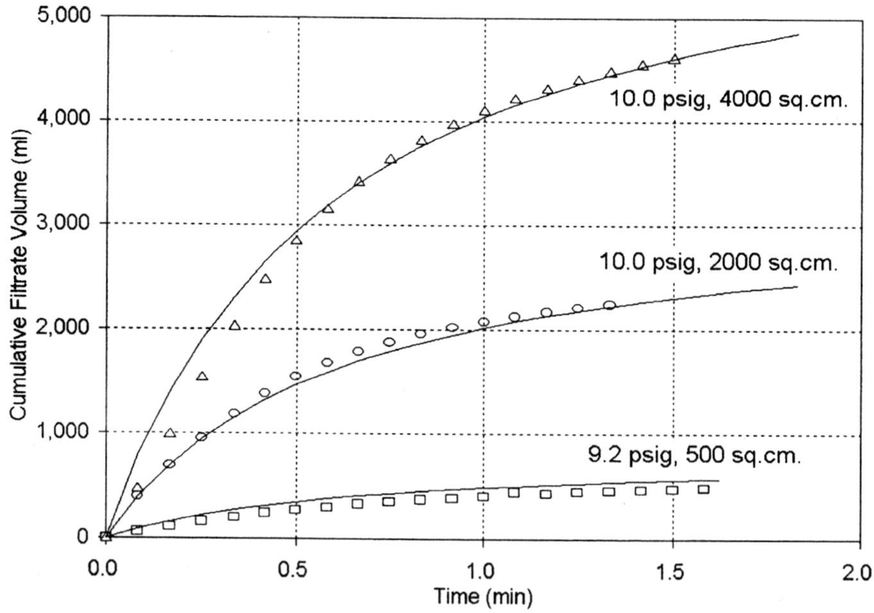

Figure 13. Comparison of actual data to scaleup prediction for raw bovine serum.

filtration train to clarify 8-L batches in 1 h or less. From a series of tests with 47-mm disks, it was determined that the optimum scheme is to prefilter with a glass-fiber filter containing 1.2-μm nominal-rated depth media, followed by final filtration with a 0.2-μm membrane filter. The prefilter should contain twice the membrane area of the final filter.

In the experiments with 47-mm disk filters, it was found that the plot of t/V versus t results in a straight line (after an initial delay), as shown in Fig. 14. From the y intercept of the dashed line in Fig. 14, J_0 was found to be 0.056 ml/min-cm^2-psi, and from the slope K_{Ads} was found to be 0.12 cm^2/ml. (In these experiments with prefiltration and final filtration, the area of the prefilter is used in the calculations. The ratio of 2:1 for prefilter area to final-filter area should be maintained throughout scale-up.) Using these measured constants and $V = 8000$ ml, $P = 5$ psi, and $t = 60$ min, the required prefilter membrane area A is calculated from Eq. (1) to be 1433 cm^2. Accordingly, a 2-ft^2 prefilter capsule should be used along with a 1-ft^2 final filter. If $A = 1858$ cm^2 (2-ft^2) is used in Eq. (1) with the above parameters, the prediction is that 8 L can be filtered in 32 min.

To test the validity of the parameters used in this scale-up prediction, 3.75 L of ascites fluid was filtered at 5 psi using a 1-ft^2 glass-fiber 1.2-μm capsule and

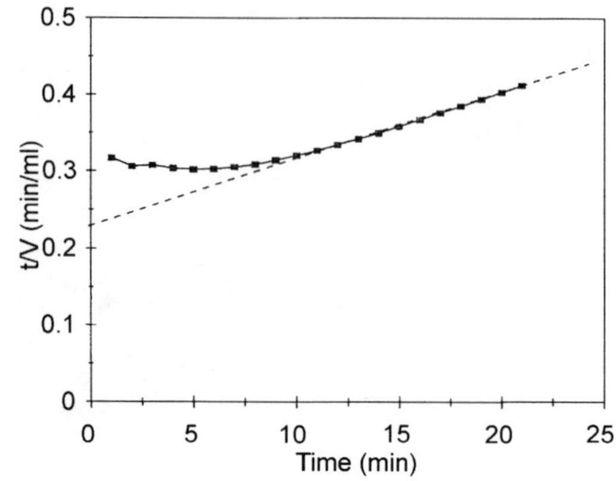

Figure 14. Ascites fluid filtration data ploted to determine J_0 and K_{Ads}.

a 0.5-ft^2 0.2-μm final-filtration capsule. The filtration took 24.5 min to complete. The predicted time to perform the filtration, based on Eq. (1) is 28 min. This is excellent agreement between experimental and predicted filtration time.

4.7. Case Studies: Sterile Filtration Scale-up

The following case studies are drawn from actual scale-up experience by the authors and their associates.

4.7.1. STERILE FILTRATION OF FERMENTATION MEDIUM

4.7.1.1. Background

The fermentation medium investigated in this study was to be sterile-filtered prior to use in the production of anti-TNF (Tissue Necrosis Factor) from CHO (Chinese hamster ovary) cells. The fermentation medium contains high concentrations of desirable soluble lipids and cholesterol, which are to be fully recovered in the filtrate stream. This medium solution quickly plugs sterilizing-grade 0.2-μm membrane filters when no prefiltration is used. A prefilter that combines layers of depth filtration can more than double the throughput of sterilizing-grade filters, making the entire process more economical.

Polypropylene depth filters are formed from a continuous monofilament of spun polypropylene, which is fused with heat. These filters are non-fiber-releasing and have low adsorptive retention of dissolved solutes. Their principal method of retention is by physical capture of particles throughout the thickness of the filter, resulting in retention of 95–99% of particles larger than the retention rating of the filter.

The objective of this study was to design a scaled-up filtration train for sterile filtration of 2000 L of fermentation medium. It was required to perform this filtration with no more than three single-element housings in series using a maximum feed pressure of 30 psi and producing a filtrate flow of greater than 30 L/min.

4.7.1.2. Materials and Methods

A 90-L batch of fermentation medium at 30°C was tested using the apparatus shown in Fig. 15. Filterability tests were performed to generate flow-decay data for process design and scale-up. The filterability test apparatus in Fig. 15 consists of three 0.5-ft^2 capsules connected in series with pressure gauges at the inlet and outlet of each capsule. Fermentation medium was fed from the feed vessel to the filter capsules using a rotary-lobe pump. The pump flow was adjusted so that the initial pressure drop across the complete filter train was 5 psi; the filtration was discontinued when the total pressure drop reached 30 psi. Filter flow rates and throughputs were determined by continuously weighing filtrate mass versus time.

4.7.1.3. Results and Discussion

A series of filterability tests was conducted using the following combination of prefilters and final filter: (1) prefilter 1: 0.5-ft^2 capsule containing Sartofine PP 0.7-μm absolute-rated depth media filter; (2) prefilter 2: 0.5-ft^2 capsule containing Sartoclean CA 0.45-μm double-membrane prefilter; and (3) final filter: 0.5-ft^2 capsule containing Sartobran P 0.2-μm sterilizing-grade membrane filter. Each of the capsules used in these tests contains the same number of membranes and support layers as is found in larger scale pleated cartridges.

Figure 16 shows filtrate flow rate versus time for the filterability test with this combination of capsules. As this figure shows, filtrate flow was constant at 3 L/min for the first 30 min of the test, after which the flow rate dropped quickly. Figure 17 shows pressure drop across prefilter 2 and the final filter during the course of this filterability study. This plot shows that after 30 min the pressure drop across the final filter increased dramatically.

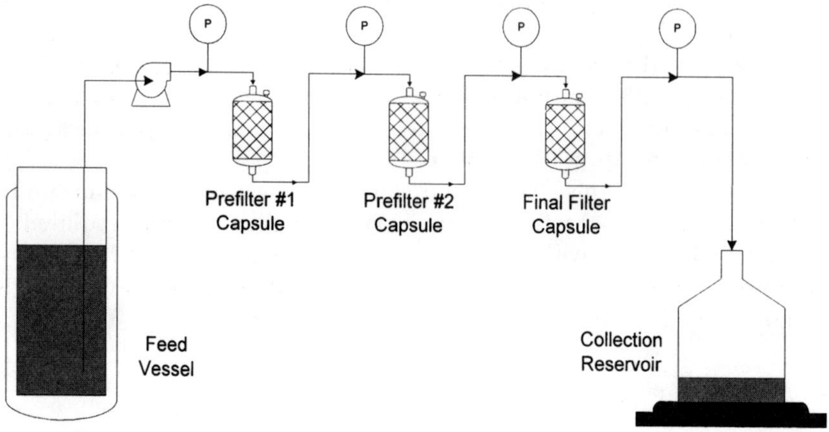

Figure 15. Filterability test apparatus.

This indicates that the final filter was fouling faster than the prefilter and is the limiting filter in this filtration train.

4.7.1.4. Conclusions

Based on the filterability test data, a scaled-up system was designed for filtration of 2000 L of media. This volume of media can be filtered at an average flow rate of 57 L/min at a feed pressure not exceeding 30 psi using the following filter train: (1) one 30-in Sartofine PP 0.7-μm absolute-rated depth media filter, (2) one 20-in Sartoclean CA 0.45-μm double-membrane prefilter, and (3) one 30-in Sartobran P 0.2-μm sterilizing-grade membrane filter.

4.7.2. CLARIFICATION OF A BACTERIAL FERMENTATION BROTH

4.7.2.1. Background

The fermentation broth tested in this study is produced during fermentation of a proprietary cell line. The broth contains cells, protein, colloidal material, polysaccharides, viral particles, and lipids. The broth is to be clarified by filtration of cells and insoluble material from the proteins and lower molecular weight broth components. Thus, the objective of this filterability study was to optimize the filter train used in a production setting for

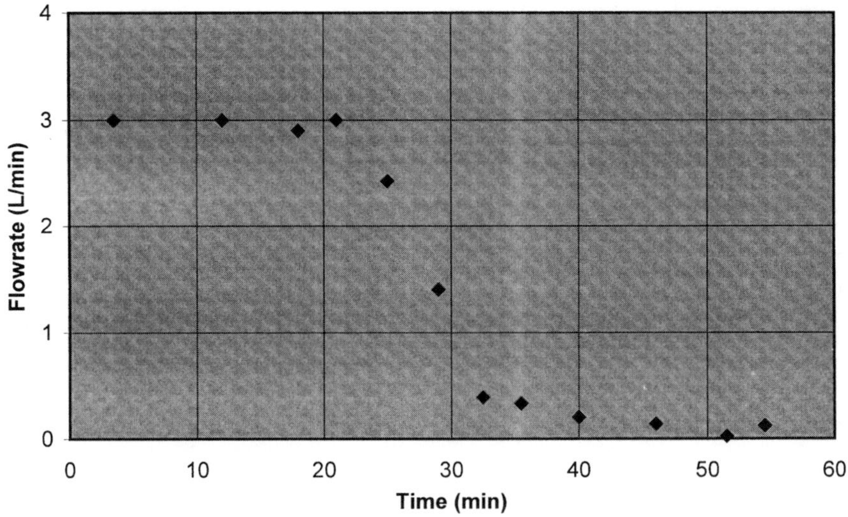

Figure 16. Filtrate flow rate versus time for filtration of HTA media.

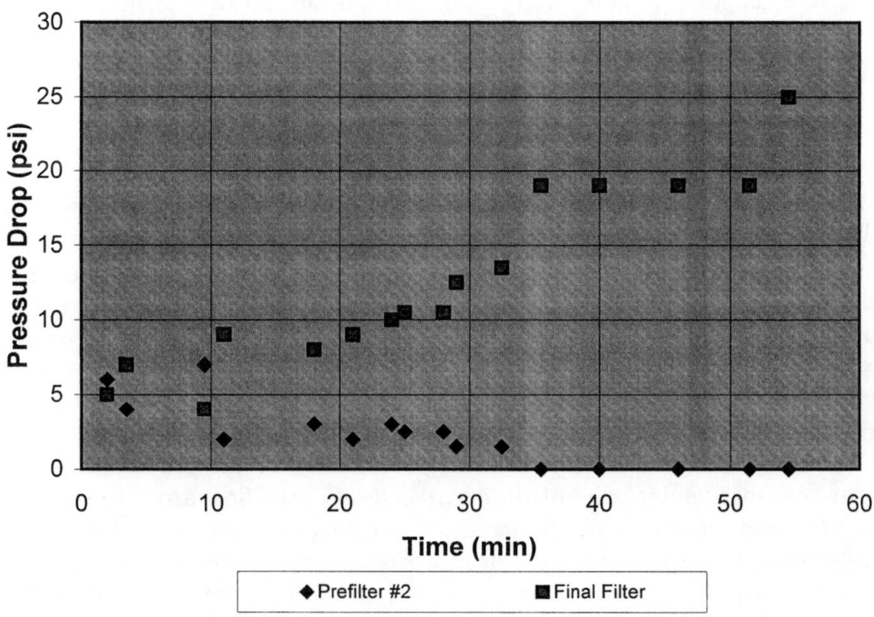

Figure 17. Pressure Drop in prefilter 2 and final filter for filtration of HTA media.

clarification of a bacterial fermentation broth. Various filtration schemes were tested to determine the effect of prefiltration prior to a 0.2-μm sterile-filtration step. Two different types of prefilters were used in this study: (1) a prefilter that contains multiple layers of progressively finer nonwoven polypropylene filter material (Sartopure PP), which has very high surface area, resulting in high capacity for filtration of particulate material from biological solutions, and (2) a glass-fiber filter preceding two cellulose acetate membrane filters, the first of larger pore size than the second (Sartoclean GF), which combines adsorptive depth filtration with membrane filtration to provide two filtration steps in one cartridge. The final membrane filter (Sartobran P) is a sterilizing-grade pharmaceutical filter that consists of a polypropylene prefilter preceding two cellulose acetate membrane filters.

4.7.2.2. Materials and Methods

In separate experiments, 100-L batches of bacterial fermentation broth were filtered at 20°C using the production setup shown in Fig. 18. Filterability tests were performed to generate pressure-drop and filter-throughput data. The setup in Fig. 18 consists of multiple 10-, 20-, or 30-in filter cartridges in stainless steel housings, which can be arranged in series or parallel as desired. Pressure gauges at the inlet and exit of each housing allow determination of which filters are plugging during the course of a run. Filtration flow rates and throughputs were determined by continuously weighing the filtrate mass versus operating time.

4.7.2.3. Results and Discussion

A series of filterability tests was conducted using various combinations of Sartorius prefilters and final filters. It was determined that the following combination of filters results in the best overall filtrate throughput: (1) prefilter 1: 10-in cartridge containing Sartopure PP 8.0-μm nominal-rated prefilter; (2) prefilter 2: two 30-in Sartoclean GF 0.80-μm double-membrane prefilters, arranged in parallel; and (3) one 10-in Sartobran P 0.2-μm sterilizing-grade membrane filter.

Figure 19 shows filtrate flow rate versus time for the production-scale tests with this combination of 10-in cartridges. Filtrate flow dropped linearly as a function of time, indicating gradual plugging of at least one filter in the train. Figure 20 shows pressure drop across the two prefilters and the final filter during the course of the described test. This plot shows that the pressure drop across the final filter is much higher than the pressure drop across either of the prefilters. This indicates that the final filter is the limiting

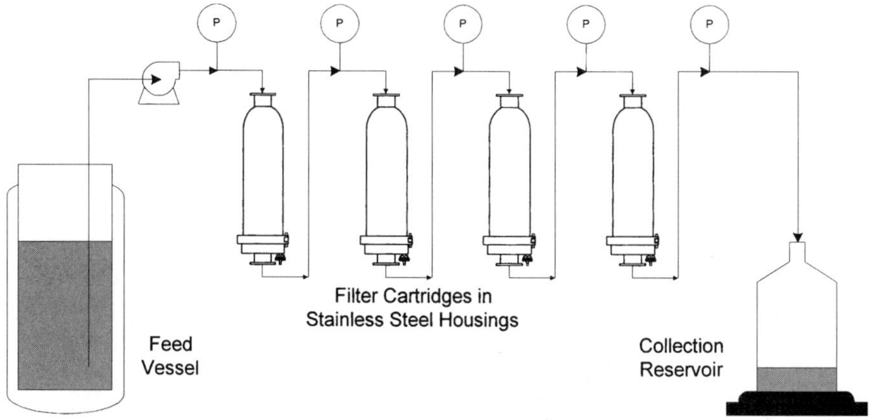

Figure 18. Filterability test apparatus for use with 10-in cartridges.

filter in this filtration train and that its surface area could be increased to increase the throughput for this filtration.

4.7.2.4. Conclusions

Based on the results of production-scale experiments for clarification of a bacterial fermentation broth, it was concluded that 80-L batches can be clarified using two stages of prefiltration followed by a final sterilizing-grade filtration step. If additional throughput is required for this filtration, this can be accomplished by increasing the number and size of prefiltration and final-filtration cartridges. Thus, it was determined that for this particular cell clarification, the optimum scheme is to perform a first prefiltration with one 10-in Sartopure PP 8.0-μm nominal-rated prefilter, perform a second filtration using two 30-in Sartoclean GF 0.80-μm double-membrane pre-filters, followed by final filtration with one 10-in Sartobran P 0.2-μm sterilizing-grade membrane filter. Using this scheme in a production setting, 80 L of fermentation broth can be clarified in 2 hr.

4.7.3. STERILIZATION OF A HUMAN FLU VACCINE

4.7.3.1. BACKGROUND

Filterability testing of concentrated human flu vaccine was conducted to evaluate the effect of prefiltration on the total throughput of a sterilizing-grade

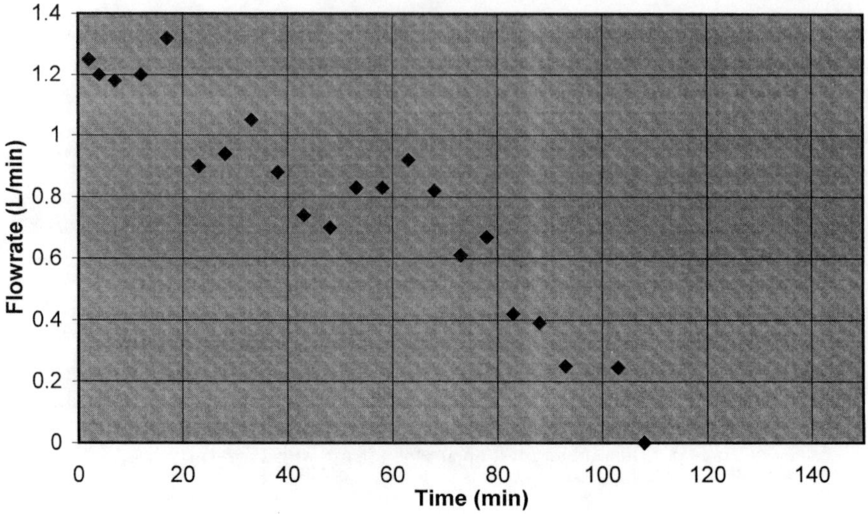

Figure 19. Filtrate flow versus time for cell-broth clarification.

final filter. This type of biological solution contains heavy loads of submicrometer particles that can quickly plug 0.2-μm membrane filters. A prefilter that combines layers of depth filtration can more than double the throughput of

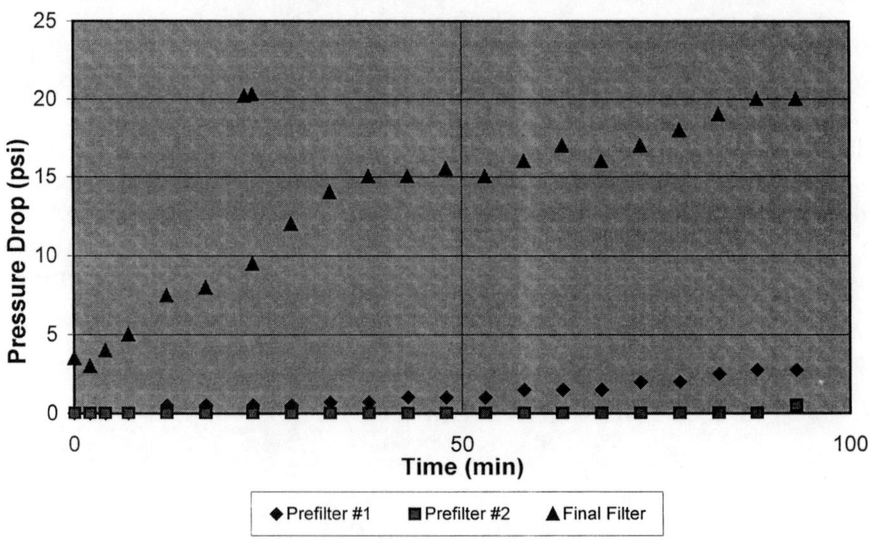

Figure 20. Pressure drop in the prefilters and final filter.

sterilizing-grade filters, making the entire process more economical. Prefilters that contain multiple layers of progressively finer glass fiber and nonwoven polypropylene filter material are very effective with this type of solution. Molecular-force interactions between colloidal particles and the glass fibers results in retention of particles smaller than the pore-size rating.

4.7.3.2 MATERIALS AND METHODS

Filterability tests were conducted using the test apparatus shown in Fig. 21. These tests were used to generate flow-decay data for proper prefilter selection and process scale-up. The filterability test apparatus consists of two 47-mm filter holders in series and pressure gauges at the inlet and the outlet of each filter holder. Feed solution was fed under air pressure from a stainless steel pressure vessel. Filter flow rates and filter throughput were determined by continuously measuring filtrate mass with a weighing balance. Several combinations of prefilter and final filter were tested to determine the optimum filtration scheme. Filtrate samples were regularly drawn for analysis of flu vaccine yield.

4.7.3.3. RESULTS AND DISCUSSION

A series of filterability tests was conducted using the following combinations of prefilter and/or final filter:

- Test 1: Baseline test using only the Sartobran P 0.2-μm absolute-rated membrane filter
- Test 2: Sartopure PP 1.2-μm nominal-rated depth filter, connected to the final 0.2-μm membrane filter
- Test 3: Sartoclean GF 1.2-μm absolute-rated membrane filter, connected to the final 0.2-μm membrane filter
- Test 4: Sartopure GF 1.2-μm nominal-rated depth filter, connected to the final 0.2-μm membrane filter.

Each of the 47-mm filters used in these tests contains the same number of membranes and support layers as is found in larger scale pleated capsules and cartridges.

Figure 22 shows total filtrate throughput and flu vaccine recovery for tests 1–4. The baseline test (test 1) and test 2 showed very low throughputs compared to tests 3 and 4, which improved throughput by threefold over the baseline. The prefilters used in tests 3 and 4 both contained glass fibers, which function to remove colloidal material by adsorptive retention. As shown in Fig. 22, in all tests the flu vaccine recovery was greater than 90%.

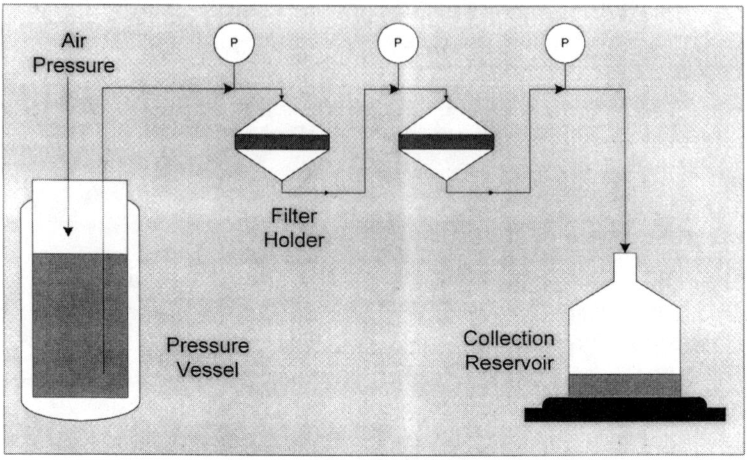

Figure 21. Filterability test apparatus.

Figure 23 shows how the addition of prefiltration to the filtration train improves the overall economics of filtration. This figure shows total filter consumable costs as a percentage of the filtration cost when no prefiltration is used (i.e., test 1). The use of Sartoclean GF for prefiltration (test 3) results

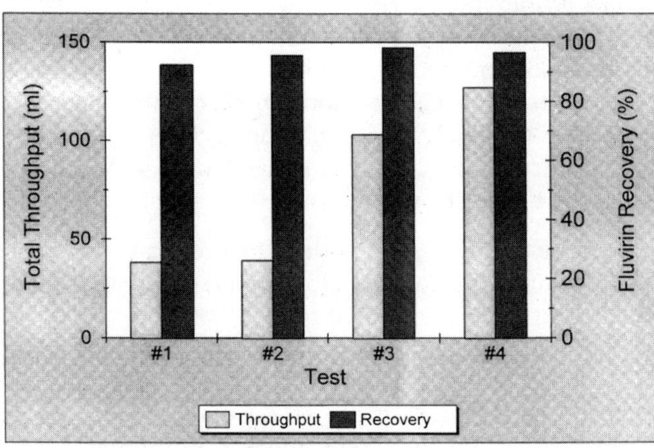

Figure 22. Total throughput and flu vaccine recovery for tests 1–4.

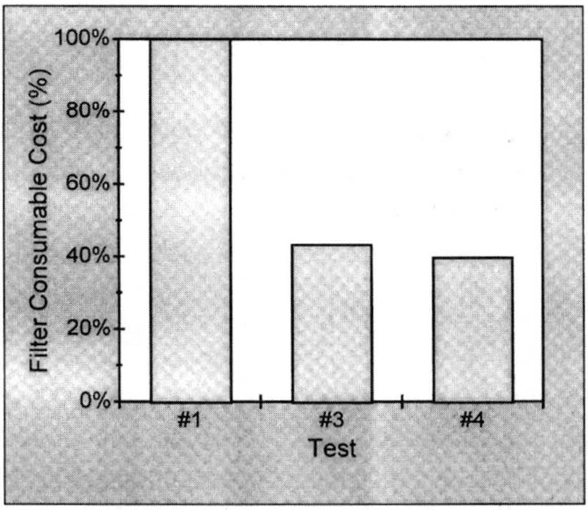

Figure 23. Total filter consumable costs as a percentage of cost with no prefiltration.

in a 57% cost savings. The use of Sartopure GF for prefiltration (test 4) results in a 60% cost savings.

4.7.3.4. CONCLUSIONS

Concentrated human flu vaccine can be rapidly and economically prefiltered using glass-fiber-based prefilters. It was determined that a two-cartridge filter train is optimum for batch sizes up to 100 L. The prefilter should be a 10-in Sartopure GF 1.2-μm nominal-rated depth medium filter. The final sterilizing-grade filter should be a 20-in Sartobran P 0.2-μm membrane filter. Use of proper prefiltration results in 60% cost savings.

4.8. Filtration-System Preparation and Sterilization

The stages for the preparation of a filtration systems may be divided into the following steps: (1) system installation and initial filter wetting, (2) sanitation and sterilization, and (3) system equilibration. All filters require initial wetting prior to use. The process of flushing a new filter with water accomplishes two tasks, first, it removes any extractables that may be present in the filter, and second, it wets all parts of an otherwise dry filter element in preparation for integrity testing. It is necessary to adequately flush a filter so as to assure both that the extractables are removed and the filter is thoroughly wetted. Failure to remove extractables present in the filter may result in the contamination of the process solution. Failure to thoroughly wet the filter may result in compromised flow rates and the possible failure of a filter integrity test.

The level of decontamination/sterilization to which a filter must be subjected depends on the stage at which the filter is used in the overall process. The highest level of sanitization is sterility. In order to assure this level of cleanliness, steam is used for the sterilization. Filter systems can be either steamed in-place or placed within an autoclave for steam sterilization. Generally, the minimum required exposure to steam is 30 min at a temperature of 121°C.

Unlike the autoclave, where the process is automated, in-line steam sterilization requires special care in the setup and execution of the steaming process. A filter system can easily be damaged by failing to run the process properly. Of particular importance is the method for starting up the steaming process. When flowing steam is applied to a cold filtration system, condensate immediately forms on the filters, filling the pores with liquid water. As a result, there is no route for steam to leave the system except by diffusion through the wetted filters (unless the bubble-point pressure has been exceeded). It is therefore critical to avoid introducing steam at a high pressure until the system has been heated to the point where all of the condensed steam has been revaporized, thereby opening the membrane pores for steam to escape as a vapor. It is only at this point that steam will freely pass through the filter.

The end of the filter-steaming process can also pose hazards to the filter element. As the system begins its cooldown, the steam begins to condense and creates a vacuum in the sealed system. Accordingly, there must be provisions to replace the condensed steam with air. If a rapid condensation vacuum occurs on only one side of the filter, a high differential pressure situation will occur, which may cause the filter element to either collapse on itself or balloon out. This may also happen in an autoclave if flow through the filter is restricted on either side.

Crossflow systems (not including ceramics) pose an additional challenge when being in-line steam-sterilized. Because the filter elements are not well supported against high-pressure differentials at elevated temperatures, steam must be introduced at both the upstream and the downstream connections of the system. Alternative methods for sanitizing filter systems may include the use of chemicals; for example NaOH or NaOCl, or water at 80°C.

After a filter system has been "sanitized" it is often recommended to preflush the system with a volume of isotonic saline or buffer similar to that of the protein solution to be filtered. Preflushing the system aids in prevention of any denaturation, precipitation, or otherwise unwanted protein instabilities.

If protein precipitates do occur, they may cause immediate fouling of the filter system.

4.9. Methods for Testing Filter Integrity

Filter integrity testing has been developed by various filter manufacturers to permit the nondestructive measurement of the integrity of a sterilizing-grade filter. These nondestructive integrity tests find their validity by a direct correlation to the ASTM bacteria-challenge test. This correlation is critical to the validation of any integrity test, because nonidealities and lack of strict applicability of the capillary model prevent methods such as the bubble-point test from being used to directly measure the largest pore and thus validate filter integrity.

The test limits for individual integrity tests are established experimentally. If a filter fails such preset limits, a nonsterile filtrate is likely. To ensure sterility of the filtered liquid, integrity tests are performed on the sterilizing-grade filters at regular intervals. Regulatory authorities (e.g., the FDA) call for an integrity test after the filtration of the product. Most membrane-filter users perform integrity tests of their sterilizing-grade filters both before and after filtration. The reasons for testing pre- and post filtration are mainly economical, due to the knowledge that if the filter is integral before and after the filtration of the product, reprocessing or even discarding of the product is avoided. There are various integrity tests in practice today, including bubble-point, diffusive-airflow, and pressure-hold tests; each of these tests is described in more detail below.

4.9.1. BUBBLE-POINT INTEGRITY TEST

The bubble-point test is based on the following formula, which predicts the pressure necessary to expel water from a wetted pore:

$$P = \frac{(4\gamma \cos \theta)}{D} \tag{8}$$

Here P is the pressure necessary to expel the water, D is the diameter of the pore, γ is the surface tension of the product that wets the pore, and cos θ is the wetting angle between the solution and the filter material, describing the adhesive forces operative between the solution and the membrane material. It is important to note that P is inversely proportional to D. This means that the smaller the pore diameter, the greater is the pressure needed to expel the wetting medium from the pores.

A water-wetted membrane retains the water within its pores. To expel the water from the pores, an upstream pressure must be applied. The bubble-point test is a measurement of the pressure needed to expel the liquid that wets the filter, and this pressure is directly correlated to the size of the largest pores within a filter matrix. If these pores are too big or a flaw exists in the filter membrane, the bubble-point pressure will be lower than the value specified by the filter manufacturer. The determination of a failed or integral filter is based on the vendor's published minimum bubble-point value. The published bubble-point supplied by the filter manufacturer is normally valid only for water-wetted membranes. Therefore, if product is used as a wetting medium, the minimum allowable bubble-point value must be empirically determined, because the surface tension of the product may be different than that for water.

4.9.2. DIFFUSIVE-AIRFLOW INTEGRITY TEST

In the diffusive-airflow test, the filter matrix is wetted with water (or another liquid) as in the bubble-point test. Pressurized air is then introduced to the upstream side of the filter, although the pressure is not increased as high as the bubble-point pressure. Instead, the pressure is maintained at a pressure of about 80% of the minimum allowable bubble-point pressure and the rate of air diffusion across the wetted membrane is measured.

In accordance with the physical laws that govern the solubility of gases in liquids, more air is dissolved in the filter-wetting liquid at higher upstream air pressures. Thus, the amount of gas dissolved in the wetted filter is a direct function of the gas pressure above the liquid. In addition, due to a pressure differential across the filter, gas diffuses through the wetted filter, flowing from the high-pressure side to the low-pressure side. As a result, the dissolved air (forced into solution by the higher pressure) diffuses to the downstream side, where it comes out of solution. The amount of gas diffusing across the wet membrane can be measured as a flow rate, in units of volume per unit time. Fick's law mathematically expresses the interrelation of factors that determine the diffusive flow of gas across the wetted filter:

$$N = \frac{NH\rho(P_1 - P_2)}{L} \tag{9}$$

where N is the gas permeation rate, D is the diffusivity of the gas in the liquid, H is the solubility coefficient of the gas in the liquid, ρ is the total porosity of the filter membrane, $(P_1 - P_2)$ is the transfilter pressure differential, and L is the thickness of the liquid layer in the membrane. As predicted by this equation, the diffusive flow N depends on the diffusivity of the gas in the liquid medium, the solubility of the gas in the liquid, the porosity of the filter, the thickness of the filter, and the differential gas pressure across the filter.

Filter integrity is determined from the amount of diffusive flow through the membrane over a specified time frame. Therefore, filter manufacturers define a maximum allowable diffusive flow in their validation documentation. Again, if a wetting medium other than water is used, the diffusive-airflow value may vary due to differences in gas solubility and diffusivity in that liquid.

Figure 24 illustrates the air flow versus differential pressure curve that characterizes wetted membrane filters. The initial, more gradual slope represents diffusive flow of air through the wetted filter. The steeper slope represents free flow of air through liquid-free pores, which commences at the bubble point that is, when surface forces between the wetting liquid and the solid filter are overcome by the hydrostatic pressure, and the largest pores are blown free of their liquid contents.

A newer method of diffusive-airflow integrity testing measures diffusive airflow over a range of pressures below the bubble-point pressure, and from these data determines, the pressure-normalized diffusive airflow. This test, termed *the multipoint-diffusion test*, is based on a linear relationship between air diffusion and air pressure, as shown by the linear portion of the curve in Fig. 24. Multipoint-diffusion testing has advantages over single-point diffusive testing, because it can more rapidly detect a pending product failure due to gradual filter degradation. In addition, multipoint-diffusion testing does not result in expelling liquid product from the membrane, and thus this integrity-test method is invaluable in the characterization of the

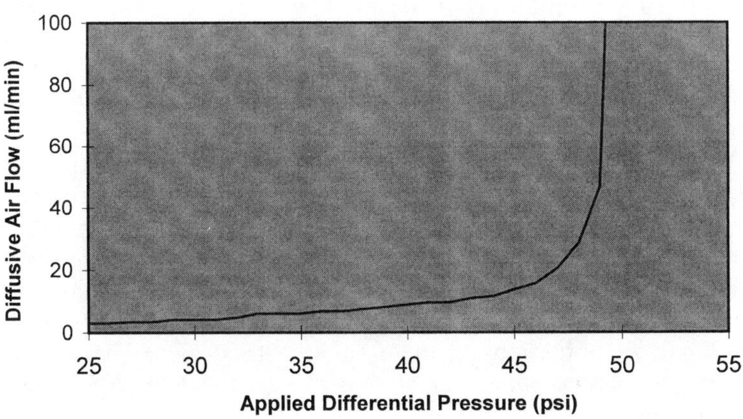

Figure 24. Diffusive air flow through a water-wetted 10-in filter cartridge.

diffusive flow of a filter when wetted with the filtered product (Jornitz *et al.*, 1998; Waibel *et al.*, 1996).

4.9.3. PRESSURE-HOLD INTEGRITY TEST

In the performance of the diffusive-airflow integrity test, a wetted filter installed into its housing is subjected to a constant inlet or test pressure on its upstream side. The pressure-hold test is a variant of the diffusive-airflow test. Test equipment is arranged as in the diffusion test, except that when the stipulated applied pressure is reached, the pressure source is valved off. The decay of pressure within the upstream side of the filter housing is then calculated as a function of time, utilizing a sensitive gauge to measure pressure changes. The resulting pressure drop can result from two sources: (1) the diffusive flow of gas across the wetted filter or (2) system leakage on the upstream side. Therefore the filter system components (i.e., housing, gaskets, and valves) need to be leak-free. The pressure-hold test measures the loss of air from the upstream side of the filtration membrane, and the ideal gas law governs the air flow measurement,

$$pV = nRT \tag{10}$$

where n is the number of moles of gas, R is the universal gas constant, T is the temperature, V is the gas volume, and p is the gas pressure.

Two factors need to be noted from the ideal gas law. First, the decrease in gas pressure measured by an upstream gauge is influenced strongly by the upstream volume of the filter holder. The larger the upstream volume, the lower is the pressure drop and therefore the lower is the sensitivity of measurement. Second, the pressure-hold test can be influenced by temperature changes during the test period. If the temperature rises, the pressure within the upstream volume of the filter holder will rise and a pressure drop cannot be accurately measured. It is important that the temperature remain constant during the pressure-hold test.

All of the integrity tests described above are conveniently performed using automatic integrity-test systems, which make the testing more reliable and less ambiguous. In addition, the tests can be performed without invasion of the sterile side of the system, making it possible to perform the test after steam sterilization. One should be aware that all of the described integrity-test methods have their advantages and disadvantages, and often a combination of these tests methods is recommended. For example, the bubble-point test is sensitive to the biggest pore within a membrane matrix, but is not capable of determining the membrane thickness, which is important in determining the sterilizing ability of an adsorptive-sequestration-type filter. The diffusive-airflow test, by comparison, does not distinguish pore sizes, but it is sensitive to the thickness of the membrane. For this reason, it may be appropriate to perform a combination of integrity tests.

The individual integrity tests are also dependent on the size of the filter system; for example, the bubble point loses some sensitivity with larger filter system due to the fact that the larger filtration area causes a higher diffusive flow, which disguises the real bubble point. On the other hand, the diffusive-airflow test cannot be used with small filter areas due to the low pore volume and resulting low airflow. The multipoint-diffusion test has significant advantages in that it can be programmed to perform the diffusion test all the way up to the bubble-point pressure, thereby determining both diffusive airflow and bubble point in the same test. The individual test methods and the implications of each are discussed in PDA (1998).

4.10. Validation Issues

The validation of a filtration systems is in principal no different than the validation of any pharmaceutical process, that is, "establishing documented evidence which provides a high degree of assurance that a specific process will consistently produce a product meeting its predetermined specifications and quality attributes" (U.S. FDA, 1987). There are numerous documents detailing the specifics of filter-system validation. For example, PDA (1992) deals specifically with validation of crossflow filtration. There are specific validation issues that may be considered to be critical to filtration processes that need to be addressed. These fall into two groups: process-specific issues and application-specific validation issues.

4.10.1. PROCESS-SPECIFIC VALIDATION ISSUES

The compatibility of a membrane system that is reused (e.g., crossflow filtration) must look at the effects of reuse on the membranes with regard to any changes in extractables after prolonged exposures to all of the fluids that the process will encounter. These include the storage, rinse, cleaning, sanitizing, and process solutions. Membrane integrity must be monitored by both direct testing of the membrane integrity and continued achievement of the desired product yield and product quality. Furthermore, the retention coefficient should also be monitored over the life of the membrane.

Retention coefficients or selectivity measurements are highly subjective. It may be best to specifically determine the retention coefficient with the products of interest by using them in the presence of normal process fluids. This then provides a practical reference by which to evaluate performance.

Membrane cleaning and sanitization must be established with experimental data and documented. It must also be demonstrated that the cleaning agents can be removed prior to reuse. The ability to demonstrate this is important, especially considering the potentially deleterious effects of residual cleaning agents on product yield and on product purity (Kirsch *et al.*, 1993). Validation of cleaning cycles can be performed through the analysis of permeate and retentate effluents after each stage of the operation. Analysis should include specific assays for compounds that have been introduced into the system. Alternative approaches are to flush the system with water for injection (WFI) and to determine if the effluents continue to meet the WFI specifications. Total organic carbon (TOC) analysis of effluent rinses offers yet another approach to this problem. Direct physical inspection of membrane systems provides further confirmation that cleaning procedures are adequate. After the membranes have been cleaned it is suggested that a water or buffer flux be determined. Clean water or buffer flux measurements can serve as a benchmark for subsequent cleanings. Generally the flux from a new membrane will never be achieved after the first production run followed by the first cleaning. It is not uncommon for the flux to drop 20%, although this reduced value should be retrievable after subsequent use and cleaning cycles.

4.10.2 APPLICATION-SPECIFIC VALIDATION ISSUES

During operation of a concentration process (e.g., in a crossflow-filtration operation), low-level contaminants may suddenly become problematic. Two specific contaminants are of particular interest. First, protein solutions may have low levels of contaminating proteases. Though the rate of proteolysis may be undetectable in the unconcentrated bulk form, after concentration the rate may be increased by the square of the concentration (PDA, 1992). The effect of proteases on product yield can be devastating to the process. Second, endotoxins may be concentrated from below-detectable limits to levels high enough to cause product-lot rejection. Apparent endotoxin content can be also influenced by the nature of the ionic content of the process fluids, which means that they may only appear after a buffer exchange step.

APPENDIX: DEFINITION OF TERMS

Adsorption	The attraction of solutes (typically proteins) to a solid surface, e.g., to the surfaces of pores in filtration membranes
Antifoam	A chemical added to a solution to prevent or minimize foaming of protein solutions, typically a surfactant
Batch process	A process where no streams flow into or out of a controlled volume, as opposed to a continuous process; in a batch filtration process, the feed solution is reduced in volume due to permeation of filtrate through the membrane; there is no continuous addition of feed solution to the feed vessel
Boundary layer	A stagnant layer of liquid against the surface of a membrane that forms due to friction between the membrane surface and the flowing liquid; higher liquid velocities reduce the thickness of the boundary layer; boundary layers are observed in crossflow filtration
Bubble point	A pressure measurement that is directly related to the size of the largest pore in a membrane; the bubble-point pressure of a membrane is dependent on the membrane polymer, the membrane wetting solution, and the largest pore in the membrane
Capsule	A dead-end filtration module, typically containing flat-sheet membrane in pleated format; membrane capsules are ready-to-use units that consist of pleated membrane encased in a molded plastic housing; the entire capsule is a disposable unit
Cartridge	A dead-end filtration module, typically containing flat-sheet membrane in pleated format; membrane cartridges are placed in stainless steel housings that come in various designs
Cassette	A crossflow filtration module, typically with flat-sheet membranes laid on top of each other in a sandwich-like configuration; crossflow cassettes are stacked one on top of another and placed between endplates
Centrifugal filtration	A filtration process where the driving force for liquid flow through the membrane is achieved by spinning the membrane device and generating a centrifugal force field, driving liquid outward and through the membrane filter

Clean-water flux	A baseline filter flow measurement performed with clean water or buffer, typically measured at 20°C
Cleaning	The process of chemically removing adsorbed, absorbed, or lodged solutes from a membrane
Concentrate	The concentrated feed solution (also known as the retentate solution) that results after removal of filtered liquid through the membrane
Concentration factor (CF)	The ratio of retained solute concentration after membrane concentration to that before concentration; this factor can be easily determined from the initial feed volume and the final feed volume, assuming that the desired solute is fully rejected by the membrane. $CF = C_f/C_i = V_i/V_f$
Concentration polarization	A situation in crossflow filtration where the rejected solute concentration is much higher at the surface of the membrane than in the bulk feed solution; this occurs when the flowing filtrate solution drives solute to the membrane faster than the solute can diffuse back into the bulk feed solution
Crossflow filtration	A membrane filtration process where the feed solution is rapidly circulated parallel to the membrane surface, resulting in minimization of the stagnant boundary layer in the liquid adjacent to the membrane surface
Dead-end filtration	A membrane filtration process where there is no crossflow of feed and an attempt is made to force all of the feed solution through the membrane
Diafiltration	A crossflow-filtration process where the bulk feed is washed, removing solutes and or solvent by continuous or batchwise addition of buffer to the feed solution, followed by filtration to reduce the total feed volume to its initial volume
Diffusion test	A test for membrane integrity that involves measuring the rate of gas diffusion through a liquid-wetted membrane
Dynamic membrane	A layer of rejected solute and/or particulate material that forms at the surface of the membrane; this dynamic membrane can have higher solute rejection than the underlying UF or MF membrane
Extractables	Substances used in the manufacture of a membrane or filter that are extracted into the filtered solution during use of the filter

Feed	The solution that is fed to the membrane process
Filter	A porous medium used for removal of material of a specific size from a solution; also known as a membrane when the pores are very small
Filtrate	The solution that permeates a filter or membrane; also known as the permeate
Flow rate	The volumetric rate of flow of a solution; expressed in units of volume per time, e.g., L/min or gal/day
Flux	The rate of filtrate (or permeate) flow, divided by the membrane area; this parameter is used to characterize membranes; sometimes "pressure-normalized" flux is reported, which is simply the flux divided by the transmembrane pressure driving force
Fouling	The process whereby solute in the feed solution blinds or blocks membrane pores, thereby reducing the filtrate flux
Integrity testing	The process of determining whether a membrane or filter has the reported pore size and solute rejection characteristics; typically the bubble-point method, diffusion method, or water-intrusion integrity test is used to determine the integrity of a membrane
Log reduction value (LRV)	A measure of the efficiency of a sterilizing membrane to retain bacteria, defined as the logarithm (base 10) of the ratio of bacterial count in the feed stream to the bacterial count in the membrane permeate; typically the LRV is reported by membrane manufacturers for retention of *Brevundimonas diminuta*
Membrane	A finely porous medium that is used for removal of materials of a specific size or solutes from a solution; also known as a filter when the pore size and retention characteristics are well defined
Membrane area	The effective surface area of a membrane that is available for filtration; not the internal pore surface area, but rather the surface of one side of a membrane filter
Microfiltration (MF)	A membrane-based filtration process where the membrane pore size is in the range of 0.02–1.0 μm; microfiltration is typically used to separate cells from solutes (e.g., proteins) in the feed solution
Module	A device that combines a large amount of membrane into a compact volume, with easily accessible feed, permeate, and possibly retentate ports

Molecular weight cutoff (MWCO)	A characteristic of membranes or filters, which specifies the average molecular weight of solutes that do not permeate the membrane; this property is dependent on the type and geometry of the solute that is used to characterize a membrane's MWCO
Nanofiltration (NF)	A membrane-based filtration process where the membrane rejects solutes with molecular weight less than 1000 Da; nanofiltration is typically used to concentrate small organic molecules or divalent ions from water
Optimum TMP	The transmembrane pressure (TMP) that maximizes the overall throughput of a membrane-filtration process; the optimum TMP should be independently determined for each filtration
Permeability	The liquid flux through a membrane which has been normalized to the pressure driving force or TMP
Permeate	The solution that permeates a filter or membrane; also known as the filtrate
Plate-and-frame	A membrane-module geometry where membranes are sandwiched together one on top of another
Polyethersulfone	A polymer used for formation of many MF and UF membranes

REFERENCES

Belter, P. A., Cussler, E. L., and Hu, W.-S., 1988, *Bioseparations: Downstream Processing for Biotechnology*, Wiley, New York.

Blatt, W. F., Dravid, A., Michaels, A. S., and Nelsen, L., 1971, Solute polarization and cake formation in membrane ultrafilters: Causes, consequences, and control techniques, in: *Membrane Processes in Industry and Biomedicine* (M. Beir, ed.), Plenum Press, New York, pp. 65–68.

Brose, D. J., and Henricksen, G., 1994, A quantitative analysis of preservative adsorption on microfiltration membranes, *Pharm. Technol.* **18:**64–72.

Brose, D. J., and Waibel, P., 1996, Adsorption of proteins in commercial microfiltration capsules, *Pharm. Technol.* **20:**48–52.

Brose, D. J., Cates, S., and Hutchison, F. A., 1994, Studies on the scale-up of microfiltration membrane devices, *J. Parenteral Sci. Technol.* **48:**184–188.

Brown, S., and Fuller, A. C., 1993, Depyrogenation of pharmaceutical solutions using submicron and ultrafilters, *J. Parenteral Sci. Technol.* **47:**285–288.

Cheryan, M., 1986, *Ultrafiltration Handbook*, Technomic, Lancaster, PA.

Datar, R., Martin, J. M., and Manteuffel, R. L., 1992, Dynamics of protein recovery from process filtration systems using microporous membranes, *J. Parenteral Sci. Technol.* **46:**35–42.

Devereaux, N, and Hoare, M., 1986, Membrane separation of protein precipitates: Studies with crossflow in hollow fibers, *Biotechnol. Bioeng.* **28**(3):422–431.

Green, L., and Duwez, P., 1951, Fluid flow through porous metals, *J. Appl. Mech.* **18:**39–45.

Hermia, J., 1982, Constant pressure blocking filtration laws—Application to power-law non-Newtonian fluids, *Trans. IChemE* **60:**183–187.

Ho, W. S. W., and Sirkar, K. K., 1992, *Membrane Handbook*, Van Nostrand Reinhold, New York.

Hu, H., Camilleri, J., and Tamashiro, W., 1993, A new membrane for biopharmaceutical filtration, *Pharm. Technol.* **17:**30–44.

Johnston, P. R., 1992a, *Fluid Sterilization by Filtration: The Filter Integrity Test and Other Topics,* 1st ed., Interpharm Press, Buffalo Grove, IL.

Johnston, P. R., 1992b, Fluids flow through filter medium: A tutorial *Pharm. Technol.*, **16**(9): 136–146.

Jornitz, M. W., Brose, D. J., and Meltzer, T. H., 1998, Experimental evaluations of diffusive airflow integrity testing, *J. Pharm. Sci. Technol.* **52:**46–49.

Kesting, R. E., 1985, *Synthetic Polymeric Membranes: A Structural Perspective*, 2nd ed. J Wiley, New York.

Kim, M., Kojima, J., Saito, K., Furusaki, S., and Sugo, T., 1994, Reduction of nonselective adsorption of proteins by hydrophilization of microfiltration membranes by radiation-induced grafting, *Biotechnol. Prog.* **10:**114–120.

Kirsch, L. E., Riggin, R., Gearhart, D. A., Lefber, D. F., and Lytle, D. I., 1993, In-process protein degradation by exposure to trace amounts of sanitizing agents, *J. Parenteral. Sci. Technol.* **47:**155–160.

Levy, P., and Shehan, J., 1991, Performance characteristics of polysulfone and cellulose membranes for the ultrafiltration of biological process streams, *BioPharm*, **14:**24–33.

Loeb, S., 1981, The Loeb–Sourirajan membrane: How it came about, in: *Synthetic Membranes* (A. Turbak, ed.), American Chemical Society, Washington, DC, Vol. 1, pp. 1–9.

Marshall, A. D., Munro, P. A., and Tragardh, G., 1993, Effect of protein fouling in microfiltration and ultrafiltration on permeate flux, protein retention, and selectivity: A literature review, *Desalination* **91:**65–108.

Martin, J. M., and Manteuffel, R. L., 1988, Protein recovery from effluent of microporous membranes, *BioPharm* **1:**20–27.

McDonogh, R. M., Bauser, H., Stroh, N., and Chmiel, H., 1990, Concentration polarization and adsorption effects in cross-flow ultrafiltration of proteins, *Desalination* **79:**217–231.

McGregor, W. C., 1990, Advances in primary separations with membranes, in *Biologicals from Recombinant Microorganisms and Animal Cells: Production and Recovery* (M. D. White, S. Reuveny, and A. Shafferman eds.), VCH, Mannheim, Germany, pp. 457–463.

Meltzer, T. H., 1987, *Filtration in the Pharmaceutical Industry*, Marcel Dekker, New York.

Meltzer, T. H. and Jornitz, M. W., eds., 1998, *Filtration in the Biopharmaceutical Industry*, Marcel Dekker, New York.

Michaels, A. S., 1986, Forward, in: *Ultrafiltration Handbook*, by Munir Cheryan, Technomic, Lancaster, PA.

Palecek, S. P., and Zydney, A. L., 1994, Intermolecular electrostatic interactions and their effect on flux and protein deposition during protein filtration, *Biotechnol. Prog.* **10:**207–213.

Parenteral Drug Association, 1992, Industrial Perspective on Validation of Tangential Flow Filtration in Biopharmaceutical Applications, Technical Report No. 15, PDA, Bethesda, MD.

Parenteral Drug Association, 1998, Sterilizing Filtration of Liquids, Technical Report No. 26, PDA, Bethesda, MD.

Pitt, A. M., 1987, The nonspecific protein binding of polymeric microporous membranes, *J. Parenteral Sci. Technol.* **41:**110–113.

Porter, M. C., 1981, The effect of fluid management on membrane filtration, in: *Synthetic Membranes* (A. Turbak, ed.), Vol. 1, ACS Symp. Ser. 153, American Chemical Society, Washington, DC, pp. 407–448.

Reif, O. W., 1998, Extractables and compatibilities of filters, in: *Filtration in the Biopharmaceutical Industry*, (T. H. Meltzer and M. W. Jornitz, eds.), Marcel Dekker, New York, pp. 204–206.

Romero, C. A., and Davis, R. H., 1988, Global model of crossflow microfiltration based on hydrodynamic particle diffusion, *J. Membrane Sci.* **39:**157–185.

Sarry, C., and Sucker, H., 1992, Adsorption of proteins on microporous membrane filters: Part I, *Pharm. Technol.* **16:**72–82.

Sarry, C., and Sucker, H., 1993, Adsorption of proteins on microporous membrane filters: Part II, *Pharm. Technol.* **17:**60–70.

Soelkner, P., and Rupp, J., 1998, Cartridge filters, in: *Filtration in the Biopharmaceutical Industry*, (T. H. Meltzer, and M. W. Jornitz, eds.), Marcel Dekker, New York, pp. 147–156.

Strathman, H., 1986, Synthetic membranes and their preparation, in: *Synthetic Membranes: Science, Engineering, and Applications* (P. M. Bungay, H. K. Lonsdale, and M. N. de Pinho, eds.), Reidel, Boston, pp. 8–34.

Tarry, S. W., 1998, Filter housings for biotechnology and pharmaceutical applications, in: *Filtration in the Biopharmaceutical Industry* (T. H. Meltzer and M. W. Jornitz, eds.), Marcel Dekker, New York, pp. 435–451.

Truskey, G. A., Gabler, R., Dileo, A., and Manter, T., 1987, The effect of membrane filtration upon protein conformation, *J. Parent. Sci. Technol.* **41**(6):180–193.

U.S. Food and Drug Administration, 1987, *Guideline on General Principles of Process Validation*, U.S. FDA, Center for Drugs and Biologics and Center for Devices and Radiological Health, Rockville, MD.

U.S. Food and Drug Administration, 1991, *Guidance for Large Volume Parenterals*, U.S. FDA, Rockville, MD.

USP, 1995, *United States Pharmacopeia*, 23rd rev., United States Pharmacopeial Convention, Rockville, MD.

Vertenstein, M., and Ronis, D., 1984, Microscopic theory of membrane transport. II. Darcy–Brinkman flow, *J. Chem. Phys.* **80:**5754–5763.

Visvanathan, C., and Ben Aim, R., 1989, Studies on colloidal membrane fouling mechanisms in crossflow microfiltration, *J. Membrane Sci.* **45:**3–15.

Waibel, P. J., Jornitz, M., and Meltzer, T. H., 1996, Diffusive airflow integrity testing, *J. Pharm. Sci. Technol.* **50:**311–316.

Wolber, P., Dosmar, M., and Banks, J., 1988, Cell harvesting scale-up: Parallel-leaf crossflow microfiltration methods, *BioPharm* **1:**38–45.

6

Fundamentals of Freeze-Drying

Steven L. Nail, Shan Jiang,
Suchart Chongprasert, and Shawn A. Knopp

1. INTRODUCTION

The development of freeze-dried injectable pharmaceutical products has traditionally been a process of trial and error, both with respect to the composition of the formulation and the process conditions used during freeze-drying. Although this approach ultimately may result in an acceptable product, it is a time-consuming and labor-intensive process, and is unlikely to result in the highest quality product attainable or in a freeze-dry process which is optimized.

The increased importance of freeze-drying as a pharmaceutical unit operation, in large part caused by widespread development of protein therapeutic agents, has spurred interest in an analytical approach to formulation and process development based on a sound understanding of the physical chemistry of freezing and freeze-drying, the material science of viscoelastic systems, and fundamentals of heat and mass transfer. A growing

Steven L. Nail • School of Pharmacy, Purdue University, West Lafayette, Indiana 47907. *Shan Jiang* • Bayer Corporation, Clayton, North Carolina 27606. *Suchart Chongprasert* • Drug Control Division, Food and Drug Administration, Muang, Nontaburi 11000, Thailand. *Shawn A. Knopp* • Genetics Institute, Andover, Massachusetts 01810.

Development and Manufacture of Protein Pharmaceuticals, edited by Nail and Akers. Kluwer Academic/Plenum Publishers, New York, 2002.

body of scientific literature has demonstrated that the scientific approach can result in improved product quality with minimum trial and error empiricism.

In addition to the need for rapid development of new pharmaceutical products, growing scrutiny of health care costs has focused more attention on minimizing cost of goods by pharmaceutical manufacturers. Among pharmaceutical unit operations, drying operations are, in general, the most expensive. Of these, freeze-drying is generally the most expensive, not only in operating cost, but also in cost of equipment. Development of overly conservative freeze-drying conditions results in long cycle times, added cost of production, and unnecessary consumption of drying capacity. Optimization of the freeze-drying cycle—minimizing drying time without a measurable adverse effect on product quality—should be integrated into the product development plan.

Freeze-drying of proteins often presents a greater challenge to the formulation scientist than that of more traditional, low molecular weight drug compounds, for several reasons. First, proteins are, in general, more susceptible to damage by the stresses involved in freezing and freeze-drying than low molecular weight drugs. Successful freeze-drying commonly requires the use of solutes intended to stabilize the protein against damage by these stresses. Second, the conventional wisdom that "the dryer, the better" with respect to both acute damage caused by freeze-drying and long-term stability does not apply to some proteins. Avoiding overdrying places added importance on careful monitoring of the process, particularly when maintenance of product quality demands keeping the dried product within a "window" of residual moisture activity. Third, the stability of protein formulations as freeze-dried solids is more uncertain than that of most low molecular weight drugs. Freeze-dried protein pharmaceutical products usually require refrigerated storage, and stability of the freeze-dried protein is more affected by seemingly subtle changes in the formulation and in processing conditions than low molecular weight drugs. Finally, the frequent high cost of bulk active material provides added incentive to maximize the yield of pharmaceutically acceptable product.

The purpose of this chapter is to review the basic physical chemistry and material science of freeze-drying, with particular reference to protein formulations, and is geared toward readers unfamiliar with the subject. We also attempt to highlight areas in need of further research. The mechanism of stabilization of proteins against freezing- and drying-induced loss of activity by cosolutes has been thoroughly reviewed by Carpenter *et al.* (1994) and Arakawa *et al.* (1993), and this body of information will not be covered in detail.

2. OVERVIEW OF THE FREEZE-DRYING PROCESS

A schematic diagram showing the major components of a modern, production-scale pharmaceutical freeze-dryer is shown in Fig. 1. The chamber contains hollow shelves through which a heat transfer fluid is circulated (usually silicone oil). This allows control of the shelf temperature from about −50°C to perhaps +50°C. A condenser, operating at temperatures from about −60°C to −75°C, collects water vapor from the product. The condenser is usually located externally to the chamber, as shown in Fig. 1, or it may be of an internal design, consisting of plates located along the side walls of the chamber. Vacuum pumps for evacuating the system are usually of the rotary, oil-sealed design, and are usually capable of maintaining a system pressure of 50 mTorr or less. Redundant compressors for the refrigeration system as well as redundant vacuum pumps are used on production freeze-dryers, and are easily justified based on the high value of the product. A valve is located between the chamber and the condenser, and the stack of shelves is hydraulically movable to allow the stack to be compressed in order to fully insert stoppers into vials after freeze-drying.

Operationally, freeze-drying of a final dosage form usually consists of filling glass vials with an aqueous solution of the solutes to be freeze-dried,

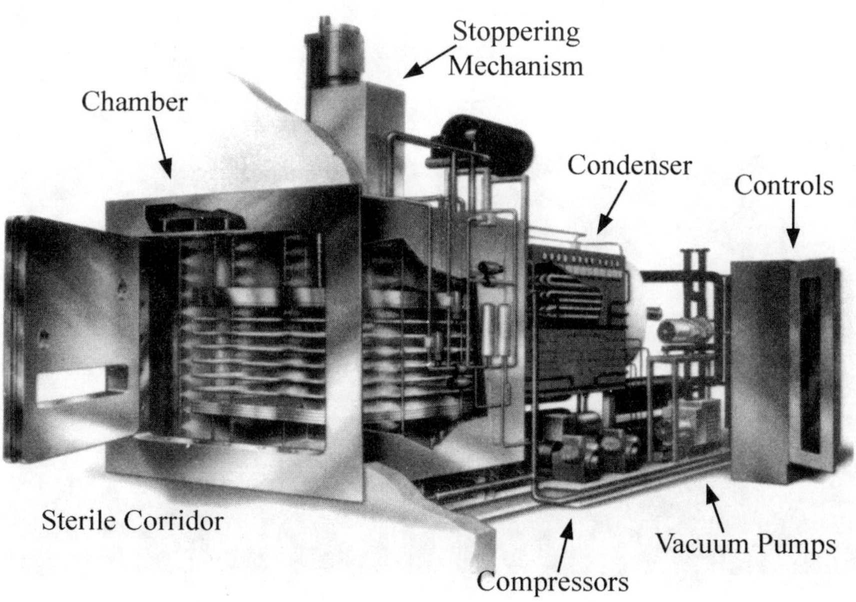

Figure 1. Schematic of a production-scale pharmaceutical freeze-dryer. Photograph courtesy of BOC Edwards Calumatic, Tonawanda, NY.

partially inserting a special rubber stopper (called a lyostopper, see Fig. 2), which allows water vapor to flow through slots in the stopper when in the partially inserted position, and transferring the vials to the shelves of the freeze-dryer.

Temperature sensors are placed in a few vials to monitor product temperature during the freeze-drying process (see Fig. 2). The shelves are then cooled to a temperature in the range of $-40°$ to $-50°C$. The vials are held long enough to approach thermal equilibrium with the shelves, usually for a minimum of 4 hr. Pressure in the freeze-dryer is then reduced to a level which is less than the vapor pressure of ice at the temperature of the product (see Table I). For example, for a product temperature of $-40°C$, the pressure in the system would be reduced to less than 0.096 mm Hg (96 μm Hg). In freeze-drying, removal of ice takes place by a process of sublimation, or a phase change from a solid to a vapor without passing through the liquid state. Ideally, this results in retention of the microstructure of the frozen material and a high specific surface area of the freeze-dried solid. This promotes rapid, complete rehydration of the solid.

In order for sublimation to take place, energy must be provided in a quantity equal to the heat of sublimation of ice ΔH_s, which is approximately 670 cal/g. This is accomplished by heating the transfer fluid and warming the

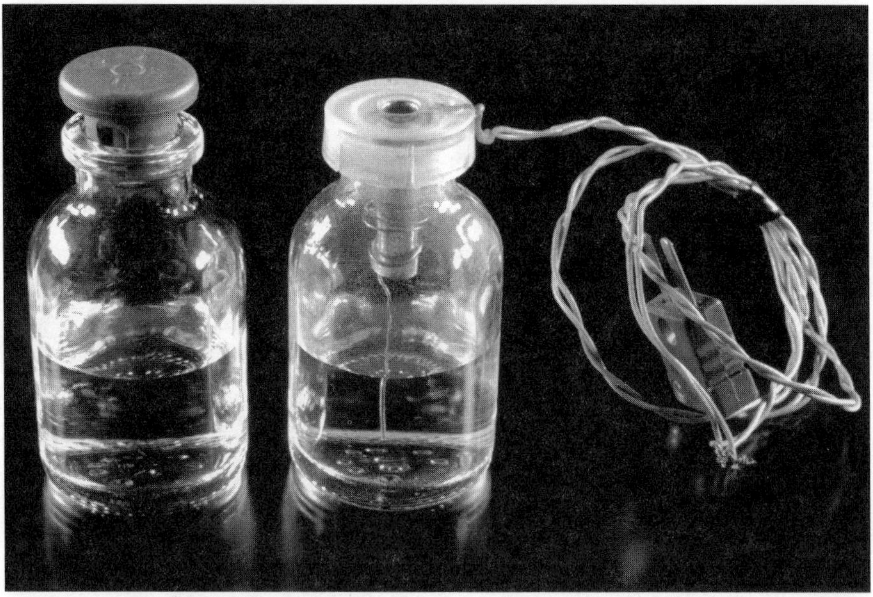

Figure 2. Filled vials with partially inserted lyostopper and a product temperature thermocouple.

Table I
Vapor Pressure of Ice

Temperature (°C)	P (mm Hg)	Temperature (°C)	P (mm Hg)
−80	0.0004	−30	0.285
−70	0.0019	−25	0.476
−60	0.0080	−20	0.776
−50	0.0295	−15	1.241
−45	0.054	−10	1.950
−40	0.096	−5	3.013
−35	0.164	0	4.579

shelves to a temperature high enough to effect sublimation, but not so high as to melt the frozen material in contact with the bottom of the vial or to collapse the partially dried product. The partial pressure of water vapor in the chamber is maintained at a low level by the condenser, and water vapor is removed from the product primarily by a process of *bulk flow* from a region of relatively high pressure (the sublimation front) to low pressure (the condenser). This phase of the process, called *primary drying*, is characterized by a visible receding boundary from the top of the frozen layer. When the ice has sublimed, the heat of sublimation is no longer needed, and the product temperature usually increases sharply toward the shelf temperature.

In general, not all of the water initially present in the product is converted to ice during freezing. The quantity of unfrozen water is a function primarily of the composition of the formulation and, to a lesser extent, the thermal history of the freezing process. Removal of this "unfrozen" water, which may be 20% or more of the weight of dry solids, is called *secondary drying*. During secondary drying, the shelf temperature is usually increased, since ice is no longer present. In contrast to primary drying, where water vapor is removed by bulk flow, water vapor removal during secondary drying is largely by a process of diffusion, or flow by molecular motion from a region of high concentration to a region of lower concentration. Because secondary drying is generally slow relative to primary drying, it often represents a larger total portion of drying time than primary drying, even though the quantity of water removed is less. At the end of secondary drying, the shelf stack is compressed together, causing insertion of the lyostoppers to the fully stoppered position. This may be done under full or partial vacuum or at atmospheric pressure. For proteins that are sensitive to oxidation in the solid state, the composition of gas in the headspace of the vial may have a critical impact on product stability (Townsend *et al.*, 1990).

Since the driving force for freeze-drying is the vapor pressure of ice, it is important from the standpoint of process efficiency to keep the product temperature as high as practical during primary drying. However, the product

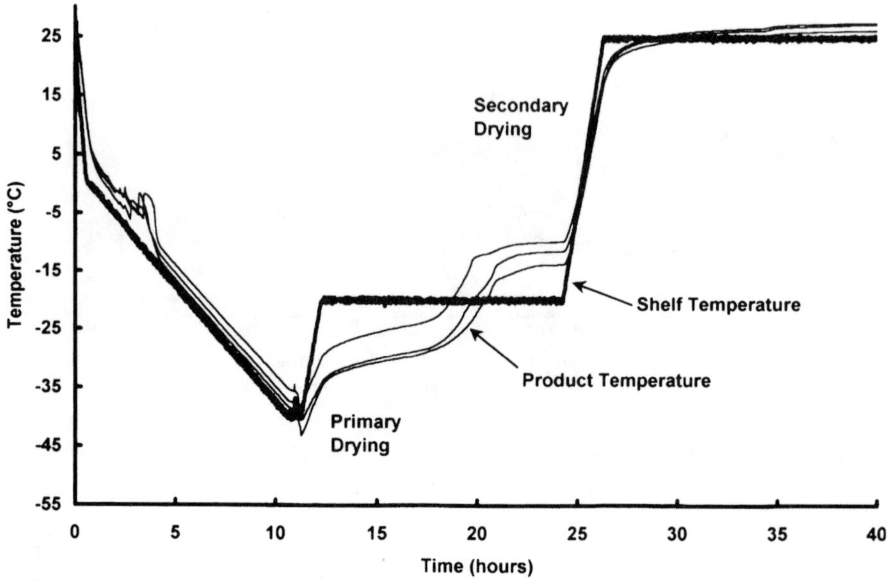

Figure 3. Plot of process variables versus time during the freeze-dry process, including shelf temperature, product temperature, and chamber pressure.

temperature must be maintained below the maximum allowable product temperature, which is either a eutectic melting temperature or a collapse temperature (see discussion below). Monitoring of product temperature is important for establishing optimum process conditions. A typical plot of process variables is shown in Fig. 3, where product temperature increases gradually during primary drying, followed by a more rapid increase at the beginning of secondary drying.

3. THE FREEZING PROCESS

An understanding of the physical chemistry of freezing is essential for understanding how both formulation composition and process conditions influence the quality of the freeze-dried product. A schematic diagram of the physical events occurring during freezing is shown in Fig. 4. Supercooling, or retention of the liquid state below the equilibrium freezing point of the solution, always occurs to some extent: It is not uncommon for aqueous solutions to supercool by 12°C or more. This is seen in Fig. 3 as a sudden "jump" in temperature during the freezing process. The effect of supercooling is to limit the ability to control the freezing rate by manipulation of shelf temperature, because the greater the degree of supercooling, the faster is the effective rate of freezing once ice crystals nucleate.

As indicated in Fig. 4, ice may not be the first component of the solution to crystallize. As the temperature is decreased, the equilibrium solubility of one or more solutes may be exceeded, allowing nucleation and crystal growth. An example of this behavior is illustrated by the drug pentamidine isethionate, where, depending on the thermal history of freezing, either drug or ice may nucleate first (Chongprasert *et al.*, 1997). Process validation studies should include determination of whether the thermal history of freezing (for example, placing vials on precooled shelves versus slowly decreasing the shelf temperature during freezing) has a measurable impact on product quality, including such factors as activity, residual moisture, appearance of solids, and reconstitution time.

As ice crystals grow in the freezing system, the solutes are concentrated. Increased ionic strength accompanying the freezing process is often an important consideration in freeze-drying of proteins as well as other biological materials, since high ionic strength may promote denaturation of proteins as well as osmotic dehydration of microorganisms. In addition, rates of some chemical reactions, particularly second-order reactions, may be accelerated by freezing through this freeze-concentration effect. Examples include reduction of potassium ferricyanide by potassium cyanide (Hatley *et al.*, 1986a), oxidation of ascorbic acid (Hatley *et al.*, 1986b), and

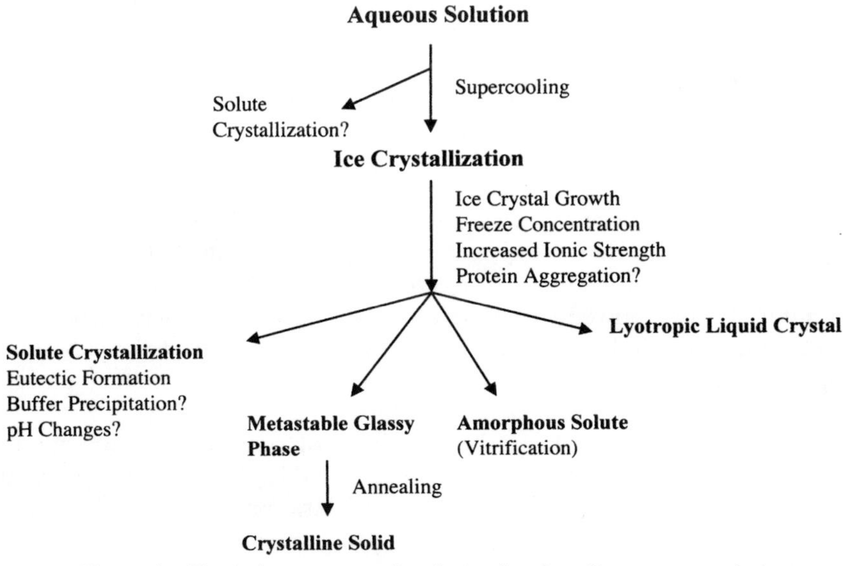

Figure 4. Physical events occurring during freezing of an aqueous solution.

polypeptide synthesis (Schuster *et al.*, 1991). The kinetics of reactions in frozen systems has been reviewed by Pincock and Kiovsky (1966).

3.1. Solute Crystallization

3.1.1. EUTECTIC SYSTEMS AND THE PHASE DIAGRAM

The fate of the solute in the freeze-concentrated solution is a key concept in understanding the material science of freeze-drying. As indicated in Fig. 4, there are several possibilities, the simplest of which is crystallization of solute from freeze-concentrated solution to form a simple eutectic mixture. A pure binary eutectic system, illustrated in Fig. 5 for the system sodium chloride/water, is not often encountered in practice, but this system is useful for a conceptual understanding of the relevance of eutectic mixtures to freeze-drying. The line *ab* represents the freezing point depression curve of water in the presence of sodium chloride, and the line *bc* represents the solubility of sodium chloride in water. The intersection of these lines is the eutectic melting temperature, which for sodium chloride/ice is −21.5°C, and the eutectic composition is about 23.3% (w/w) sodium chloride. The freezing of a 5% solution of sodium chloride in water is represented by the line *defgh* (called an *isopleth*). At room temperature (point *d*), the system is entirely liquid. As the solution cools, ice appears at point *e* (in the absence of supercooling). As the system cools, ice continues to crystallize and the remaining solution becomes more concentrated in sodium chloride. At point *f*, two phases are present, ice and a freeze concentrated solution of sodium chloride in water. The freeze concentrate has the composition given by point *i*, which is in equilibrium with ice. At point *g*, the freeze concentrate is saturated with respect to sodium chloride, and (again, in the absence of supercooling) sodium chloride begins to precipitate. It is only below the eutectic temperature that the system is completely solidified. The relevance of the eutectic temperature to freeze-drying is that it represents the maximum allowable product temperature during primary drying, because eutectic melting would introduce liquid water and destroy the desirable properties of a freeze-dried solid. Melting of the frozen system during primary drying is sometimes called *meltback*.

The phase diagram illustrates the degree of concentration of sodium chloride caused by the freezing process. For example, if a solution of normal saline (0.9% w/v, or 0.15 *N*) is frozen, the concentration reaches 23.3% before sodium chloride crystallizes and forms a eutectic mixture, or just over 4.0 *N*. This increase in ionic strength associated with freezing of salt solutions is a

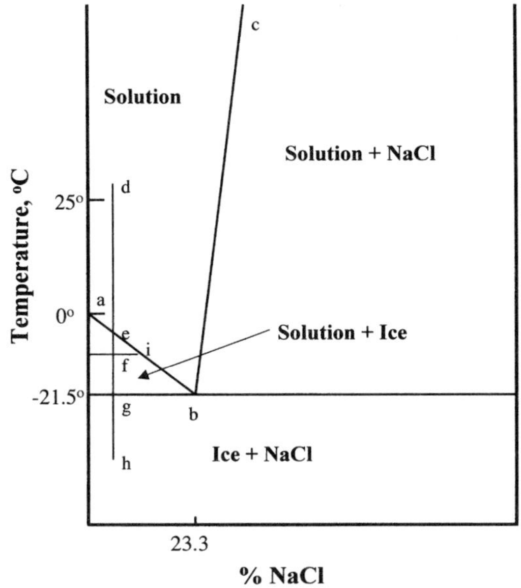

Figure 5. Phase diagram of sodium chloride/water.

contributing factor in osmotic dehydration of cells during freezing and has been thought to contribute to protein denaturation during freezing.

The phase diagram of glycine/water, illustrated in Fig. 6, is more representative of the freezing characteristics of aqueous solutions of organic compounds that crystallize from a freezing system. The eutectic melting temperature of the glycine/ice eutectic mixture, −3.6°C, is much closer to the melting temperature of ice than is that of the sodium chloride/ice eutectic mixture, and the amount of glycine present in the eutectic mixture is lower, about 12%. In general, the lower the solubility of a given solute in water, the higher is the eutectic melting temperature. Figure 6 is probably an oversimplification, however, since glycine is known to exist in three polymorphic forms. Since different polymorphs of a compound have different physical properties, it might be expected that the solubility curve would be different. This, in turn, would change the eutectic melting temperature. This aspect of the physical chemistry of freeze-drying has not been investigated in depth, and more work is needed to better understand the impact of polymorphism on the physical chemistry of freeze-drying.

3.1.2. pH SHIFTS DURING FREEZING

Crystallization of phosphate buffers during freezing is a special case of eutectic mixture formation during freezing, and is worthy of mention not

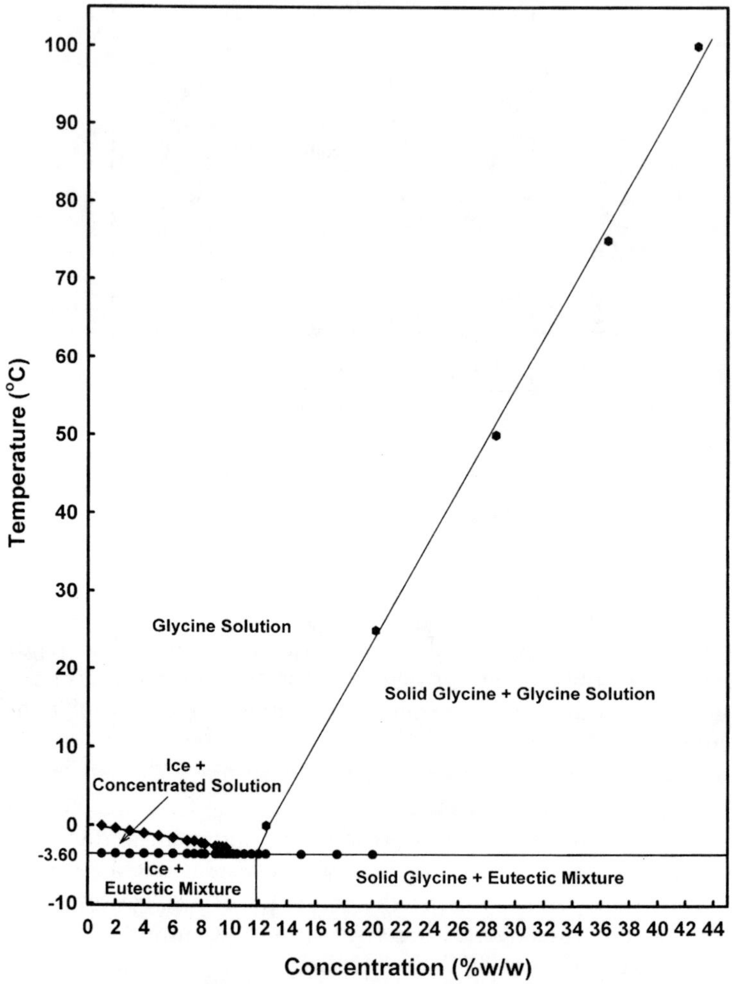

Figure 6. Phase diagram of glycine/water.

only because phosphate salts are the most common buffers used in protein formulations, but also because the crystallization process can cause significant shifts in pH during freezing. Such shifts are an illustration of LeChâtelier's principle, which states that, when a disturbance is imposed on a system at equilibrium, the equilibrium shifts so as to compensate for the disturbance. If a component precipitates from solution, the equilibrium shifts so as to compensate, as follows:

$$H_2PO_4^- + H_2O \leftrightarrow HPO_4^{2-} + H_3O^+$$

The effect of freezing on the pH of sodium and potassium phosphate buffers has been extensively studied by van den Berg and co-workers (van den Berg, 1959; van den Berg and Rose, 1959). These investigators used a technique where precipitated solids were removed from unfrozen solution by a sintered glass filter, and the pH was measured above 0°C. The pH was found to decrease with temperature during the precipitation of ice alone, as would be expected simply from freeze concentration. The pH shift was shown to depend on the salt used. The monobasic salt of potassium phosphate, being less soluble than the dibasic salt, tends to precipitate during freezing and causes an increase in pH. For sodium phosphate, the dibasic salt is less soluble than the monobasic salt, and the opposite effect is observed. Changes in pH of as much as three units toward the acid side have been reported for the sodium phosphate system. Changes toward more alkaline pH for the potassium phosphate system are less pronounced. Eutectic temperatures of the phosphate buffer salts are listed in Table II, including ternary eutectic mixtures consisting of ice and monobasic and dibasic salts. It is important to remember, however, that a eutectic mixture is only formed as a result of crystallization of one or more solutes, and the presence of a phosphate buffer in a formulation does not mean that the pH will shift during freezing, other than the small (about 0.5 unit) shift which would result from freeze concentration of the solution.

Later work by van den Berg reported development of a calomel and glass electrode modified to make it suitable for use at low temperatures, allowing direct measurement of pH in frozen systems (van den Berg, 1960). This was accomplished by use of reference solutions containing roughly equal weight proportions of water and ethylene glycol or glycerol, saturated with potassium chloride. A glass membrane with a high electrical conductivity was also used to decrease the electrode response time (electrodes suitable for this type of measurement are available from Mettler-Toledo, Inc., Wilmington, MA). This technique was used by

Table II
Eutectic Temperatures of Phosphate Buffers[a]

Solute	T_e (°C)
Na_2HPO_4	−0.5
NaH_2PO_4	−9.7
NaH_2PO_4–Na_2HPO_4	−9.9[b]
K_2HPO_4	−13.7
KH_2PO_4	−2.7
KH_2PO_4–K_2HPO_4	−16.7[b]

[a] From Murase and Franks (1989).
[b] Ternary eutectic.

Gomez *et al.* (2001) in a further study of crystallization during freezing of aqueous solutions of sodium phosphate buffers. This work extended previous work by Larsen (1973) by examining the influences of initial buffer solution pH and concentration on subsequent pH changes during freezing, as well as the influence of other species in solution on buffer salt crystallization. Gomez *et al.* (2001) reported that the pH changes associated with crystallization of a sodium phosphate buffer solution initially at pH 7.4 are directly related to the initial concentration of buffer in the range of 8–100 m*M*. Further, the lower the initial pH of the buffer, the higher is the observed pH at −10°C. These investigators further reported that addition of NaCl increases the ion product of dibasic sodium phosphate, thereby leading to larger pH changes. Solutes such as sucrose and mannitol inhibited crystallization of buffer species, resulting in smaller pH shifts upon freezing. The presence of sucrose and mannitol at concentrations above 3 mol per mole of dibasic sodium phosphate completely prevented salt crystallization. In this case, the pH change upon freezing was only 0.5 unit, which was attributed to the effect of freeze concentration.

Murase and Franks (1989) used differential scanning calorimetry to study freezing of phosphate buffer solutions. This study differed from previous work in that precipitation was studied under conditions far from equilibrium using small volumes of solution. Precipitation was found to be both concentration- and freezing-rate-dependent, with fast freezing rates inhibiting crystallization. Whereas dibasic sodium phosphate was found to precipitate readily, monobasic sodium phosphate did not precipitate at all under the experimental conditions examined. Monobasic potassium phosphate was found to precipitate readily. Dibasic potassium phosphate also precipitated readily and had the lowest eutectic point of any of the phosphate salts. Ternary eutectics were not observed in any of the systems studied.

Other pharmaceutically relevant buffer systems have not been as well characterized as phosphate with respect to pH changes accompanying freezing. Larsen (1973) reported that acetate, citrate, glycine, and Tris show only small pH shifts upon freezing.

Orii and Morita (1977) reported the use of pH indicators such as phthaleins, sulfonaphthaleins, and universal pH indicators to study pH changes with freezing. Although this method is subject to errors caused by changes in p*K* of indicators with temperature, as well as effects of increased ionic strength on dissociation constant, it can provide a convenient method of estimation of pH changes in a semiquantitative way. Of the 30 buffers they studied, Orii and Morita reported that over half exhibited pH changes with freezing.

There are very few published studies which examine the practical significance of pH shifts on the integrity of pharmaceutically relevant proteins, primarily because very few studies of the relationship between formulation or processing conditions and protein integrity after freeze-drying attempt to measure pH changes with freezing. The most direct evidence of a pH shift-induced loss of protein activity during freeze-drying was reported by Anchordoquy and Carpenter (1996) for lactate dehydrogenase (LDH) in a sodium phosphate buffer system. Dissociation of LDH was attributed to a large decrease in pH in the frozen state, from 7.5 to 4.5. Addition of polymers such as bovine serum albumin (BSA) and polyvinylpyrrolidone (PVP) to the solution inhibited crystallization of buffers, thereby attenuating the pH shift and stabilizing LDH against dissociation. Stabilization could not be attributed solely to prevention of buffer crystallization, however, since stabilization of the protein via Timasheff's preferential exclusion hypothesis (Carpenter *et al.*, 1994) would also be expected.

Williams-Smith *et al.* (1977) detected pH changes in phosphate, pyrophosphate, and Tris buffers containing three different metalloenzymes, using visual pH indicators, and measured pH-dependent changes in the electron paramagnetic resonance spectra of these proteins. Zwitterionic buffers resulted in relatively small pH changes upon freezing. Lam *et al.* (1996) used Fourier-transform infrared (FTIR) spectroscopy and circular dichroism to study the influence of succinate and glycolate buffers on stability of freeze-dried interferon γ. It was hypothesized, based on electrical resistance measurements, that the monosodium form of succinate crystallized during freezing, resulting in an effective pH in the freeze concentrate of 1–2 pH units lower than the original solution pH of 5.0 (the pH during freezing was not actually measured). This appeared to promote unfolding of the protein and cause loss of bioactivity. Use of glycolate buffer instead of succinate minimized the pH shift during freezing. The freezing-associated conformational change was also reflected in the solid-state stability of the lyophilized protein, with the glycolate formulation showing superior stability after 4 weeks of storage at 25°C.

Although it was not directly related to buffer-induced pH changes accompanying freezing, Costantino *et al.* (1997b) reported that lyophilized organic compounds containing protein functional groups (amino, carboxylic, and phenolic) exhibit "pH memory", that is, the ionization state of the solid, as reflected by the FTIR spectrum, is similar to that of the aqueous solution from which the compound was freeze-dried.

Tarelli and Wheeler (1994) reported that freeze-drying from aqueous phosphate buffers can promote O-phosphorylation of hydroxyl groups on serine and threonine as well as saccharides and alditols. Primary hydroxyl groups are more reactive than secondary hydroxyl groups, and the reaction occurs more readily at acid pH. Residual moisture in the dried solid also affects this reaction, with the most rapid reaction occuring in the range of 3–

10% residual moisture. The potential for pH shifts with freezing of phosphate buffers was not mentioned in this study.

Sarciaux *et al.* (1999) reported substantially higher levels of insoluble aggregates in formulations of bovine immunoglobulin G (IgG)-containing sodium phosphate as compared with potassium phosphate at the same concentration. This effect was not due to a pH shift upon freezing, however. It was hypothesized that the difference in degree of aggregation observed for the two buffer systems was attributable to Hofmeister series effects.

Regardless of the relative scarcity of published data linking pH changes with freezing to loss of protein integrity, the potential for significant changes exists, particularly with phosphate buffers. Given the impact of pH on many mechanisms for loss of protein activity, both chemical and physical, the formulation scientist who is alert to the potential effects of these changes is wise.

The list of buffers used in approved products is relatively small and includes, in addition to sodium phosphate, citrate (p*K* of 3.15, 4.8, and 6.4), Tris-hydroxymethylaminomethane (THAM, p*K* 8.1), succinate (p*K* of 4.2 and 5.6), histidine (p*K* 1.8, 6.0, and 9.0), glycine (p*K* 2.35 and 9.8), and arginine (p*K* 2.18 and 9.1). It should be noted that glycine and arginine, although often listed as buffers, would only be effective well above or well below neutral pH. The physical state of L-histidine after freeze-drying and long-term storage was studied by Osteberg and Wadsten (1999). The tendency of L-histidine to crystallize from frozen solutions was dependent upon the pH of the formulation, and was found to be maximum at pH 6.0, corresponding to the p*K* of the imidazoline function. Histidine was also reported to crystallize from the amorphous solid state upon exposure to water vapor.

As a general formulation guideline, it is good practice to keep the concentration of buffers to a minimum. This minimizes the chance of precipitation during freezing, thereby minimizing the potential for pH shifts.

3.2. Vitrification during Freezing

Many solutes do not crystallize from a freeze-concentrated aqueous solution. In this case, the freezing behavior is illustrated by Fig. 7, which describes sucrose, a common lyoprotective solute, which remains amorphous during freezing and freeze-drying. This type of representation was first suggested by MacKenzie (1977) and termed a *supplemented* phase diagram, whereas other investigators have used the term *state* diagram (Blond *et al.*, 1997). The freezing curve and the solubility curve are the same as presented in Fig. 5 for sodium chloride/water, and the composition of the freeze concentrate is given by the freezing curve; however, sucrose is kinetically inhibited from crystallizing at T_{eut}. Instead, the composition of the freeze concentrate continues to be represented by the dashed line, which represents a supercooled liquid. Ice crystallization continues, and the freeze-concentrated system becomes more viscous, until ice crystal growth (on a practical time scale) ceases and the degree of freeze concentration of the solute reaches a maximum. The glass transition of the maximally freeze-concentrated solution is designated as T_g'. Note that T_g' is a point on the *glass transition curve*, which represents a reversible change of state between a viscous liquid and a rigid, glassy system. The viscosity of this freeze concentrate changes by three or four orders of magnitude over a temperature range of a few degrees in the region of the glass transition. The point T_g represents the glass transition of amorphous sucrose (about 65°C), and the intersection of the glass transition curve with the y-axis represents the glass transition temperature of amorphous solid water, at about −135°C. Extremely fast freezing rates, on the order of 10^5 K/min, are required to form solid amorphous water when starting with pure water (Johari *et al.*, 1987).

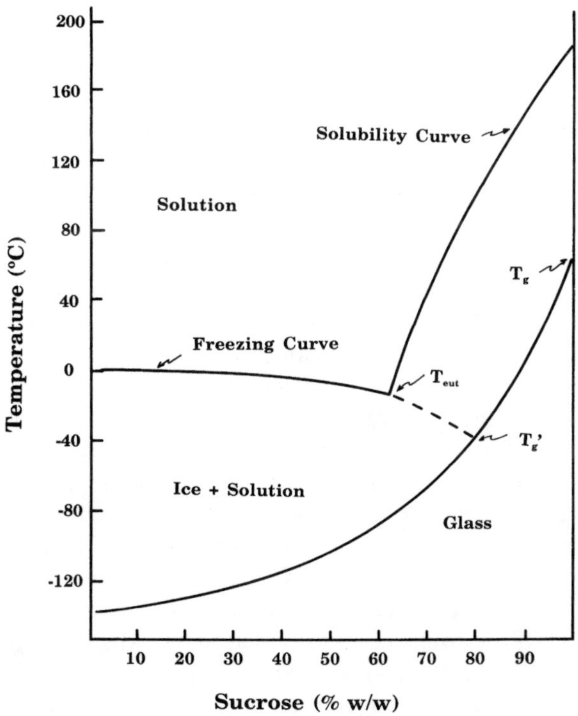

Figure 7. State diagram of sucrose/water.

The glass transition curve in Fig. 7 illustrates that only a small amount of water in amorphous sucrose causes a sharp drop in the glass transition temperature of the solid. Thus, water is acting as a *plasticizer*. The point designated as T_g', the glass transition temperature of the maximally freeze-concentrated solute, corresponds to about 20% water/80% sucrose. This glass transition is generally accepted as the physical basis for *collapse* in freeze-drying. Low-temperature thermal analysis of the sucrose/water system and the uncertainty concerning the relationship between T_g' and collapse are discussed below. Collapse is caused by viscous flow of the freeze-concentrated solute on the time scale of freeze-drying, with resultant loss of the microstructure that was established by freezing, decreased surface area of the freeze-dried solid, and loss of pharmaceutical elegance. Thus, the onset of collapse represents the maximum allowable product temperature during freeze-drying of systems in which the solute(s) remain amorphous during freezing.

3.3. Differences in Freeze-Drying Behavior between Crystalline and Amorphous Solutes

If the solute crystallizes completely during freezing, the microstructure of the system is illustrated by the sketch in Fig. 8, where the interstitial material between the ice crystals consists of an intimate mixture of small crystals of ice and solute. For an amorphous solute, the interstitial material consists of a solid solution of solute and unfrozen water. For a eutectic-forming system, the upper product temperature limit during primary drying is the eutectic melting temperature, whereas for an amorphous system the upper temperature limit is the collapse temperature. Although eutectic melting temperatures of ice/inorganic salt mixtures can be quite low (the eutectic melting temperature of calcium chloride/water is about −52°C), eutectic melting temperatures of ice and most organic compounds are in the range of just below 0°C to perhaps −12°C. These relatively high melting temperatures allow maintenance of high product temperatures during primary drying, resulting in a more efficient drying process. Collapse temperatures are generally much lower than eutectic melting temperatures. Collapse temperatures below −40°C are not uncommon, particularly for freeze-dried protein formulations containing sugars as protective additives. These formulations often contain added salts, which further decrease the collapse temperature (Her *et al.*, 1995). Given that most production-scale freeze-dryers are unable to consistently maintain product temperatures below about −40°C, freeze-drying is not feasible for all formulations. Even

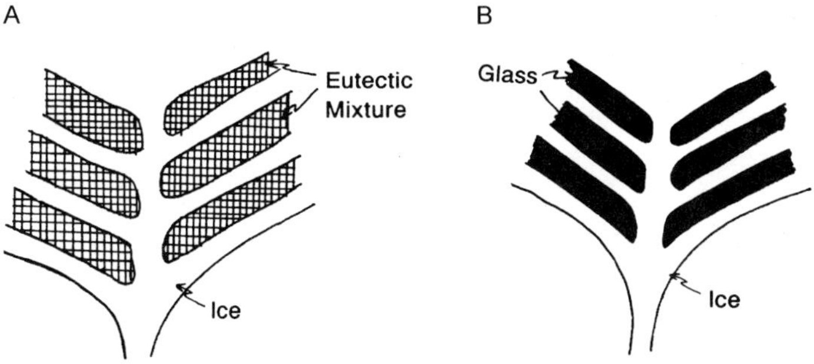

Figure 8. Microstructure of frozen system when (A) solute crystallizes versus (B) when the solute remains amorphous upon freezing.

if the product temperature can be controlled at temperatures below −40°C, the vapor pressure of ice at such low temperatures, and therefore the driving force for freeze drying, are relatively low. Because the vapor pressure of ice at −40°C is only 0.096 Torr, the system pressure must be kept well below this pressure, which can be a significant demand on the vacuum system. As a general formulation guideline, formulations with collapse temperatures of −40°C or lower should be avoided if possible.

As illustrated in Fig. 8, nearly all water is present as ice (either pre-eutectic or eutectic) when the solute crystallizes. As the sublimation front moves through the frozen system, all of the ice can be removed by a process of *bulk flow*, or transport of water vapor from a region of high pressure to lower pressure, through the porous bed of dried solid. There is essentially no secondary drying. For the sucrose/water system, representing an amorphous solute, there is 20% water associated with sucrose, which must be removed during secondary drying. This water must be removed by a process of *diffusion*, or flow by molecular motion from a region of high concentration to a region of lower concentration. This combination of a significant amount of water to be removed by secondary drying and the slower transport mechanism often means that secondary drying is a significant part of the total drying time needed.

In addition to these influences on the drying process itself, the physical state of a drug as a freeze-dried solid may be important with respect to solid-state stability of the compound. It is well known that freeze-drying of some cephalosporins is not feasible because the drug does not crystallize during freeze-drying, and the amorphous form of the drug is not sufficiently stable, despite being thoroughly dried, to support a reasonable shelf life of a freeze-dried product at room temperature (Pikal *et al.*, 1978).

Of course, it is common (and often desirable) to have both amorphous and crystalline phases present in a freeze-dried formulation. This is particularly relevant to freeze-dried proteins, where the lyoprotectant is present in the amorphous state and another component, such as glycine, is present as a crystalline solid in order to impart mechanical integrity and pharmaceutical elegance to the lyophilized solid.

3.4. Other Types of Freezing Behavior

3.4.1. METASTABLE GLASS FORMATION

In addition to eutectic crystallization and glass formation (vitrification), other types of freezing behavior may be observed (see Fig. 4). A metastable glass may form, which, with subsequent heating, undergoes crystallization. The most common example of metastable glass formation is mannitol, which crystallizes when heated above its T_g' (see discussion of thermal analysis below). This type of behavior is the basis for *annealing* during freeze-drying (see below), which refers to warming the product after an initial freezing step, generally to a temperature above T_g', but below the onset of melting, holding for a period of perhaps 1–4 hr, then cooling the material again before starting drying.

Mannitol is one of the most commonly used excipients in freeze-dried formulations; however, the formulation scientist should be aware of problems associated with the use of mannitol. It is well known that mannitol is associated with vial breakage during freeze-drying. This appears to be related to crystallization of mannitol from the metastable amorphous state. Vial breakage associated with mannitol can be minimized by keeping the depth of fill in a vial to no more than about 30% of the overflow capacity of the vial. The thermal expansion coefficient of the glass and the thermal history of freezing may also be significant factors affecting vial breakage.

3.4.2. LYOTROPIC LIQUID CRYSTAL FORMATION

States of matter which have degrees of order intermediate between amorphous and crystalline are called liquid crystals. Liquid crystals are broadly categorized as *thermotropic*, which are formed by heating, and *lyotropic*, which are formed by addition of solvent to a solid. Compounds which form lyotropic liquid crystals are surface-active, and the liquid crystal represents a more ordered structure than a micelle. These higher ordered structures are a result of freeze concentration, and may be either lamellar or rod-shaped.

There have been few reports of lyotropic liquid crystal formation in aqueous solutions of drugs, and even fewer which are relevant to freeze-drying. Powell *et al.* (1994) reported peptide liquid crystal formation by the luteinizing-hormone-releasing hormone deterelix and the effect of added salts on thermodynamic stability of the liquid crystal phase. Vadas *et al.* (1991) reported that a leukotriene D_4 receptor antagonist forms lyotropic liquid crystalline phases when lyophilized from aqueous solution. Bogardus (1982) studied the phase equilibria of nafcillin sodium–water and reported a lamellar mesophase in aqueous solutions containing more than 55% nafcillin sodium. Milton and Nail (1996) extended this work by characterizing the low-temperature differential scanning calorimetry (DSC) thermogram of frozen aqueous solutions of nafcillin as well as the freeze-dried solid. Freeze-drying of frozen systems containing lyotropic mesophases appears to result in a unique X-ray diffractogram consisting of a single sharp peak at low angle (less than about 5° 2θ) in addition to the "halo" characteristic of amorphous solids. Herman *et al.* (1994) reported a similar X-ray powder diffraction pattern in methylprednisolone sodium succinate. The influence of liquid crystal formation during freezing on critical quality attributes of freeze-dried products is a subject which remains largely unexplored.

3.4.3. LIQUID–LIQUID PHASE SEPARATION DURING FREEZING

Crystallization of one or more solutes during freezing is the most extreme form of phase separation. However, another type of phase separation is possible: liquid–liquid phase separation into phases which are enriched in one component. This type of behavior has been reported for aqueous solutions of polymers such as poly(ethylene glycol)/dextran, PVP/dextran, and Ficol/dextran as well as polymer/phosphate buffer systems such as poly(ethylene glycol) (PEG)/potassium phosphate and PVP/sodium phosphate (Walter *et al.*, 1991). This type of behavior arises when enthalpically unfavorable interactions between solutes overcome the entropy gained by complete miscibility. A critical concentration of the solutes must be reached in order for the enthalpic contribution to be dominant. For example, the critical concentration of PEG 3350 and dextran T500 to form a two-phase system at 0°C is about 5% for each component (Heller *et al.*, 1997). Alternatively, phase separation has been proposed to arise because of differing effects of solutes on water structure (Zaslavsky *et al.*, 1989).

Izutsu *et al.* (1996) studied the effect of selected sugars and polymers on crystallization of PEG in frozen solutions by differential scanning calorimetry and pulsed NMR. Whereas some solutes are effective inhibitors of PEG crystallization, certain polymers, such as PVP and dextran, have virtually no influence on PEG crystallization. Izutsu and co-workers concluded that this is caused by phase separation between the incompatible polymers during freezing. Some possible consequences of liquid–liquid phase separation with respect to protein stability have been the subject of a review by Randolph (1997), and are discussed further below.

3.5. Effect of Freezing Rate on Recovery of Protein Activity

Many biochemists involved in protein purification hold the view that rapid freezing promotes high recovery of biological activity. There are scientifically sound reasons for this point of view, including minimization of buffer crystallization with resulting pH shifts, minimization of exposure of the protein to the high ionic strength which results from freeze concentration if salts are present, and reduction of the time during which cold denaturation could take place. Although the body of published literature on the subject is not large, there is growing evidence that rapid freezing adversely affects recovery of protein activity after freeze-thawing and freeze-drying, and may adversely affect the stability of the lyophilized solid during storage. Eckhardt *et al.* (1991) studied the effect of freezing rate (from 0.5 to about 50°C/min) on the formation of soluble and insoluble aggregates of human growth hormone using combinations of phosphate buffer and mannitol as an excipient system in the pH range of 7.4–7.8. Formation of soluble aggregates was not measurably affected by freezing rate; however, the effect of freezing rate on formation of insoluble aggregates (measured turbidimetrically) was dramatic, with considerably more aggregation at higher freezing rates. The effect of the excipient system was not clear in this study, except that formation of insoluble aggregates seemed to be promoted by a lower pH prior to freezing. pH of the system during freezing was not measured, and the effect of freeze-drying on aggregation was not studied.

Chang *et al.* (1996) reported that there is a strong correlation between the tendency of a protein to denature at surfaces and its tendency to denature during freezing, and their data support the conclusion that the ice–water interface can play a significant role in protein denaturation during freezing. Proteins examined were phosphofructokinase, lactate dehydrogenase, glutamate dehydrogenase, interleukin-1 receptor antagonist, tumor-necrosis-factor-binding protein, and ciliary neurotropic factor. Slow freezing (3°C/hr) resulted in markedly less turbidity after freezing than dipping vials in liquid nitrogen (quench-freezing), presumably because the relative area of the ice–liquid interface is smaller after slow freezing. Furthermore, addition of 0.01% Tween 80 resulted in essentially complete protection of all of the model proteins after freeze-thawing, presumably because surfactants can compete with proteins for the available ice surface. These findings are consistent with a report by Strambini and Gabellieri (1996), where measurements of phosphorescence lifetime of tryptophan residues demonstrated that freezing is accompanied by loosening of the native fold and loss of secondary and tertiary structure. Although this process is largely reversible, a fraction of the protein molecules do not recover either the original phosphorescence or the original activity. Proteins examined were alcohol dehydrogenase, alkaline phosphatase, glyceraldehyde 3-phosphate dehydrogenase, lactate dehydrogenase, ribonuclease T_1, and apoazurin. Cryoprotectants such as sucrose and glycerol were found to reduce or eliminate the perturbation caused by freezing. Neither freezing rate effects nor surfactant effects were reported by Strambini and Gabellieri.

Jiang and Nail (1998) measured recovery of enzymatic activity of lactate dehydrogenase and β-galactosidase as a function of freezing rate, and found that rapid freezing of solutions of these proteins by placing vials in liquid nitrogen consistently resulted in decreased recovery of activity relative to slower freezing methods, and that this effect was more pronounced at low protein concentration. The effects of shelf freezing methodology were examined by comparing recovery of activity after (1) "slow" freezing by placing vials on the shelf of the freeze-dryer and slowly decreasing the shelf temperature at a rate of 0.5°C/min to −45°C versus (2) "fast" freezing by placing vials on shelves which had been precooled to −45°C. Recovery of activity of the model proteins was significantly higher when vials were placed on precooled shelves compared to the "slow" freezing method. Although further study is needed, these results are not necessarily inconsistent with faster freezing rates resulting in loss of protein integrity after freezing, because the degree of supercooling and, consequently, the effective freezing rate actually may be faster when more time is allowed for thermal equilibrium to be established throughout the volume of the supercooled solution.

Sarciaux *et al.* (1999) reported lower levels of insoluble aggregate in bovine IgG after freeze-drying following slow cooling relative to quench cooling in liquid nitrogen. Annealing of quench-cooled samples resulted in a significant reduction in the level of insoluble aggregates after freeze-drying, along with a corresponding decrease in the specific surface area of the freeze-dried solids. The data in this study are consistent with a mechanism for aggregation involving denaturation of the protein at the ice/freeze-concentrate interface which is reversible upon freeze-thawing, but becomes irreversible after freeze-drying and reconstitution.

Although not widely investigated, freezing rate may affect the storage stability of freeze-dried protein formulations. Hsu *et al.* (1995) reported that the rate of increase in turbidity during storage of reconstituted formulations of recombinant tissue-type plasminogen activator is correlated with the specific surface area of the lyophilized solid. This finding was the result of investigation into the reason for the observation that fast freezing consistently produced a less stable product than slow freezing. Hsu *et al.* hypothesized that denaturation takes place at the solid–void interface during storage, which is not reversible upon reconstitution of the solid.

Millqvist-Fureby and colleagues (1999) used electron spectroscopy for chemical analysis (ESCA) to examine freeze-drying of proteins with various carbohydrates by measuring the surface composition of the amorphous solid. The proteins were found to be enriched at the solid surface relative to the bulk concentration of protein in the powders, and the mechanism was believed to be preferential adsorption of the proteins to the ice/freeze-concentrate interface, where the degree of surface accumulation depends on the carbohydrate used. The surface accumulation of protein was increased in annealed samples compared with nonannealed controls.

Regardless of the mechanism of loss of protein integrity after freezing and freeze-drying, the published literature points to the need to consider the thermal history of freezing when identifying critical process variables during validation studies.

4. MATERIAL CHARACTERIZATION

Minimizing empiricism in formulation and process development for freeze-dried products requires characterization of the formulation intended to be freeze-dried. The result of such characterization should be information on the upper product temperature limit during the primary drying stage of freeze-drying, knowledge of the physical state of the solute(s), and an assessment of the degree to which the characteristics of the frozen system are affected by changes in composition of the formulation. Characterization of the freezing behavior of a formulation is an important part of process validation for a freeze-dried product. Failure to generate this information systematically results in an approach to both formulation and process development which is strictly trial and error. Besides being inefficient, this approach has come under increased regulatory scrutiny, particularly since the advent of preapproval inspections for new pharmaceutical products. Characterization of the formulation is also needed for process optimization. For the sake of this discussion, an optimized freeze-drying process is one in which freeze-drying is carried out in the minimum amount of time while meeting all specifications for quality attributes of the final product.

A frequently overlooked aspect of development of freeze-dried products is characterization of the freeze-dried solid. This includes determination of the physical state of the dried product, that is, crystalline, partially crystalline, or amorphous. It may also include identification of the polymorph of a crystallizing component which exhibits polymorphism and determination of whether the crystal form observed is affected by changes in formulation and processing conditions. For amorphous systems, the glass transition temperature of the amorphous solid as well as the extent to which T_g changes with residual moisture may be critical attributes of the product with regard to both physical and chemical stability, both of which are discussed in more detail below.

Because the methods used for characterization of frozen systems and freeze-dried solids are similar, both will be discussed together. Aspects unique to either frozen systems or freeze-dried solids will be pointed out as needed.

4.1. Thermal Analysis

Changes in any material which accompany a change in temperature are almost always accompanied by absorption or release of energy in the form of heat. Thermal analysis is used to measure the temperatures at which these changes occur, determine whether the transitions are endothermic (absorption of heat) or exothermic (release of heat), and quantitate the amount of energy involved in the transition. Experimentally, a small quantity (typically 5–20 mg) of sample is placed in an aluminum pan and a lid is crimped into place. An empty pan and lid is generally used as a thermally inert reference. Subambient capability is needed for analysis of frozen solutions, which is accomplished either by using a mechanically refrigerated cooling accessory or by liquid nitrogen cooling.

There are two types of differential scanning calorimetry instruments. In *power compensation* DSC (for example, Perkin-Elmer, Norwalk, CT) the sample and reference pans are placed in isolated furnaces. The sample and reference are maintained at the same temperature, and the power required to do so is measured. In *heat flux* DSC (for example, TA Instruments, Newcastle, DE) the sample and reference are placed in the same furnace and heated from the same source. The temperature difference between the sample (T_s) and reference (T_r) is measured and converted to heat flow (ΔQ) using the relationship

$$\Delta Q = (T_s - T_r)/R_T \qquad (1)$$

where R_T is the thermal resistance of the cell. In either case, the result of the experiment is a plot of heat flow versus temperature. Thermal transitions which are commonly observed in frozen systems are illustrated in Fig. 9, where, for the sake of this discussion, a deflection upward indicates an endothermic transition. The glass transition is seen as a shift in the baseline toward higher heat capacity. Crystallization during the DSC experiment is observed as an exothermic event, and eutectic melting is an endothermic transition which precedes melting of ice. The convention for assignment of onset temperature is the intersection of the extrapolated baseline with the tangent at the steepest slope of a transition. Glass transitions are commonly reported as either the onset temperature or the midpoint of the transition.

4.1.1. EXPERIMENTAL CONSIDERATIONS IN THERMAL ANALYSIS

Thermal analysis is a very valuable tool for development of freeze-dried formulations and processing conditions, yet it is too often regarded as a "quick and dirty" screening technique. As a result, valuable information about the system under study is too often lost. The development scientist should recognize that the results obtained by thermal analysis reflect the

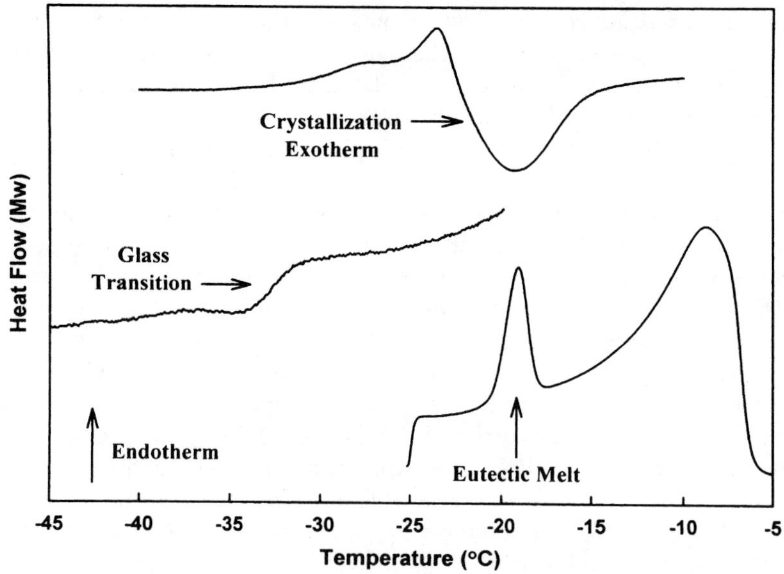

Figure 9. Typical thermal transitions in frozen solution, including eutectic melting, glass transitions, and crystallization exotherms.

experimental conditions used and that useful information can be gained by using more than one experimental condition, such as sample concentration, cooling rate, and heating rate.

A purge gas should always be used when carrying out a DSC experiment. For subambient thermal analysis, it is important to avoid condensation of atmospheric moisture between the sample and the surface of the sample well. The purge gas also helps to transfer heat from the sensor to the sample pan. Either helium or nitrogen is usually acceptable as purge gas. Nitrogen has a wider operating temperature range, and helium has a higher thermal conductivity and usually reduces baseline curvature relative to nitrogen.

The calorimeter must be calibrated for temperature measurement, and it is often desirable to calibrate for enthalpy measurement. Common temperature calibration standards for low-temperature operation are listed in Table III.

Liquid sample sizes of 10–20 mg are typical. Larger sample sizes are not necessarily better, because thicker samples will have a larger temperature gradient across the sample, which can decrease the sharpness of observed transitions. This is particularly important for temperature calibration.

For analysis of freeze-dried solids, sample sizes of 5–10 mg are typical. A larger mass of sample is not necessarily better, for the reason mentioned above. We have found preparation of a thin powder compact using a punch-and-die with a diameter slightly smaller than the inside diameter of the sample pan to be useful for maximizing the quantity of material in the sample pan while minimizing sample thickness. A representative thermogram of freeze-dried maltodextrin as a powder compact versus the same quantity of sample as a loose powder is shown in Fig. 10, where the glass transition region is better defined when the sample is presented as a powder compact.

Because aqueous systems are prone to supercooling, the most common practice is to record the thermogram during the heating cycle after freezing of the sample. This avoids supercooling-related artifacts. However, the cooling

Table III
Common Standards for Thermal Analysis

Material	Melting temperature (°C)	Heat of fusion (J/g)
Indium	156.6	28.71
Mercury	−38.8	11.44
n-Octane	−56.7	182
n-Decane	−29.6	202.1
n-Dodecane	−9.7	216.7

rate of an aqueous solution can have a significant influence on the thermogram recorded during subsequent heating of the sample, particularly for solutes which tend to crystallize during the time course of freezing. The term "critical cooling rate" usually refers to a cooling rate above which crystallization of solute during freezing is prevented. For example, at cooling rates less than about 2°C/min, crystallization of mannitol from a freezing solution takes place during the cooling cycle, whereas, at faster cooling rates, crystallization is inhibited. In this case, complex thermal behavior is observed in the subsequent heating thermogram, which consists of a glass transition followed by a melting endotherm and an exotherm indicating crystallization of the solute (Her and Nail, 1994).

Heating rates used for thermal analysis are typically 5–10°C/min, but it is often useful to carry out experiments at lower rates in order to maximize resolution of thermal events taking place at nearly the same temperature. The choice of a heating rate for conventional DSC involves a trade-off between sensitivity and resolution, where sensitivity refers to the ability to detect a given transition, and resolution is the ability to separate two thermal events taking place at nearly the same temperature. In thermal analysis of frozen systems, particularly those in which the solute crystallizes, resolution is often a more important issue, because the eutectic melting temperature may be very close to the melting temperature of ice. An example of the use of slow heating rates to maximize resolution is shown in Fig. 11 for the glycine–water system. In this case, three thermal events—

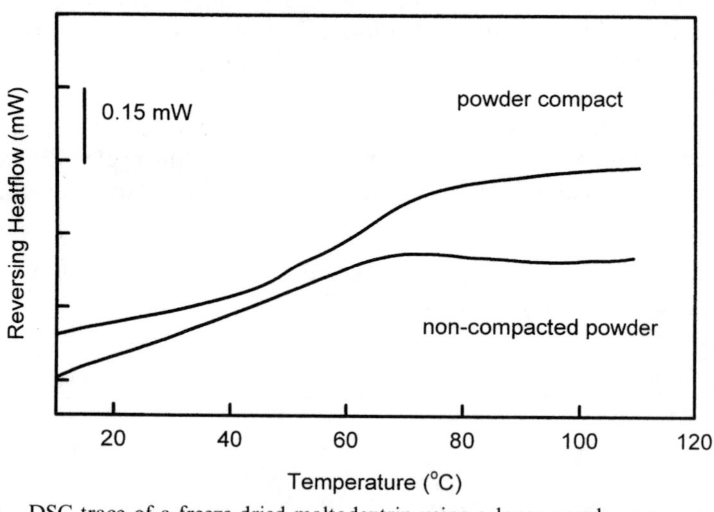

Figure 10. DSC trace of a freeze-dried maltodextrin using a loose powder versus a powder compact.

readily apparent at a very slow heating rate—could be interpreted as simply the melting of ice if only rapid heating rates are used. In the glycine–water system, the complex thermal behavior in the region of the melting endotherm of ice arises from the polymorphism of glycine (Chongprasert *et al.*, 2001).

For detection of glass transitions in both frozen systems and freeze-dried solids, sensitivity may be more important than resolution, because glass transitions are often difficult to detect. In this case, relatively rapid heating rates are recommended. However, it should be remembered that heating rate influences the observed glass transition temperature, where faster heating rates cause observation of the transition at higher temperatures. For example, for a frozen solution of 10% dextran, the observed

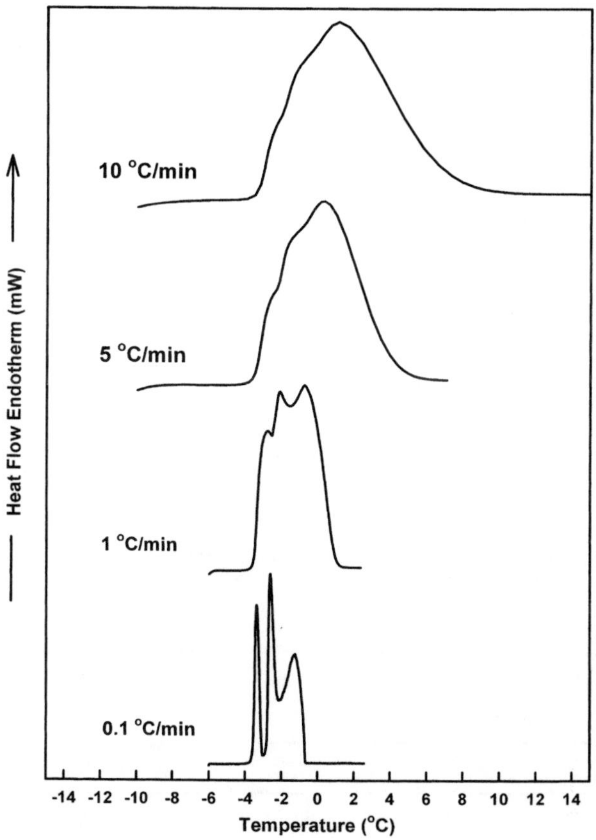

Figure 11. DSC trace of frozen glycine solution at different heating rates in the region of ice melting.

midpoint of the glass transition increases from about −13.5°C to about −9.5°C as the heating rate is increased from 2°C/min to 20°C/min.

Another reason to use more than one heating rate when characterizing frozen systems by DSC is that, depending on the heating rate, a solute may crystallize during the DSC experiment. This is illustrated in Fig. 12 for a sucrose/glycine system containing 10% (w/v) total solids at a 1:1 weight ratio of sucrose to glycine. At the fastest heating rate, the only notable feature of the thermogram other than melting of ice is the glass transition region in the temperature range of about −45°C to about −50°C. At slower heating rates, an exotherm arising from crystallization of glycine is observed (note also that the glass transition region is not apparent at very slow heating rates). This is important information, given that the time scales of thermal analysis and freeze-drying may be more than an order of magnitude different. Using a too-rapid heating rate can result in misleading information regarding the physical state of the formulation after freeze-drying.

For thermal analysis of frozen solutions, it may be useful to increase the solute concentration above the intended formulation concentration in order to maximize sensitivity, particularly for glass transitions. As the state diagram predicts, the glass transition temperature is essentially independent of solute concentration. Her and Nail (1994) demonstrated this effect for solutions of lactose and related compounds.

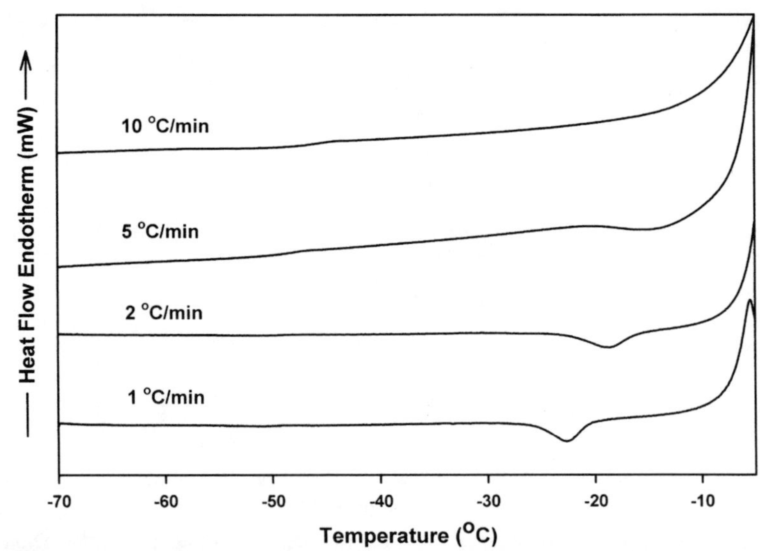

Figure 12. DSC thermograms of sucrose:glycine (7:3 weight ratio) at different heating rates.

4.1.2. SOURCES OF UNCERTAINTY IN INTERPRETATION OF DSC THERMOGRAMS

Although the transitions illustrated in Fig. 9 are easily interpreted, uncertainties often arise. For many organic compounds which crystallize from freezing aqueous solution, the eutectic melt occurs very close to the melting temperature of ice. In this case, the eutectic melting endotherm may be partially, or completely, obscured by the ice melting endotherm even at very slow heating rates. Examples are the ice/mannitol and ice/dibasic sodium phosphate eutectic melts at about −1°C and −0.5°C, respectively. Although resolution of such closely spaced endotherms is enhanced by slow heating rates, this is by no means assured.

Glass transitions, both in frozen systems and freeze-dried solids, can be difficult to detect. This may be caused by the small heat capacity change associated with the glass transition, a broad glass transition region, or both. Interpretation is made more uncertain by baseline drift or other noise. In addition, other thermal events at temperatures close to the glass transition, such as enthalpy recovery or crystallization, may disguise the heat capacity change associated with the glass transition. Sensitivity to such transitions is enhanced by rapid heating rates.

Thermal analysis of frozen solutions is a relatively small subset of the thermal analysis literature, most of which is directed toward thermal analysis of polymer systems. As a result, some phenomena which are commonly observed are poorly understood, including lyotropic liquid crystal formation during freezing and apparent multiple glass transitions in frozen systems. A pharmaceutically important and relatively widely studied example of uncertainty in interpretation of a DSC thermogram is the system sucrose/water, illustrated for six different concentrations of sucrose in Fig. 13. The glass transition which is commonly reported, illustrated by the thermogram of 5% sucrose, is in the range of −32°C to −35°C. Examination of higher concentrations, however, reveals another transition at a lower temperature. This apparent multiple glass transition has given rise to uncertainty regarding both the physical significance of both of these transitions and assignment of the T_g' temperature for the purpose of characterizing frozen solutions of amorphous solutes. Other investigators have reported results similar to the data in Fig. 13 for sucrose and related solutes, and the bulk of the evidence supports the conclusion that the lower temperature transition is the glass transition of the maximally freeze concentrated solution. Ablett *et al.* (1992a) reported that, in sucrose and glycerol solutions which are sufficiently concentrated that ice formation is inhibited (bottom thermogram in Fig. 13), a single glass transition is observed at the lower temperature. It is only when ice is present that the complex transition is observed. Ablett and

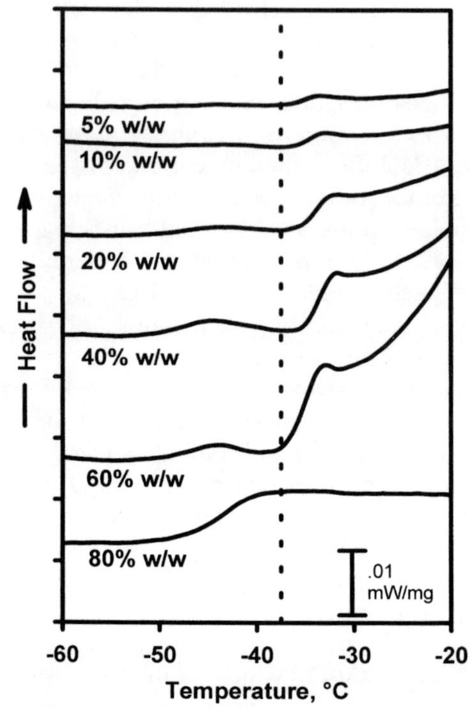

Figure 13. DSC thermograms of sucrose solutions in the glass transition region.

co-workers argue convincingly that the higher temperature transition is not a glass transition at all, because (1) maximum freeze concentration was observed at −40°C, (2) the heat capacity change at the higher temperature transition is well in excess of the heat capacity change for totally vitrified samples, and (3) the higher temperature transition is considerably narrower than the glass transition region for totally vitrified samples. The physical significance of the second transition is interpreted by Ablett and co-workers as the temperature at which the mobility of the solute has increased to the point where ice dissolution can begin. Shalaev and Franks (1995) also reported a double transition in frozen solutions of sucrose/water and fructose/water. The lower temperature transition is attributed to the glass transition, whereas the higher transition is interpreted as a softening process. Izutsu *et al.* (1996) also refer to the higher temperature transition as a softening temperature, T_s, which is distinct from the "real" glass transition temperature. The physical significance of the higher temperature transition is attributed to a relaxation process between ice crystals and the freeze-concentrated solute phase.

From the formulation scientist's point of view, it is important to know which of these transitions, if either, is predictive of collapse in freeze-drying.

Although a comprehensive answer is not available yet, a study by Knopp *et al.* (1998), using both thermal analysis and microscopic observation during freeze-drying of sucrose solutions, concluded that the onset of collapse occurs approximately midway between the end of the lower temperature transition and the onset of the higher temperature transition. This is denoted by the vertical dashed line in Figure 13. Since the higher temperature transition is usually the only "glass transition" observed in thermal analysis of disaccharide-containing protein formulations, this transition is generally accepted as the thermal event that is predictive of collapse, and it is often very close to the observed collapse temperature. However, more data are needed for a wide variety of solutes in order to better determine the degree to which the higher temperature transition is predictive of collapse during freeze-drying.

Another source of uncertainty in thermal analysis of frozen systems relates to glass transition temperatures of protein solutions. When the concentration of protein is low, the freeze-drying characteristics of the formulation are generally determined by added solutes, or excipients. When the protein is the predominant solute, as in the case of some immunoglobulin preparations, for example, it is common to observe no transitions other than the melting of ice. Although the reason for this is not clear, possible explanations include the following: (1) The transition temperature is too close to the onset of ice melting for the two events to be resolved, (2) the heat capacity change at the transition is too small to be measured by the instrumentation, or (3) the transition region is so broad that the transition is "lost" in baseline drift or noise in the thermal analysis experiment. This behavior is probably related to the type of glass formed, that is, strong versus fragile, which is discussed below. In any event, other methods are often needed in order to adequately characterize the system.

4.1.3. MODULATED-TEMPERATURE DSC

A recent advance in thermal analysis, first described by Reading *et al.* (1994), that is finding application in pharmaceutical research and development is modulated-temperature DSC (MTDSC). Briefly, the total heat flow observed in conventional DSC consists of the sum of two components, a heat capacity component, $C_p(dT/dt)$ and a kinetic component $f(T, t)$:

$$dH/dt = C_p(dT/dt) + f(T,t) \qquad (1)$$

where the heat capacity component is heating-rate-dependent and the kinetic component is a function of time at constant temperature. In

MTDSC a sine wave modulation is superimposed on the underlying linear temperature program, where the amplitude and the frequency of the modulation are controllable. A discrete Fourier transform algorithm is applied to the resulting sample response to deconvolute the signal into a heating-rate-dependent, or reversing, component and a time-dependent, or nonreversing, component. Examples of reversing transitions are the glass transition and some melting phenomena. Nonreversing transitions include enthalpy recovery, evaporation, crystallization, decomposition, curing of polymers, and some melting phenomena. Melting can be observed in both components to differing degrees, depending upon experimental conditions.

MTDSC offers a solution to the trade-off between sensitivity and resolution by providing, in principle, both the high resolution provided by a slow underlying heating rate and the high sensitivity provided by a high *instantaneous* heating rate from the modulation. Other advantages include the possibility of separating overlapping thermal phenomena, detection of metastable melting phenomena, and greater ease of measuring heat capacity. We have observed that conventional DSC is more suitable for detection of glass transitions in frozen solutions, because the upper limit of the heating rate for modulated DSC is about 3°C/min. However, modulated DSC appears to provide a significant improvement over conventional DSC for analysis of freeze-dried amorphous solids, where detection of the glass transition is often uncertain due to overlapping thermal events. An example is shown by the thermogram in Fig. 14 for a freeze-dried maltodextrin, where the dominant feature of the total heat flow thermogram is the enthalpy recovery endotherm, which is superimposed on the glass transition region.

Using modulated DSC, an experimental technique referred to as stepwise isothermal heat capacity measurement may provide added enhancement of sensitivity for glass transition measurement. With scanning DSC, even in the modulated mode, detection of glass transitions becomes more difficult as the glass transition region becomes broader and the heat capacity change at the glass transition becomes smaller. This is illustrated in Fig. 15 for maltose and a series of maltodextrins of increasing molecular weight. Note that the glass transition region is uncertain in curves *C* and *D*. In stepwise isothermal heat capacity measurement, the sample is held at a constant underlying temperature while the temperature is modulated for a period of about 5 to as much as 15 min. This allows calculation of the heat capacity of the sample. The temperature is then increased by perhaps 1°C, and the experiment is repeated. The use of stepwise isothermal heat capacity measurement is shown in Fig. 16 for a maltodextrin with an average molecular weight of 3600 after equilibration at 23% and 11% relative humidity. The heat capacity change over the relatively broad glass transition

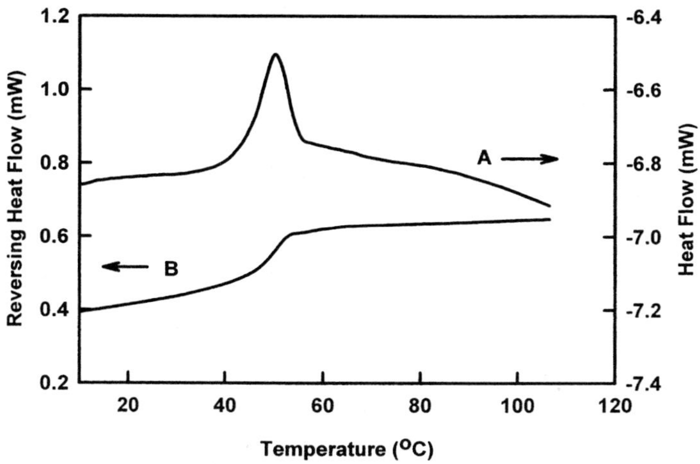

Figure 14. DSC thermograms of freeze-dried maltodextrin showing total heat flow (A) versus reversing heat flow (B) in the glass transition region.

region is apparent in both thermograms. Although stepwise isothermal heat capacity measurements are time-consuming, this technique appears to offer enhanced sensitivity for glass transition temperature measurement relative to scanning calorimetry.

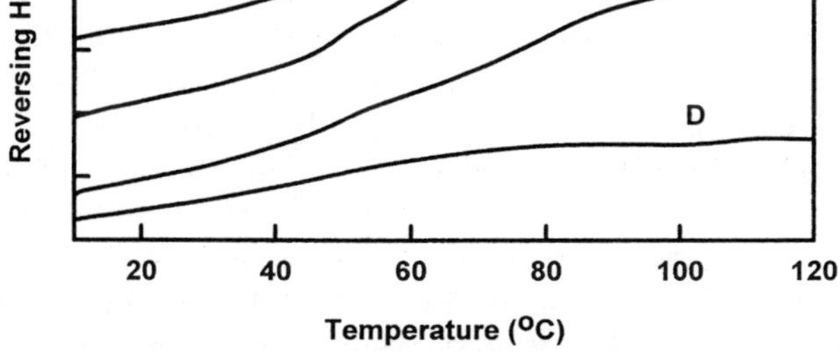

Figure 15. DSC thermograms of maltose and different molecular weights of freeze-dried maltodextrins in the glass transition region. As molecular weight increases, the heat capacity change at the glass transition decreases and the width of the glass transition increases. (A) Maltose; (B) maltodextrin, MW 900; (C) maltodextrin, MW 1800; (D) maltodextrin, MW 3600.

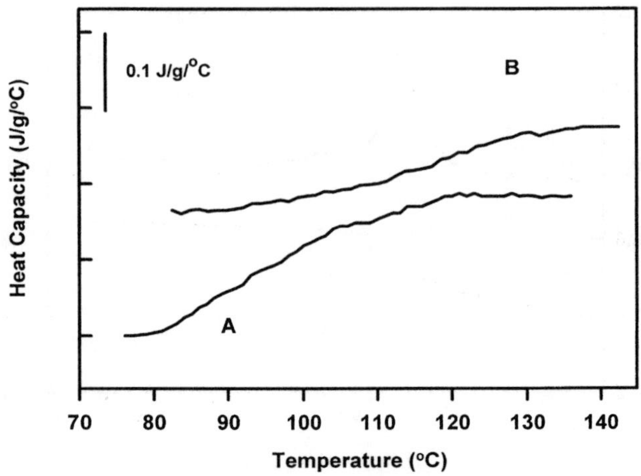

Figure 16. DSC thermograms of maltodextrin (MW 3600) in the glass transition region by stepwise isothermal heat capacity measurement. (A) Equilibrated at 23% relative humidity; (B) equilibrated at 11% relative humidity.

4.2. Optical Microscopy

Optical microscopy using a special low-temperature stage which can be evacuated to operate at pressures representative of freeze-drying allows direct observation of materials during freezing and freeze-drying, generally under a low magnification of 50–100×. The technique is not new. Rey (1957) was perhaps the first to describe a cryomicroscope, to which he later added vacuum capability. MacKenzie (1964) later built a microscopy system suitable for the study of freeze-drying. Other devices for freeze-drying microscopy were subsequently described, and the designs varied widely (Flink and Gejl-Hansen, 1978; Flink *et al.*, 1973; Freedman *et al.*, 1972; Kochs *et al.*, 1989; Pikal *et al.*, 1983). However, growing interest in the technique has resulted in at least two types of commercially available instruments. One type uses a two-stage Peltier cooling device (Physi-Temp Instruments, Clifton, NJ) capable of cooling the top plate of the device to about −50°C. A second type of freeze-drying stage uses liquid nitrogen vapor circulated through a silver block along with a heater to control the surface temperature (Linkam Scientific Instruments, Tadworth, Surrey, UK). This allows cooling the sample to well below −150°C. Both instruments (see Fig. 17) offer the ability to program the temperature profile, which allows studies such as the effect of thermal history of freezing as well as annealing on the morphology of the frozen and freeze-dried material.

It is important in freeze-drying microscopy to avoid vertical temperature gradients, because it is not practical to measure the sample temperature by placing a thermocouple directly in the sample. Calibration of temperature measurement is easily done by comparing observed transitions of standard materials with reference temperatures. In addition to the standards used for DSC listed above, eutectic melting temperatures for frozen aqueous solutions of inorganic salts can also be used, such as ice/NaCl (−21.5°C), ice/KCl (−11.1°C), ice/NaBr (−28.0°C), and ice/NaI (−31.1°C). Observed transitions should be within about 0.5°C of published values for standards. Temperature calibration should be checked both at atmospheric pressure and under vacuum. Temperature accuracy may be improved by using a heat sink compound or silicone oil to improve thermal contact between the sample and the surface of the stage.

In order to minimize vertical temperature gradients in the system, it is not appropriate to use traditional microscope slides. Cover slips should be used instead. Some investigators use quartz cover slips in order to further reduce vertical temperature gradients, because the thermal conductivity of quartz is significantly higher than that of glass. A disadvantage of quartz cover slips is the expense and fragility of the material.

Experimentally, freeze-drying microscopy is usually carried out by placing a small quantity, perhaps 10 μl, of sample on a cover slip, freezing the sample, evacuating the system, then systematically varying the sample temperature and observing the resulting morphology in the dried material. This technique is particularly useful for measuring the *collapse* temperature in freeze-drying, which may be somewhat different from glass transition temperatures measured by thermal analysis. Glass transitions are measured on closed systems, whereas collapse is a dynamic phenomenon taking place during freeze-drying. An example of collapse phenomena is shown in the photomicrograph in Fig. 18a. In the authors' experience, the correlation between collapse observed microscopically and collapse in a vial is within a few degrees Celsius, although several exceptions have been observed.

Freeze-dry microscopy is useful for examining the influence of formulation composition and freezing history on the morphology of ice. Dendritic ice is normally observed (see Fig. 18b); however, at high solute concentrations, spherulitic ice can form (Fig. 18c). The significance of ice morphology is that, when dendritic ice forms, open channels are created by prior sublimation of ice, providing a relatively low resistance to transfer of water vapor through the partially dried solids. Spherulitic ice, on the other hand, is associated with a high resistance to mass transfer and inefficient freeze-drying.

Another application of freeze-drying microscopy is for observation of secondary, or solute, crystallization during freezing. Crystallization of

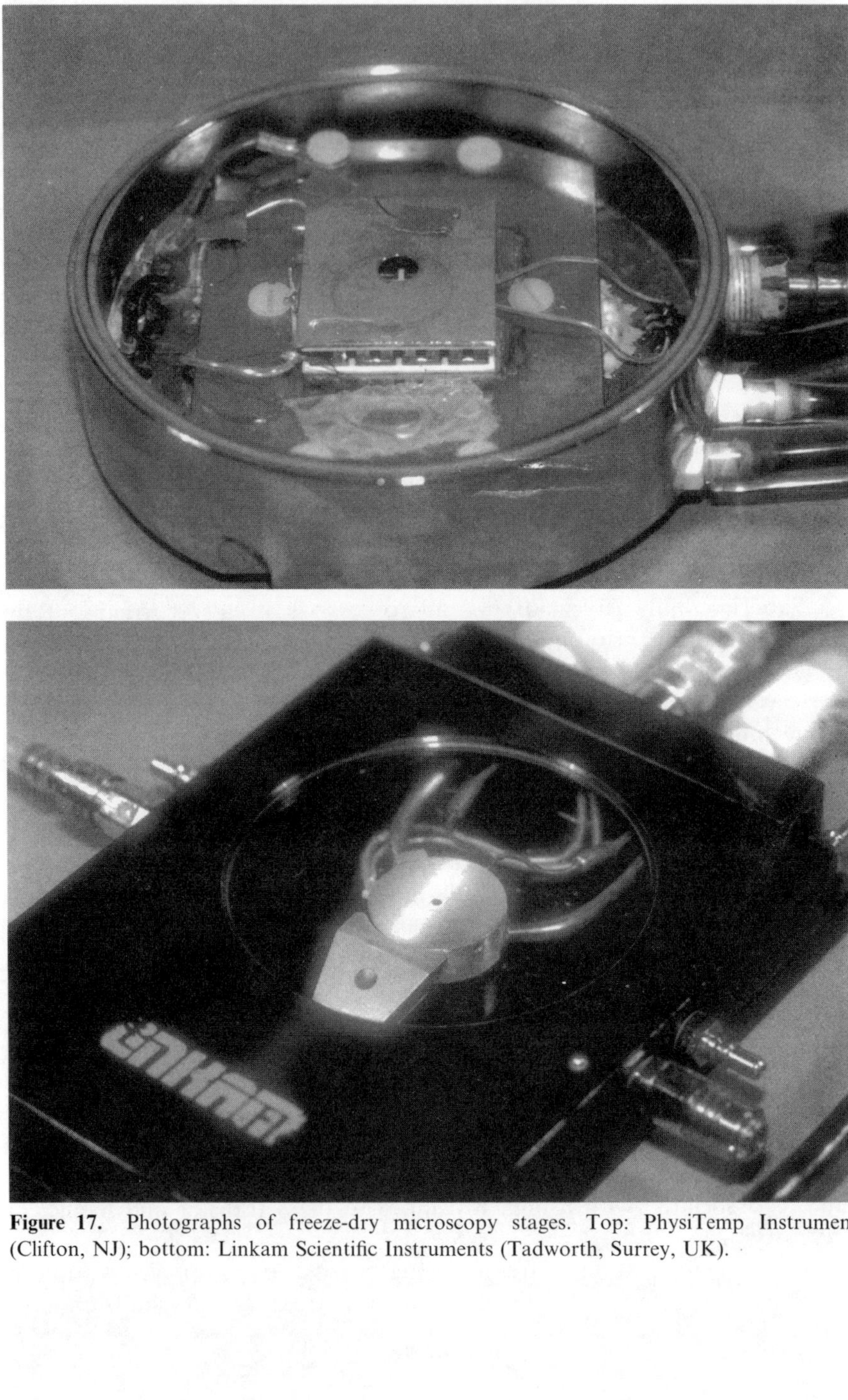

Figure 17. Photographs of freeze-dry microscopy stages. Top: PhysiTemp Instruments (Clifton, NJ); bottom: Linkam Scientific Instruments (Tadworth, Surrey, UK).

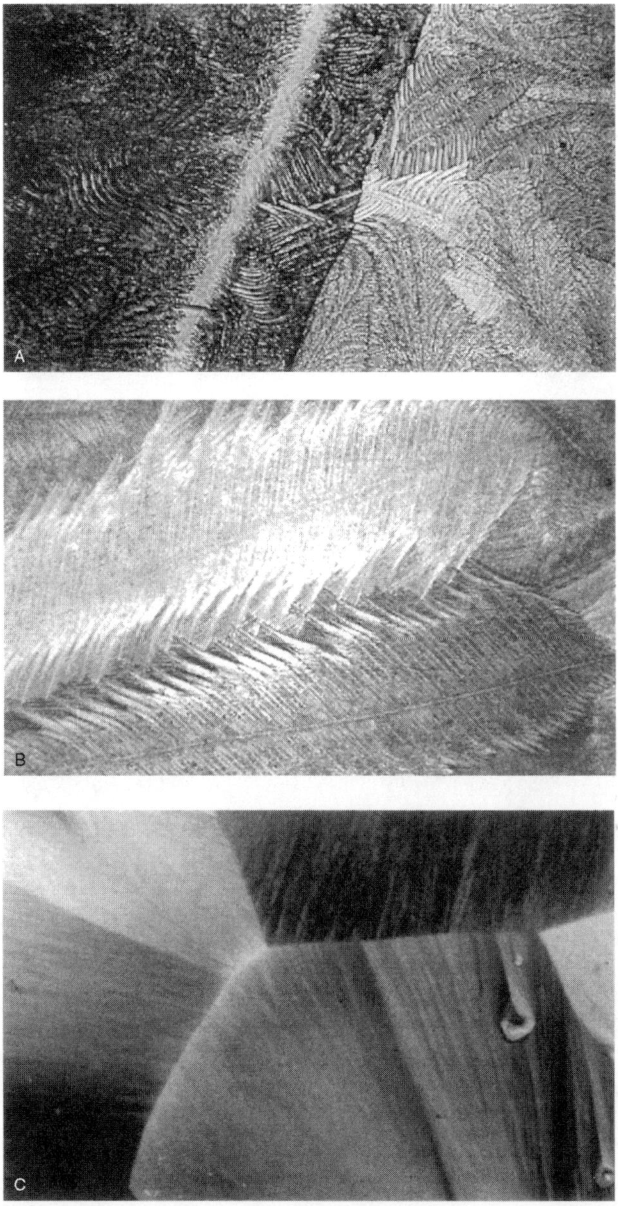

Figure 18. Freeze-drying photomicrographs illustrating (A) collapse versus retention behavior of recombinant human hemoglobin, (B) dendritic ice morphology, (C) spherulitic ice morphology at high solute concentration, (D) crystallization of nafcillin sodium during annealing, and (E) liquid crystalline behavior of methylprednisolone sodium succinate.

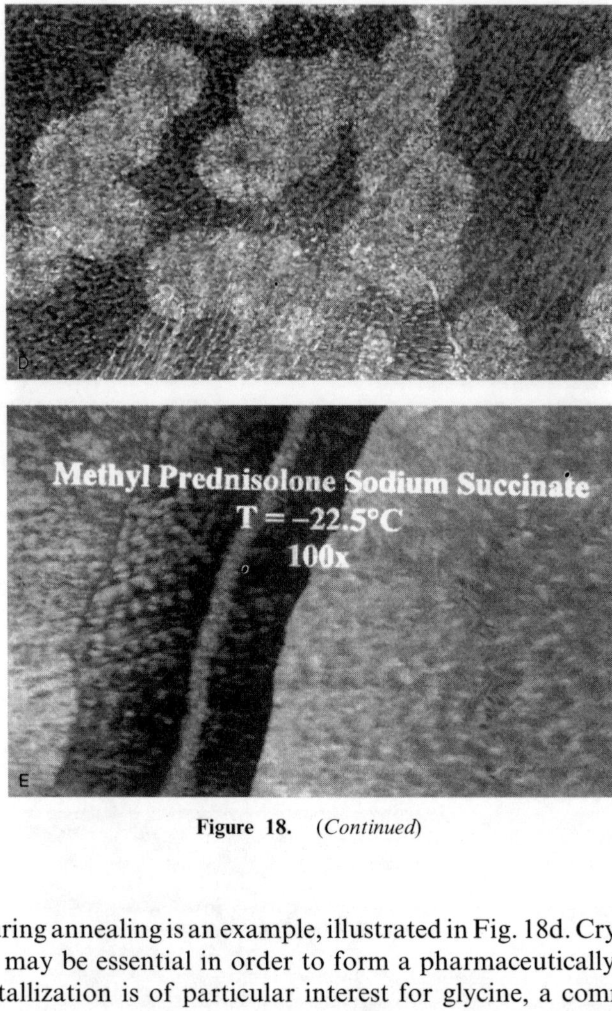

Figure 18. (*Continued*)

nafcillin during annealing is an example, illustrated in Fig. 18d. Crystallization of a solute may be essential in order to form a pharmaceutically acceptable cake. Crystallization is of particular interest for glycine, a commonly used formulation excipient, because the glass transition of amorphous glycine/water is well below the temperature range available in a commercial freeze-dryer, whereas the eutectic melting temperature of glycine/ice is about −3.6°C.

Eutectic melting and collapse phenomena are usually easy to observe microscopically and are the most common types of freeze-drying behavior. However, another type of freeze-drying behavior, illustrated in for methyl-prednisolone sodium succinate in Fig. 18e, is not uncommon, yet is not well recognized. The freeze-dried material is highly birefringent, indicating order in at least two dimensions; however, this material appears to collapse as the product temperature exceeds a certain temperature. Microscopic observation suggests a crystalline material that collapses, which is not realistic for materials that are crystalline in the traditional sense. It is more likely that these compounds are examples of lyotropic liquid crystalline behavior. The significance of this behavior is that it makes determination of the maximum allowable operating temperature during primary drying uncertain. As stated above, the influence of lyotropic liquid crystal formation in freeze-drying on critical product attributes is not well understood.

4.3. Scanning Electron Microscopy

Scanning electron microscopy (SEM), either conventional SEM or environmental SEM, can be particularly useful in characterization of freeze-dried solids. Two examples are illustrated in Fig. 19. Figures 19A and 19B are examples of two different polymorphs of pentamidine isethionate formed under different freezing conditions (Chongprasert *et al.*, 1998). Figures 19C and 19D illustrate an amorphous material (lactose) dried under conditions of retention of microstructure (Fig. 19C) and after the onset of collapse (Fig. 19D). The characteristic "Swiss cheese" morphology can be observed before macroscopic collapse is apparent macroscopically in the freeze-dried solid, and is associated with decreased resistance of the dried material to flow of water vapor during the primary drying phase of freeze-drying (Milton *et al.*, 1997).

4.4. X-Ray Powder Diffraction

X-ray powder diffraction is the method of choice for characterization of the physical state of a freeze-dried solid. Documenting the physical state of a product both immediately after freeze-drying and periodically during stability studies is often overlooked during product and process development, but it can provide valuable insight into the processes underlying physical or chemical stability problems.

For protein formulations requiring a lyoprotectant such as a disaccharide, the freeze-dried solid will be at least partially amorphous, because it is generally accepted that such lyoprotectants must be amorphous in order to be effective. However, it is common practice to include a crystallizing component such as glycine or mannitol in order to promote a physically stable freeze-dried cake, that is, one that retains the size and shape of the original frozen solution (see stability discussion below) throughout the shelf life of the product. A particular concern is crystallization of a

Figure 19. SEM photomicrographs illustrating different polymorphs of pentamidine isethionate (A, B) observed after different cooling histories and freeze-drying with retention (C) versus freeze-drying at the onset of structural collapse (D) of lactose.

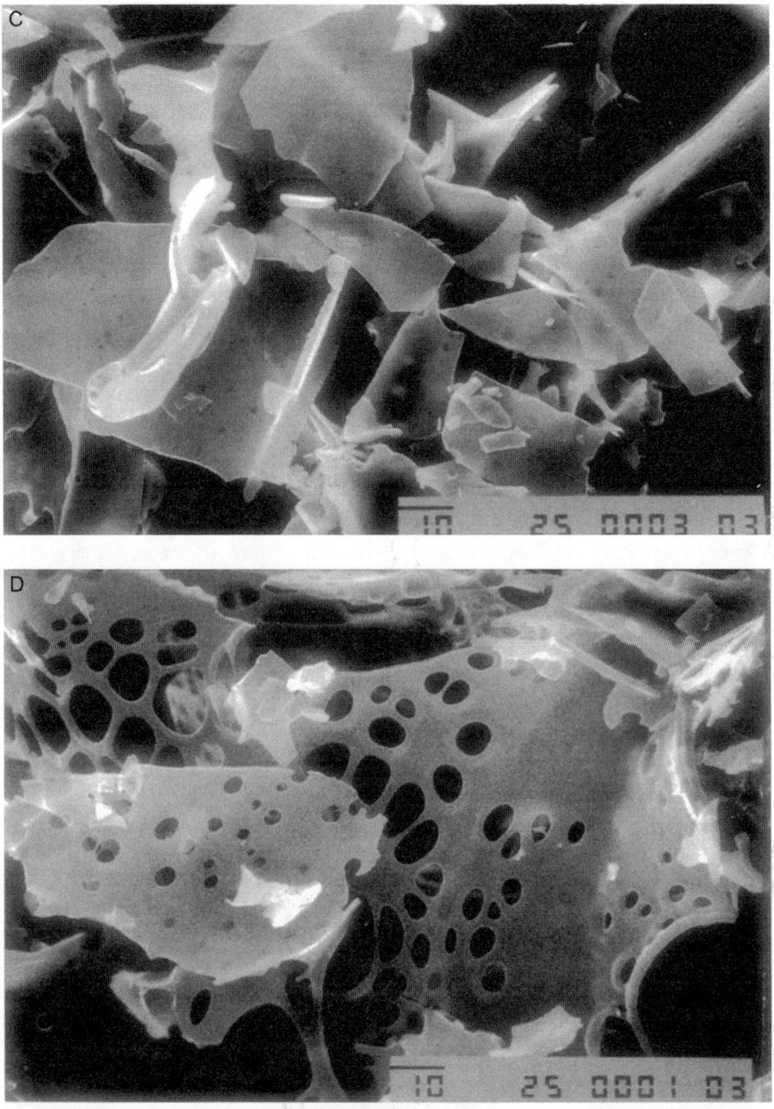

Figure 19. *(Continued)*

component of the formulation from the solid state during storage, illustrated for a mannitol-containing formulation in Fig. 20. This can lead to a redistribution of water within the solid, and the increased water activity in the microenvironment of the drug can lead to increased rates of degradation either through the participation of water as a reactant or through the effect of water on glass transition-associated mobility (Herman *et al.*, 1994).

Polymorphs of crystalline materials have different physical properties, such as differences in melting point, dissolution rate, and solubility. As mentioned above, the influence of polymorphism of both drugs and excipients that crystallize during freezing or during subsequent annealing processes (see below) has not been investigated extensively, but there is a large enough body of literature to support the idea that formulation scientists should be aware of the effects of seemingly subtle changes in formulation and processing conditions on the crystal form in the freeze-dried solid. Kim *et al.* (1998) demonstrated that different polymorphs of mannitol form during freezing depending both on the initial concentration of mannitol and the rate of freezing. Chongprasert *et al.* (1998) studied the effects of freezing conditions on crystallization of pentamidine isethionate and established that at least two polymorphs are observed in the freeze-dried solid, depending on the thermal history of freezing. In this example, different photostability is observed in the freeze-dried solid for product frozen under different conditions.

X-ray powder diffraction can also be used to study crystallization processes directly in frozen systems by using a low-temperature stage. Cavatur and Suryanarayanan (1998) applied this technique to the *in situ* crystallization of nafcillin sodium and sodium phosphate dodecahydrate from frozen solutions.

For a comprehensive treatment of the theory and practice of X-ray diffraction, the reader is referred to Klug and Alexander (1976).

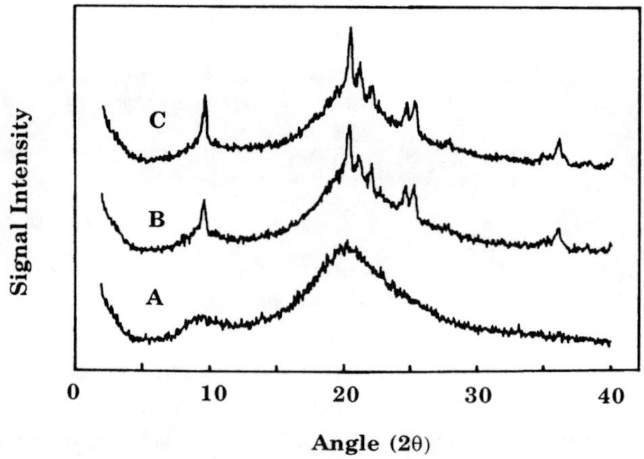

Figure 20. X-ray powder diffractograms showing crystallization of mannitol during storage at 40°C. (A) Initial; (B) 3 weeks of storage; (C) 5 weeks of storage.

Table IV
Characteristic Amide I Absorption Frequencies for Common Secondary Structures[a]

Structure	Amide I Frequency (cm^{-1})
Antiparallel β sheet/aggregated strands	1675–1695
3_{10} helix	1660–1670
α-helix	1648–1660
Unordered	1640–1648
β Sheet	1625–1640
Aggregated strands	1610–1628

[a] Data from Jackson and Mantsch (1995).

4.5. Infrared Spectroscopy

It has been known since the early 1950s that a correlation exists between absorption frequencies in the amide regions of the infrared spectrum of proteins and the predominant secondary structural motif (α helix, β sheet, and so forth) of proteins as determined by other methods (Elliot and Ambrose, 1950). Advances in instrumentation and data processing software as well as the relative ease of experimental procedures, small quantity of protein required (about 100 μg), and moderate cost of the instrumentation have spurred interest in the application of infrared spectroscopy to characterization of freezing and drying effects on protein structure.

The amide I absorption, from about 1600 to about 1700 cm^{-1}, arises primarily (about 80%) from C=O stretching, with a smaller contribution (about 20%) from C–N stretching. The amide II absorption (about 1600–1500 cm^{-1}) is due to both N–H bending vibrations and C–N stretching vibrations in roughly equal proportions, and the amide III region (about 1330–1220 cm^{-1}) consists of C–N stretching vibrations coupled to N–H in-plane bending vibrations with contributions from C–C stretching and C=O in-plane bending (Jackson and Mantsch, 1995). Several other amide absorption regions are discussed by Dong *et al.* (1995). Most investigators using IR spectroscopy to examine protein structure use the amide I region, probably because most of the absorption arises from one functional group. However, Costantino *et al.* (1996a, 1996b, 1997a) have used the amide III region in several studies, pointing out that, unlike the amide I frequency range, water absorption does not interfere in the amide III region.

The utility of infrared spectroscopy for the study of protein structure results from the fact that each type of secondary structure has a characteristic hydrogen bonding pattern between the amide C=O and N–H groups. The strength of hydrogen bonding influences the absorption frequency, where stronger hydrogen bonding causes decreased electron density around the carbonyl group, resulting in absorption at a lower frequency than for a carbonyl group which is not hydrogen-bonded. Table IV summarizes approximate frequency assignments for different secondary structures.

Of course, proteins generally contain more than one type of secondary structure, where the width of individual absorptions is greater that the separation between bands, often giving rise to featureless spectra in the range of interest. Several techniques for processing of the data in order to enhance resolution are available, including Fourier self-deconvolution and calculation of derivative spectra. A detailed discussion of these techniques is beyond the scope of this chapter, and the reader is referred to reviews by Jackson and Mantsch (1995), Surewicz *et al.* (1993), and Dong *et al.* (1995). All of these reviews advise caution when attempting to apply FTIR quantitatively to determine the amounts of different types of secondary structure within a protein. Most pharmaceutically relevant studies use FTIR to compare the effects of different stabilizers on the spectrum of the dried protein relative to the same protein in aqueous solution. In this case, calculation of a spectral correlation coefficient between second derivative spectra has been reported (Dong *et al.*, 1995; Prestrelski *et al.*, 1995), which provides a quantitative estimate of the relative effects of formulation and processing parameters on protein structure.

The O–H bending absorption of water at 1644 cm^{-1} will obscure the amide I region of the spectrum. Most modern instruments have software for digital subtraction of the water or buffer spectrum from that of a protein solution. However, problems associated with the high molar absorptivity of water can cause problems with detector linearity. As a result, very short pathlength cells (not more than about 10 μm) are required for aqueous solutions of proteins, which means that relatively high concentrations of protein are needed (10–50 mg/ml). Alternatively, the protein can be dissolved in D_2O, because the O–D bending absorption is shifted to a frequency below the amide I region. Longer pathlength cells and lower concentrations of protein can then be used. Deuterons will exchange with the amide nitrogen, but this phenomenon can be exploited to gather information concerning the degree of solvent exposure within secondary structures. Residual water in dried proteins does not seem to be a significant source of interference in second derivative spectra up to about 10% residual water (Dong *et al.*, 1995), because the water absorbance is broad relative to the protein absorbances. This reduces the intensity of water absorbance in the second derivative spectra.

The flexibility of infrared spectroscopy with regard to the physical state of the sample makes it well suited to the characterization of freeze-dried proteins. In addition to examining the protein as a solution in water or D_2O, solutions in organic solvents can be used as well as colloidal dispersions and dried solids, either as a powder or as a film. Typically, freeze-dried powders are pressed into KBr pellets. General features of drying-induced changes in the amide I region include broadening of individual bands as a result of a general disordering of secondary structure (Prestrelski *et al.*, 1994). However, changes between the spectrum of a dried protein relative to the same protein in solution are very protein-specific. In general, drying-induced changes fall into three broad categories: (1) essentially no change in the spectrum upon drying, (2) conformational changes which are restored to the original solution conformation upon rehydration, and (3) conformational changes which are not restored upon hydration. A new band near 1615 cm^{-1} is commonly observed upon drying, which is consistent with intra- or intermolecular β sheets resulting from unfolding and aggregation (Prestrelski *et al.*, 1994). Prestrelski *et al.* (1995) studied the influence of formulation pH on lyophilization-induced secondary structural changes and on the solid-state stability of freeze-dried interleukin-2 (IL-2). Drying from a pH 7 solution resulted in significant changes in the spectrum in the amide I region, whereas IL-2 remained native if dried from a pH 5 solution. The spectra are shown in Fig. 21. Note the appearance of the band at 1617 cm^{-1}, which is consistent with a β sheet structure formed by unfolding and aggregation of the protein during lyophilization. Additional unfolding was observed upon incubation at elevated temperature, and a strong correlation was observed between retention of nativc structure after freeze-drying and stability as a freeze-dried solid.

Remmele *et al.* (1997) described an internal reflection IR technique for *in situ* monitoring of protein structure during freeze-drying. In this study, the amide I band was used to monitor the effect of sucrose in the stabilization of lysozyme.

4.6. Nuclear Magnetic Resonance

The application of nuclear magnetic resonance to the characterization of frozen solutions and freeze-dried solids has been primarily by use of low-field pulsed NMR, or relaxometry. This NMR technique does not provide chemical shift information, but is used to measure molecular mobility for solids, liquids, and frozen systems through the spin–lattice (T_1) and spin–spin (T_2) relaxation times. The reader is referred to Harris (1986) or other books on theoretical and experimental aspects of NMR spectroscopy for background information.

Izutsu *et al.* (1995, 1996) used low-field pulsed NMR to measure mobility of water and solutes in frozen solutions of various solutes in both water and D_2O, and used the term “softening temperature” to describe the

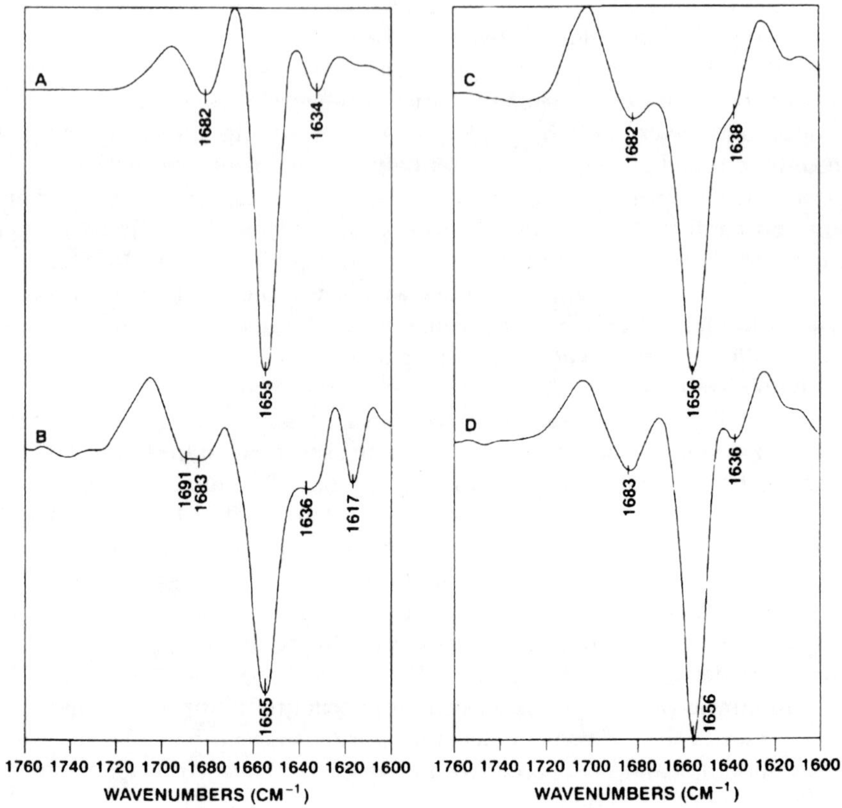

Figure 21. FTIR spectra of freeze-dried interleukin-2: (A) aqueous solution, pH 5; (B) lyophilized, pH 7; (C) lyophilized, pH 6; (D) lyophilized, pH 5. Reproduced with permission from Prestrelski *et al.* (1995). Copyright 1995 Plenum Press.

higher temperature transition such as that shown in Fig. 13. Izutsu *et al.* reported that water molecules gain mobility at significantly lower temperatures than solutes, where water molecules become mobile at around the "real" glass transition temperature and solute molecules become mobile at the softening temperature.

Yoshioka *et al.* (1996) measured molecular mobility of BSA and bovine IgG (BGG) as a function of water content by measuring T_2 and correlating this measurement with aggregation susceptibility. Increased aggregation susceptibility appeared to be directly related to T_2. In later work by the same research group (Yoshioka *et al.*, 1997a), the free induction decay (FID) signal of freeze-dried BSA and BGG was analyzed as a function of temperature at different water contents. At low temperature, the FID is composed of a single

Lorentzian decay arising from water protons and a single Gaussian decay due to protein protons. As the temperature increases, however, a temperature is reached where the FID is no longer described by these two decays and an additional Laurentian decay is needed. This is illustrated in Fig. 22 for bovine IgG containing 0.212 g H_2O/g protein at 10°C and 90°C. This was interpreted as the appearance of a population of protein protons with higher mobility, and the temperature at which this population appeared was defined as the softening temperature, or $T_{s(T_2)}$. Yoshioka *et al.* proposed that this softening temperature is an important property of protein formulations, particularly because T_g of freeze-dried protein formulations as determined calorimetrically is often uncertain, as discussed above. An abrupt increase in the rate of aggregation of both BSA and BGG was observed above the softening temperature, although significant aggregation was observed below this temperature, particularly at low moisture levels. The term $T_{s(T2)}$ was later replaced by T_{mc}, or the "mobility changing" temperature (Yoshioka *et al.*, 1997b), and applied to study the stability of BGG in model formulations containing varying molecular weights of dextran and poly(vinyl alcohol) (PVA) (Yoshioka *et al.*, 1998) as an excipient. Both T_{mc} and protein stability increased as the molecular weight of dextran increased, and the temperature dependence of the protein aggregation rate changed at T_{mc}. Above T_{mc}, the temperature dependence of the aggregation rate was related to T_{mc} by the Williams–Landel–Ferry equation, where T_g was replaced by T_{mc}. Formulations containing PVA were less stable than those containing dextran, despite having lower residual water contents. This was attributed to the lower T_{mc} of the PVA formulations.

4.7. Dielectric Analysis

Measurement of the electrical properties of frozen systems by monitoring electrical resistance as a function of temperature has been used as a characterization tool since the early days of freeze-drying. The principle of the technique is that, when an aqueous solution is thoroughly frozen, the mobility of charge-carrying species is restricted. As the frozen system warms and begins to thaw, however, the mobility of charge-carrying species increases markedly and causes a large decrease in electrical resistance. This is a sensitive method for the detection of eutectic melting. However, the interpretation of electrical resistance data as a function of temperature is uncertain in frozen systems containing amorphous solutes, and widespread use of this technique has been hampered by the frequency dependence of the response and, more generally, by a lack of standardized instrumentation and methodology.

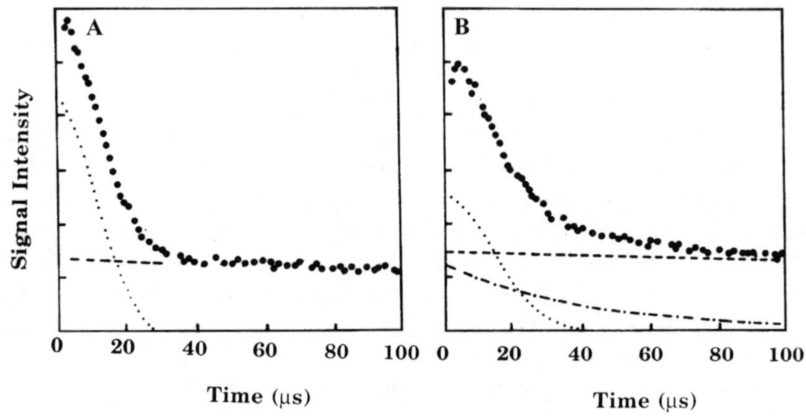

Figure 22. NMR free induction decay signal for freeze-dried bovine IgG containing 0.212 g H_2O/g protein at 10°C (A) and 90°C (B). Reproduced with permission from Yoshioka *et al.* (1997a). Copyright 1997. American Chemical Society and American Pharmaceutical Association.

A more recent advance in electrical thermal analysis is dielectric relaxation spectroscopy (DRS). The reader is referred to Smith *et al.* (1995) for a comprehensive review of the theory and applications of this method. Briefly, DRS measures the rate and extent of polarizability of materials when placed in an oscillating electromagnetic field. At low frequency, the molecular dipoles reorient with the applied field so as to minimize the potential energy of the system. As the frequency is increased, a frequency is reached where the molecular dipoles can no longer reorient fast enough to stay in phase with the applied field. The frequency response of a given material depends on the rotational freedom of the molecule, which is a reflection of its environment and its interaction with neighboring molecules. The method covers a broad frequency range (from about 10^{-5} to about 10^{11} Hz), thus enabling investigation of a wide range of processes. Since multiple frequencies are used, DRS can distinguish between first-order transitions, which show no frequency dependence, and second-order transitions such as the glass transition, which are frequency-dependent (Evans *et al.*, 1995). DRS can be applied to characterization of both frozen systems and dried solids, although there have been few studies reported in the field of freezing and freeze-drying. Morris *et al.* (1994) applied DRS to the prediction of collapse temperatures for binary glass-forming systems, including sucrose, trehalose, an antibiotic, and sorbitol in water. A "take-off" frequency model was used to establish good correlation between the predicted onset of viscous flow as measured by DRS and collapse as directly observed by microscopy.

5. THE DRYING PROCESS

5.1. Annealing in Freeze-Drying

Annealing actually precedes the drying process, and refers to warming the product to a subzero temperature—while avoiding melting the product—and holding for a period of perhaps 2–6 hr in order to improve the freeze-drying characteristics of the formulation. This is generally done in order to facilitate crystallization of a component of the formulation, either the drug substance or an excipient. The rationale for annealing is based on glass transition-associated mobility, where mobility increases dramatically above the glass transition temperature. Holding the system above the glass transition temperature of the metastable glass may then allow crystallization of components which would otherwise not crystallize. Oguchi *et al.* (1990) used annealing in the temperature range of $-13°$ to $-18°C$ to prepare crystalline inclusion complexes of α-cyclodextrins with a series of benzoic acid derivatives as guest molecules. These investigators pointed out the marked temperature dependence of annealing, where decreasing the annealing temperature to $-21°C$ resulted in no crystallization of the complex even when held for up to 24 hr. Gatlin (1991) studied the kinetics of crystallization of cefazolin sodium during annealing at a range of temperatures above the glass transition. Analysis of the data using the Johnson–Mehl–Avrami equation yielded kinetic data that were consistent with a crystallization mechanism of heterogeneous nucleation of a two-dimensional growth habit.

In addition to promoting crystallization of solutes, annealing can also promote crystallization of unfrozen water in glassy systems. Annealing even a few degrees below the glass transition temperature can thus deplasticize the amorphous phase, resulting in a readily measurable increase in the glass transition temperature.

Sarciaux *et al.* (1999) studied the effect of annealing on aggregation of bovine IgG after freeze-drying, and observed decreased aggregation following annealing along with a significant reduction in the specific surface area of the freeze-dried solid. This observation is consistent with a mechanism involving denaturation at the ice–freeze-concentrate interface, which becomes irreversible after freeze-drying.

An important example of annealing with respect to protein formulations is the sucrose/glycine system. Figure 23 shows how the thermogram of the frozen system changes with composition, where the total solids concentration is 10% (w/v), and the ratios of the two solutes are indicated in the figure. Addition of glycine to a sucrose solution at relative concentrations too low for glycine to crystallize causes a decrease in the apparent glass transition of the frozen system. Figure 24A shows the appearance of freeze-dried solids as a function of the composition of this system, where samples were frozen at −45°C and freeze-dried at a shelf temperature of −30°C. Under these drying conditions, the samples containing 7:3 and 6:4 weight ratios of sucrose-to-glycine exhibit collapse. As the relative concentration of glycine increases, glycine is able to crystallize. Glycine crystallization allows formation of a freeze-dried solid which retains its macroscopic structure during freeze-drying and results in a pharaceutically acceptable cake. If the frozen solutions with sucrose-to-glycine ratios of 7:3 and 6:4 are annealed at −20°C for 3 hr prior to drying, glycine crystallizes from these systems. Freeze-drying of the annealed samples results in a pharmaceutically acceptable cake, shown in Fig. 24B.

Searles, Carpenter, and Randolph (2001a, 2001b) reported that the main determinant of primary drying rate is the ice nucleation temperature, where a higher degree of supercooling results in smaller ice crystals, a higher resistance to mass transfer, and a slower drying rate. However, ice nucleation is not directly controllable in normal pharmaceutical freeze-

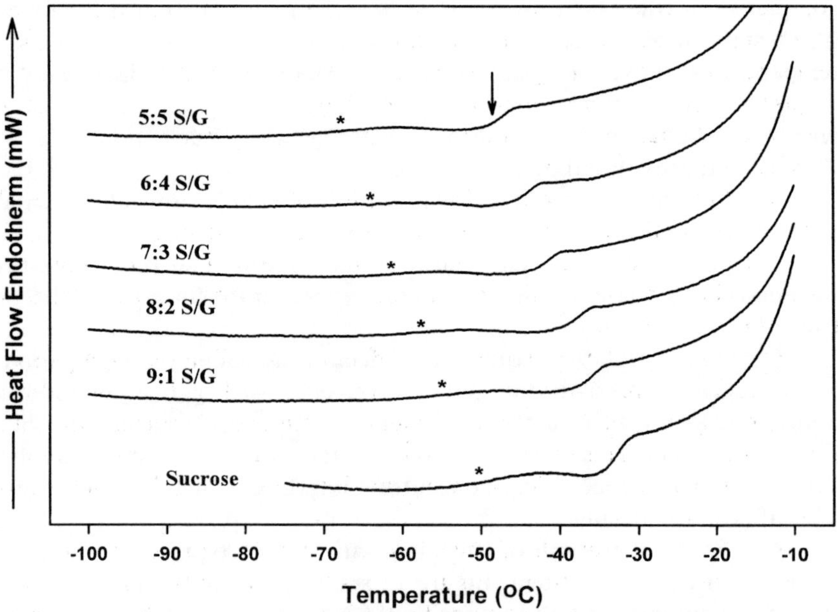

Figure 23. DSC thermograms of sucrose/glycine as a function of ratio of sucrose to glycine. Asterisk indicates lower temperature transition and arrow indicates higher temperature transition.

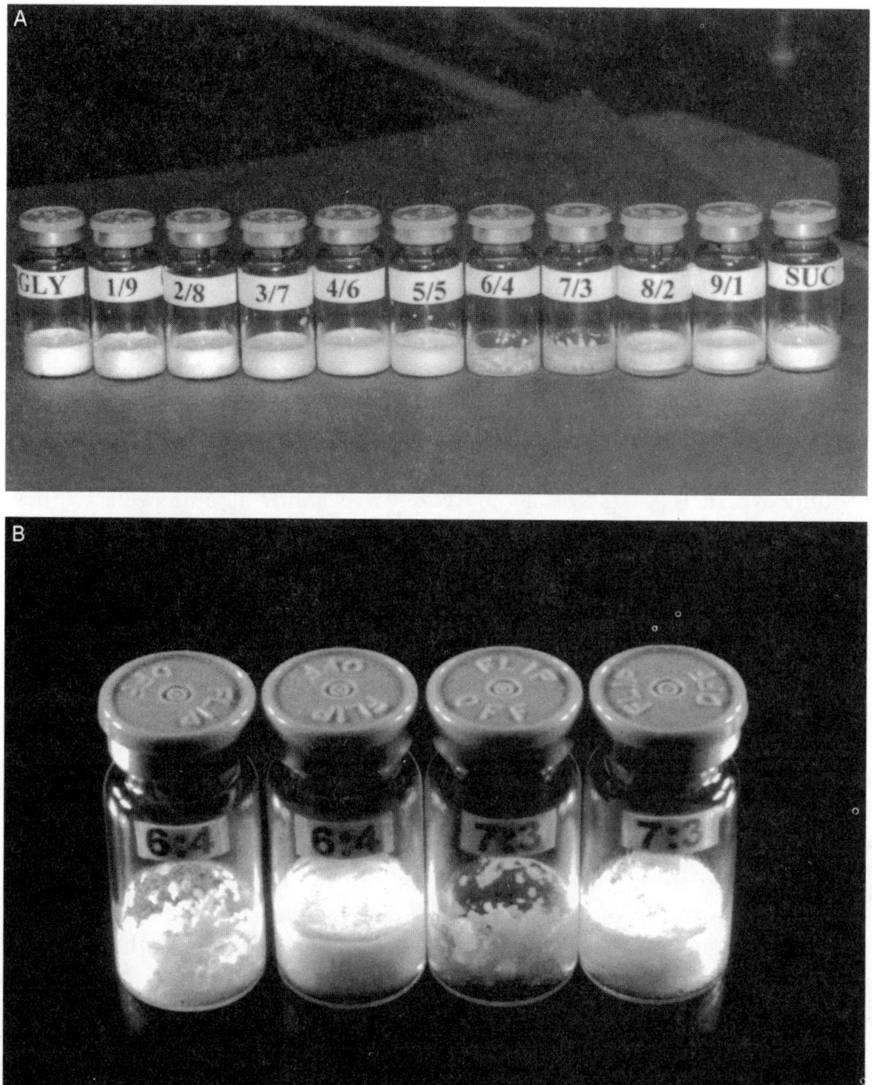

Figure 24. Freeze-dried samples of sucrose glycine at different weight:weight ratios of the two solutes (A) and the effect of annealing on structural collapse during freeze-drying (B) of 7:3 and 6:4 sucrose/glycine systems.

drying. Annealing at temperatures above the T_g' of the formulation was found to increase ice crystal size. This not only reduces the heterogeneity in drying rate caused by variation in ice nucleation, but also results in increased primary drying rate by as much as a factor of 3.5. Annealing below T_g' did not result in increased drying rates.

5.2. Collapse during Freeze-Drying

If for no other reason than pharmaceutical acceptability, drying conditions must be developed which avoid collapse during freeze-drying. However, a more fundamental question must be asked: Does collapse during freeze-drying cause a marked decrease in recovery of activity after freeze-drying? Furthermore, does collapse during freeze-drying result in a product which is less stable as a freeze-dried solid? The answers to these questions have important product quality implications, particularly in the widely observed case of partial or localized collapse in a few vials within a batch of product. Total collapse can be easily observed visually, and these vials can be culled from the batch during inspection. However, areas of collapse may be present within the freeze-dried solid which are not apparent during inspection. Since measurement of protein activity after freeze-drying is based on statistical sampling, a small percentage of partially collapsed vials could easily go undetected. If collapsed regions within the lyophilized solid were associated with significantly lower recovery of activity or significantly decreased stability, the consequences could be subpotent vials, which is far more significant than pharmaceutical elegance.

There has been little published literature on the effect of collapse during freeze-drying on stability of proteins against damage during freeze-drying, although there is widespread acceptance of the idea that collapse can cause a sharp drop in recovery of activity. Jiang and Nail (1998) measured the profile of activity loss during freeze-drying of the model proteins catalase, β-galactosidase, and lactate dehydrogenase under conservative drying conditions resulting in complete retention of microstructure versus aggressive conditions which resulted in collapse during primary drying. Results are shown in Table V. The only protein for which a large decrease in recovery of activity was observed is LDH. This may be due at least in part to the sensitivity of LDH to freeze–thaw damage, since some melting of frozen solution during primary drying could not be ruled out. These results may also be partially confounded with the effect of residual moisture in the freeze-dried material. It was found in this study that the most important determinant of activity recovery after freeze-

Table V
Recovery of Activity of Model Proteins after Freeze-Drying under Conditions of Retention of Structure and Collapse of Cake

Protein	Structure retention	Collapse
Catalase	92	88
β-Galactosidase	95	85
Lactate dehydrogenase	95	50

drying was the amount of residual moisture in the lyophilized solid, and collapsed material, having a lower specific surface area, tends to have a high residual moisture level.

5.3. Heat and Mass Transfer in Primary Drying

The heat of sublimation of ice must be supplied by transfer of heat through a series of resistances from the freeze-dryer shelf to the ice at the sublimation front. Of course, this must be done without melting the frozen material in contact with the bottom of the container. The rate of the solid-to-vapor phase change is, relative to the time course of freeze-drying, instantaneous. Following the phase change, water vapor must be removed by flow through the porous matrix of partially dried product. A knowledge of both heat and mass transfer is important for the development of optimum processing conditions. In particular, it is important to understand the dominant, or controlling, resistance in both of these transfer operations. There is a small, but important body of literature on heat and mass transfer in freeze-drying of pharmaceuticals, most notably by Pikal and co-workers (Pikal, 1985; Pikal *et al.*, 1983, 1984), and the reader interested in a detailed analysis is referred to these papers. The discussion below is intended to summarize this work and to provide broad guidelines pertaining to freeze-dry cycle development.

There are three mechanisms by which heat can be transferred:

- *Conduction*, the flow of heat by molecular motion between one differential volume element of a material and the next
- *Convection*, exchange of heat caused by bulk flow of a fluid
- *Radiation*, flow of heat by thermal excitation which gives rise to electromagnetic radiation

Because freeze-drying takes place in an evacuated system, heat transfer by convection is probably not a significant mechanism. This has been demonstrated by measuring the rate of heat loss from a heated filament as a function of orientation of the filament at pressures representative of freeze-drying (Dushman and Lafferty, 1962). Because of the modest temperatures used in freeze-drying, it has been shown that thermal radiation probably does not account for more than about 5% of the total heat flow (Nail, 1980), where the total heat flow is given by

$$Q = \Delta H_s \cdot m \tag{3}$$

where Q is the heat flow, which is linked to the rate of sublimation m by the heat of sublimation of ice ΔH_s. However, thermal radiation can be significant in some situations, particularly in scaleup from laboratory to production-scale equipment.

The dominant mechanism of heat transfer in freeze-drying is conduction. Analysis of this heat transfer process (Nail, 1980) has demonstrated that the controlling resistance to heat transfer by conduction is the poor thermal contact between the surface of the freeze-dryer shelf and the product. This contact resistance is particularly important when freeze-drying from vials. The rate of heat transfer is thus determined by the thermal conductivity of the vapor phase, which is a function of the pressure in the chamber. Under conditions of free molecular flow of the vapor phase (mean free path is larger than the characteristic dimension of the system), the thermal conductivity of the vapor increases linearly with pressure. As the system pressure increases to above about 0.1 mm Hg, the vapor flow regime becomes transitional, and thermal conductivity increases more slowly with increased pressure (at still higher pressure, where the flow regime is viscous, thermal conductivity of a gas is independent of pressure). Analysis of heat transfer leads to the conclusion that, contrary to earlier belief, very low chamber pressure during freeze-drying slows the rate of drying by inhibiting heat transfer.

Selection of an optimum system pressure involves a trade-off between heat transfer and mass transfer considerations. The pressure in the chamber, which consists almost entirely of water vapor during primary drying (Nail and Johnson, 1992), must be lower than the vapor pressure of ice in the product in order for drying to take place. However, a pressure which is too low unnecessarily retards the process. The optimum chamber pressure depends on the desired product temperature during primary drying. A reasonable guideline for selection of an optimum chamber pressure during primary drying is that the chamber pressure should be no more than about one-half and not less than about one-fourth of the vapor pressure of ice at the desired product temperature. For example, assume that the target product temperature during primary drying is −20°C. From Table I, the vapor pressure of ice at this temperature is 0.76 mm Hg. An optimum pressure for primary drying would be in the range of 0.25–0.35 mm Hg

(250–350 μm Hg). If the product temperature must not exceed, for example, −40°C during primary drying, then the viewpoint that "the lower the pressure, the better" is appropriate, because the optimum system pressure would be less than 0.05 mm Hg.

When freeze-drying from vials, the vial itelf is the main reason for the poor thermal contact between the shelf and the product. Seemingly insignificant variations in vial geometry can have a measurable impact on heat transfer and, consequently, drying rate. It is often useful to measure the vial heat transfer coefficient. These data can be valuable in vial selection for new products and evaluation of the impact of a change in vial supplier, vial size, or vial type (molded vs. tubing glass) with respect to impact on freeze-drying characteristics. A simple procedure for measuring the vial heat transfer coefficient is as follows (Pikal, 1985). First, the vials to be measured are filled to no more than about one-third full with water, marked for identification, and weighed. Thermocouples are carefully placed in several vials such that the thermocouple is touching the bottom center of the vial (see Fig. 1) and at least one other thermocouple is placed on the surface of the shelf. Vials are then transferred to the freeze-dryer and frozen. Freeze-drying is carried out at the desired shelf temperature and at constant chamber pressure. After a period of time during which at least half of the ice is sublimed, the process is stopped, the drying time is noted, and the vials are weighed. The mass loss per unit time is calculated, and the total rate of heat transfer is calculated from Eq. (3). The vial heat transfer coefficient is then calculated from

$$Q = A_v K_v (T_s - T_b) \tag{4}$$

where A_v is the cross-sectional area of the vial based on the outside vial diameter, T_s is the shelf surface temperature, and T_b is the temperature measured at the bottom center of the vial. Since the vial heat transfer coefficient is a function of pressure, K_v should be calculated at several pressures in the range of 0.05 to perhaps 0.4 mm Hg. Representative vial heat transfer coefficient data are shown in Fig. 25, which illustrates that the vial heat transfer coefficient is strongly dependent on both the pressure and the type of vial.

Analysis of mass transfer follows the same general approach as analysis of heat transfer, where

$$\text{Sublimation rate} = \text{Pressure difference/Resistance}$$

and the total resistance to mass transfer consists of three components: (1) the dried product layer, which increases in thickness as drying proceeds, (2)

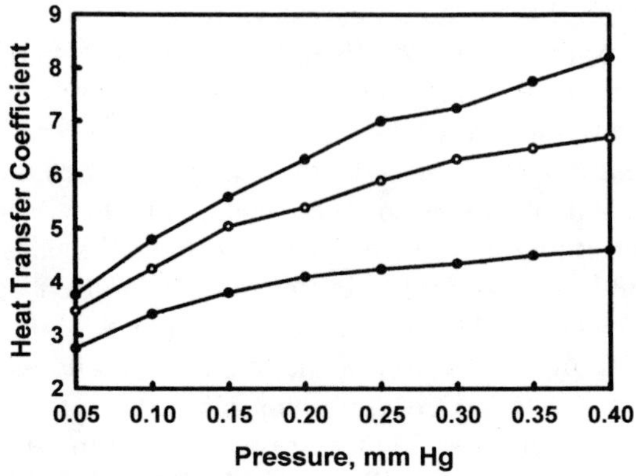

Figure 25. Representative vial heat transfer coefficient data illustrating dependence on type of vial and chamber pressure. Reproduced with permission from Pikal (1985). Copyright 1985 PDA Journal of Pharmaceutical Science and Technology.

the partially inserted stopper, and (3) the resistance of the freeze-dryer itself to flow of vapor from the drying chamber to the condenser. Mass transfer in freeze-drying has been rigorously analyzed by Pikal *et al.* (1983; Pikal, 1985), and the limiting resistance has been identified as the dried product layer, which usually accounts for more than 90% of the total resistance to mass transfer. Measurement of dried product resistance is more involved than measurement of heat transfer coefficients. The most precise method is the microbalance technique reported by Pikal, which requires specialized equipment. The "vial" method, also reported by Pikal, also requires special equipment in order to measure the pressure in the headspace of the vial. Pikal accomplished this by a sidearm connected to a pressure gauge. A more recent method, reported by Milton *et al.* (1997) and discussed below, uses analysis of the transient pressure response when the valve between the chamber and condenser is briefly closed during primary drying. Although this method requires only an external condenser design with a valve that closes quickly (about 0.5 sec), further study is needed in order to verify the accuracy of this method for resistance of the product to water vapor transport.

Given that the product resistance is the controlling resistance to mass transfer and that this resistance increases with product thickness, the depth of fill should be minimized as much as practical. In general, it is good practice to avoid filling a vial more than about one-third full. This also helps to prevent vial breakage during freezing, as discussed above for mannitol-based formulations. There is a trade-off here, however, because minimizing the depth of fill also means that less product can be placed in the freeze-dryer.

5.4. Secondary Drying

Secondary drying is removal of water which did not freeze. This is a particularly important aspect of drying of protein formulations because, unlike traditional drugs of low molecular weight, proteins can be subject to loss of activity by overdrying. For example, Hsu *et al.* (1992) studied the effect of residual moisture in an excipient-free freeze-dried formulation of tissue-type plasminogen activator (tPA) on both turbidity upon reconstitution (insoluble aggregates) and stability of the biological activity upon storage. This was related to the water vapor adsorption isotherm of the protein. Samples that were dried to a residual water level below monolayer coverage of the solid exhibited greater turbidity upon reconstitution than samples containing a higher residual moisture level, and the turbidity of the driest material increased during storage at 50°C more rapidly than samples containing a higher residual moisture content. However, samples that were dried to a residual moisture level corresponding to monolayer or multilayer coverage showed more rapid loss of biological activity upon storage. The results suggest that dryer is not always better, and that the optimum moisture level must be chosen based on both physical stability and chemical stability considerations.

As with heat and mass transfer during primary drying, it is useful to know the rate-limiting process during the secondary drying phase of freeze-drying. The possibilities are (1) diffusion in the glassy matrix to the surface of the solid, (2) evaporation at the solid surface, and (3) water vapor transport through the porous bed of solids. Pikal *et al.* (1990) studied the kinetics of secondary drying as a function of temperature and pressure, although protein formulations were not used for these studies. Mannitol was used as a model crystalline solid along with two amorphous materials, moxalactam disodium and polyvinylpyrrolidone. It was reported that residual moisture decreases rapidly during the first few hours of secondary drying, then slows considerably, and reaches a plateau level which exceeds the equilibrium moisture content of the solid calculated from the desorption isotherm and the partial pressure of water vapor in the chamber. This plateau level is very temperature-sensitive, and higher secondary drying temperatures cause a significant reduction in the plateau level. The secondary drying rate increases with specific surface area of the dried solids, and the rate of secondary drying was observed to be independent of the chamber pressure during secondary drying. The rate of secondary drying was found not to be dependent on the depth of the dried product layer. This, along with the independence of secondary drying rate on pressure, supports the conclusion that the rate-limiting step is not water vapor transport through the porous bed of solids. The rate-limiting step must therefore be either diffusion of water through the solid or evaporation at the solid surface.

Spieles *et al.* (1995) studied secondary drying using aqueous solutions of hydroxyethyl starch, which were concentrated to the point where ice formation was prevented during freezing. In this case, the entire drying process consists of secondary drying. Drying kinetics was studied by measuring weight loss at constant temperature in the range of 254–293 K and pressure in the range of 5–44 Pa until equilibrium was reached after 5–8 days. The main focus of this study was the relationship between water activity and residual moisture content, which was described by a Freundlich relationship

$$W = q(a_w)^r \tag{5}$$

where q and r are empirical constants. The relationship between water content and water activity was independent of temperature over the range studied. This finding is not consistent with Pikal's data, although the experimental conditions and, in particular, the time scale of the drying experiments were quite different.

When small changes in residual moisture have a significant impact on a critical product quality attribute such as stability, the distribution of residual moisture levels between vials and within a lyophilized cake may be as important as the average residual moisture level. Vial-to-vial variability should be determined by extensive sampling during process validation, ideally by "mapping" residual moisture level as a function of position in the freeze-dryer. "Edge" effects can be significant, where vials at the edge of a shelf dry faster and have lower residual moisture levels than the bulk of the lot because of lateral heat transfer from the sides, the door, and the back of the chamber. Pikal and Shah (1997) studied the intravial distribution of moisture. Theoretical analysis predicted that residual moisture near the top of the lyophilized cake is lower than at the bottom, particularly for large fill depths. Experimental data demonstrated that the top is drier than the bottom of the cake, but that the "outer" section of the cake—that nearest the vial wall—is consistently the driest.

Better definition of the role of residual moisture in the stability of freeze-dried proteins and better control of the secondary drying process are subjects of significant practical importance where more research is needed.

5.5. Process Monitoring

Optimization of freeze-drying conditions requires control of shelf temperature and chamber pressure such that the product temperature is just below the maximum allowable temperature during primary drying. Measurement of product temperature is therefore necessary for cycle development and scale-up. In order to obtain the most meaningful data from product temperature measurement, it is important to pay attention both to use of an appropriate temperature-measuring device and placement of the sensor within individual vials. The best place to measure product temperature is in the center of a vial, with the sensor touching the bottom of the vial. The reason for this placement is that, whereas the sublimation front is planar at the beginning of primary drying, it becomes concave (as viewed from the bottom of the vial) as the sublimation front recedes, and the curvature increases with time. This effect can be easily demonstrated by observing freeze-drying of ice, where the last ice to sublime is at the bottom center of the vial. Thermocouples are better for measurement of temperature at a precise location than resistance-type temperature detectors (RTDs), because the measuring element for a RTD is larger, and the temperature measured is an average temperature over the surface of the measuring element. Thin-gauge thermocouples are recommended not only because they facilitate precise location of the thermocouple, but also because the light weight and better flexibility of the wire minimizes problems with the knocking down of vials by the thermocouple wire. A device similar to that shown in Fig. 1 is useful for precise placement of thermocouples, because the thermocouple is held in place by a cap which fits over the top of the vial, and the wire is flexible enough to be located accurately, yet rigid enough to stay in position during the drying cycle.

Although these measures can minimize inaccurate process data arising from sensor placement, there are other sources of uncertainty in product temperature measurement which are not a function of placement. The temperature sensor facilitates nucleation of ice, so that monitored vials generally undergo less supercooling during freezing than nonmonitored vials. This results in a slower effective freezing rate for monitored vials, a larger average ice crystal size, larger pores in the dried solid, less resistance to mass transfer during primary drying, and faster drying. Monitored vials are generally considered to dry about 10% faster than the nonmonitored vials. For this reason, soak times are used when product temperature data are used to sequence a cycle automatically from freezing to freeze-drying.

The trend in the pharmaceutical industry is toward larger production freeze-dryers, and scaleup to large equipment brings about more problems with product temperature measurement, where an important issue is location of monitored vials within the batch. If vials are placed in metal trays, it is best to monitor vials in the center of the tray, because warpage (with resulting poor thermal contact with the shelf) of a tray tends to be maximum at the center, and vials in the center receive less lateral heat transfer from the sides of the trays. Both of these factors contribute to vials in the center of a tray drying slower than vials near the perimeter. Likewise, the dryer has its own "edge effect" due to heat transfer from the side walls and the door of the chamber. This means that the most appropriate trays to monitor are in the center of a given shelf. However, easy access to the center of a shelf can be problematic, particularly for large freeze-dryers. Since placement of temperature probes is largely a manual process requiring close proximity of an operator to open vials of product, placement of product temperature probes compromises sterility assurance to some extent.

For these reasons, there is a need for technology which enables monitoring of product temperature without placing temperature probes in individual vials or product. Although there is currently no way to do this directly, Milton *et al.* (1997) explored measurement of product temperature by a manometric method. In this technique, which requires an external condenser design, the isolation valve between the chamber and the condenser is closed for between 10 and 20 sec during primary drying, and the transient response of the chamber pressure is monitored. The data are fitted to an equation based on a theoretical model of the transient pressure response, which takes into account (1) continued sublimation of ice while the valve is closed, which stops when the chamber pressure reaches the vapor pressure of ice at the temperature of the sublimation front, (2) continued heating of the product during the measurement, and (3) equilibration of temperature across the frozen layer during the measurement. Least squares fitting of the data to the model results in estimates of the value of three variables: (1) product temperature at the sublimation front, (2) the resistance of the dried product layer to transport of water vapor, and (3) the vial heat transfer coefficient. Although further refinement of this method is needed in order to explain the effects of batch size and edge effects on accuracy of temperature and product resistance data, it does represent a step toward developing an approach to measuring product temperature while avoiding the problems associated with monitoring the temperature of individual vials.

Another reason for monitoring individual vials of product is to detect the end of drying. The end of drying (see Fig. 2) is approached as the product temperature approaches the shelf temperature during secondary drying. Again, this information is necessary for cycle development, even though the nonrepresentative nature of monitored vials must be recognized and compensated for by the use of "soak times" at the end of the process. When it is desired to eliminate product temperature probes for routine monitoring of production lots, it is helpful to have another way to detect the end point of drying, preferably a method that is based on monitoring of the composition of the vapor phase in the chamber. There have been a few published reports of such methods for detection of drying end point. Nail and Johnson (1992) described a comparative pressure measurement technique, where a capacitance manometer pressure gauge (which measures total pressure, independent of vapor composition) is used to control the pressure, and a thermal conductivity-type pressure gauge (which is preferentially sensitive to water vapor) is used to monitor pressure. At the end of drying, the apparent pressure as determined by the thermal conductivity gauge decreases in response to the decreased partial pressure of water vapor in the chamber. Roy and Pikal (1989) described use of an electronic hygrometer to measure the partial pressure of water vapor in the chamber and demonstrated that this is a sensitive method for detecting the end of drying. A limitation of this type of instrumentation is that, at this writing, it cannot be steam-sterilized. A "pressure rise" technique can be used, similar to the manometric temperature measurement method described above, except that the valve is closed during secondary drying. When drying ceases, the pressure rise due to continued evolution of water vapor from the product ceases, and the pressure rise is only due to leaks in the system. This can be an effective method, provided that the "background" leak rate is low and approximately constant. Use of a residual gas analyzer (RGA) for monitoring the lyophilization process has been described by Connelly and Welch (1993) and Nail and Johnson (1992). The RGA is a small, quadrupole mass spectrometer with an upper limit of mass-to-charge ratio m/e of perhaps 200. Monitoring of the composition of the vapor in the freeze-dry chamber provides a sensitive method for endpoint detection as the partial pressure of water vapor decreases. Connelly and Welch (1993) present data on partial pressure of water vapor versus time for freeze-drying of bovine serum albumin showing a gradual decrease in partial pressure of water vapor during secondary drying. Nail and Johnson (1992), using a crystalline freeze-dried solid (mannitol), demonstrated a sharp drop in partial pressure of water vapor at the end of the process. As discussed above, this would be expected, because there is virtually no secondary drying when the solute crystallizes during freezing. In addition to end point detection, the ability to monitor vapor phase composition during the process has other applications, such as (1) monitoring oxygen partial pressure in the chamber during stoppering of protein formulations which are sensitive to oxygen in the headspace of the vial, (2) leak detection, either from an external leak or from an internal leak of refrigerant, and (3) monitoring evolution of solvents from the material being freeze-dried, either residual solvents from the purification process or organic solvents, such as ethanol or *t*-butanol, used in the formulation.

6. STABILITY OF FREEZE-DRIED FORMULATIONS

Stability assessment and shelf-life prediction is usually a major focus of a pharmaceutical scientist's attention in the development any freeze-dried dosage form, and this is especially true of freeze-dried protein formulations. Ideally, the product would demonstrate at least a 2-year shelf life during storage at room temperature. However, approximately 90% of all freeze-dried protein pharmaceuticals sold in the United States require refrigerated storage. This can be a major logistical problem, particularly in underdeveloped countries, where a suitable "cold chain" cannot be maintained. In addition, the predictive value of accelerated stability testing for freeze-dried products is questionable, and the Food and Drug Administration (FDA) does not recognize accelerated stability data in support of the requested expiration dating; that is, the expiration date must be based on actual data at the intended storage temperature. For these reasons, there is a need for further development of the application of pharmaceutical materials science to assist in designing more stable freeze-dried formulations as well as development of the scientific basis for meaningful interpretation of accelerated stability data.

6.1. Molecular Motion in Amorphous Systems and Relevance of the Glass Transition Temperature to Stability

Freeze-dried protein formulations are at least partially amorphous. Whereas a crystalline fraction is usually desireable in maintaining the mechanical integrity of the solid, it is the amorphous phase which contains the protein and predominates in determining both physical and chemical stability. The concepts of a glass transition and the temperature dependence of molecular motion in the region of the glass transition are central to understanding the behavior of amorphous materials. Given that molecular motion slows dramatically below T_g, there is a body of opinion which holds that all molecular motion relevant to storage stability ceases a few degrees below T_g (Franks, 1990; Hemminga *et al.*, 1993) and that storage below T_g is necessary and sufficient for stability. This point of view, sometimes known as the *vitrification hypothesis*, is an oversimplification, however, and a better understanding of the relationship between the temperature dependence of the molecular mobility and stability of pharmaceutical systems is a subject for which much more research is needed. This has been the subject of several excellent reviews, particularly by Hancock and Zografi (1997) and Angell (1995). The reader is referred to these reviews for a more in-depth treatment.

The temperature dependence of molecular motion in amorphous systems is described by the empirical Vogel–Tammann–Fulcher (VTF) equation,

$$\tau = \tau_0 \exp(B/T - T_0) \tag{6}$$

where τ is the molecular relaxation time, T is the temperature, τ_0 and B are constants, and T_0 is the temperature at which the configurational entropy of the system reaches zero. T_0 is also known as the Kauzman temperature, and is believed to be roughly 50°C below the experimentally measured glass transition temperature, depending on the nature of the system (see discussion below). Note that, when T_0 is zero, the VTF equation reduces to the Arrhenius equation, where B is analogous to the activation energy. The Williams–Landel–Ferry equation, originally developed to describe the temperature dependence of viscosity of solutions above the glass transition temperature, is another special case of the VTF equation:

$$\eta = \eta_g \exp\left[C_1(T - T_g)/(C_2 + (T - T_g))\right] \tag{7}$$

where η_g is the mean viscosity at T_g and C_1 and C_2 are constants equal to B/$(T_g - T_0)$ and $(T_g - T_0)$, respectively. The constants have been found to be universal for a variety of polymer systems (-17.4 and 51.6 for C_1 and C_2, respectively). Roy *et al.* (1991) tested the applicability of the WLF equation to the prediction of the temperature dependence of the stability of a freeze-dried monoclonal antibody–vinca conjugate above the T_g of three different formulations, all of which were below 40°C. Three degradation processes were monitored: aggregation, hydrolysis of the hydrazone linkage, and decomposition of the vinca moiety. Stability data at 25°C and 40°C were plotted as ln R/R_g versus $(T - T_g)$ for each of the processes. All were observed to be linear, which indicates that $C_2 > (T - T_g)$. The slope of the line, C_1/C_2, was constant for all of the degradation reactions.

Glasses have been classified as "strong" or "fragile," depending on the temperature dependence of the activation energy for molecular motion in the region of T_g (Angell, 1995; Hancock and Zografi, 1997). Strong glasses exhibit an Arrhenius temperature dependence of molecular mobility and have small changes in heat capacity at T_g and broad glass transition regions (when they can be detected at all). Proteins are good examples of strong glass-formers. Fragile glasses, on the other hand, undergo much larger changes in heat capacity at T_g and have narrower T_g ranges. The molecular mobility around T_g of a fragile glass is much more temperature-dependent than that described by the Arrhenius equation: The molecular mobility may change by an order of magnitude with a 10 K change in temperature, as opposed to roughly a factor of two over the same temperature range for a mobility which is described by the Arrhenius relationship. The values of B and T_0 in the VTF equation are related to glass fragility. Fragile glasses are characterized by low values of B (<10), and $(T_g - T_0)$ is generally less than 50. Strong glasses, on the other hand, have high values of B (>100), and $(T_g - T_0)$ is usually greater than 50. Another method for estimating the fragility of a glassy system simply uses the ratio of the melting temperature to the glass transition temperature (T_m/T_g, in K); strong glasses have ratios greater than 1.5 (Hancock and Zografi, 1997).

Hancock *et al.* (1998) described a method for determining fragility of some amorphous pharmaceutical materials based on measuring the width of the glass transition as measured by DSC. Fragility m has been defined as

$$m = \Delta H/(2.3RT_g) \tag{8}$$

where ΔH is the activation energy for structural relaxation and R is the gas constant. Further,

$$\Delta H = CR/(1/T_g^{\text{onset}} - 1/T_g^{\text{offset}}) \tag{9}$$

where C is a constant. Hancock and co-workers examined 11 pharmaceutical materials by this method, finding values of m in the range of 36–60, which corresponds to intermediate fragility. However, values of m were found to vary considerably with the experimental conditions used, particularly on whether measurements were made during heating or cooling. It also appears that the relationship between the width of the glass transition region and fragility is dependent upon the material being examined, which makes measurement of the width of the glass transition as a means of measuring fragility somewhat suspect.

When a glassy solid is allowed to "age" below its glass transition temperature, its normal molecular motions combined with the thermodynamic driving force toward a lower enthalpy state causes it to "relax," or assume a more ordered structure. This exothermic process is termed *enthalpy relaxation.* When the glass is reheated through the glass transition region, energy must be added to the system in order to reestablish the degree of disorder present in the viscous liquid state above the glass transition. This process is called *enthalpy recovery.* The rate at which the material relaxes is a measure of the molecular mobility below the glass transition, and the limited published data support the idea that this relaxation takes place at about the same time and temperature conditions for both polymeric species and low molecular weight glass-formers (Hancock *et al.*, 1995). This mobility is expressed quantitatively as the relaxation time constant, which is the average time required for a motion of a particular type. The time constant is calculated from enthalpy recovery data as follows. First, the maximum enthalpy recovery at a given temperature ΔH_∞ is calculated using the equation

$$\Delta H_\infty = (T_g - T)\Delta C_p \tag{10}$$

where T_g is the glass transition temperature, T is the experimental temperature, and ΔC_p is the change in heat capacity at T_g. The degree to which a material relaxes at any time and temperature is then given as

$$\phi_t = 1 - \Delta H_t/\Delta H_\infty \tag{11}$$

where ΔH_t is the measured enthalpy recovery at a given time and temperature of annealing. The structural relaxation time is then calculated, typically by using the Williams–Watts equation, where the ϕ_t data are fit to a stretched exponential function,

$$\phi_t = \exp(-t/\tau)^\beta \tag{12}$$

where τ is the relaxation time constant and β is a relaxation time distribution parameter, with a value between 0 and 1. Values of β and τ are calculated by nonlinear least squares regression. Values of β close to unity are representative of strong glasses, and the glass becomes more fragile as the value of β decreases. This approach was used by Duddu *et al.* (1997) to study sucrose and trehalose formulations of a monoclonal antibody in order to determine whether physical stability (aggregate formation in this case) is related in a predictable way to molecular mobility below the glass transition temperature. Values of β calculated from enthalpy recovery data were 0.38 and 0.44 for sucrose and trehalose formulations, respectively, and these values were confirmed by dielectric relaxation studies. Based on these data, the sucrose formulation was more fragile than the trehalose formulation. Although the sucrose formulation had the lower glass transition temperature (59°C and 81°C for the sucrose and trehalose formulations, respectively), the greater temperature dependence of the mobility of the sucrose formulation results in the relaxation time constant τ of the sucrose formulation being less than that of the trehalose formulation below about 12°C. It was not possible to test the significance of this observation with respect to rate of aggregation due to the very slow rate of aggregation at low temperatures. Although much more research is needed, these results suggest that both the glass transition temperature and fragility data are needed in order to correctly interpret accelerated stability data.

The data for the nonenzymatic browning reaction between xylose and lysine in a carboxymethylcellulose/lactose matrix as a function of $(T - T_g)$, shown in Fig. 26, are consistent with the idea of molecular motion relevant to storage stability ceasing below T_g. There is as yet insufficient published literature on actual rates of degradation reactions in protein formulations as a function of $(T - T_g)$ to allow broad generalizations, particularly at temperatures below T_g. Duddu and Dal Monte (1997) investigated the role of the glass transition temperature in the stability of a monoclonal antibody in either sucrose or trehalose. The sucrose formulation ($T_g = 59°C$) was observed to undergo aggregation during storage of the freeze-dried solid at 60°C, whereas the trehalose formulation ($T_g = 80°C$) was stable. Increasing the protein/sucrose ratio increased the T_g of both formulations. For example, increasing the protein concentration from 5 to 50 mg/ml at a constant sucrose concentration of 62.5 mg/ml resulted in an increase in T_g from 59°C to 89°C. Increasing the protein/sugar ratio also significantly improved the stability of the formulation. It should be remembered, however, that increasing the concentration of protein, in general, improves stability, and this observation may not be totally explained by changing the glass transition temperature. Duddu and Weller (1996) measured the stability of a freeze-dried aspirin formulation in the region of T_g and pointed out the significant error introduced by measuring the degradation rate above T_g and extrapolating to temperatures below T_g. For a system with a T_g of about 36°C, extrapolation of 40°C, 45°C, and 50°C data yielded a predicted rate constant of 0.009 day^{-1} at 22°C, whereas the observed value was 0.003 day^{-1}.

In studying the relationship between T_g and stability, it is important to recognize that water can serve not only as a plasticizer of the amorphous phase, but also as a reactant. Bell and Hageman (1994) used a freeze-dried preparation of aspartame in different molecular weights of polyvinylpyrro-

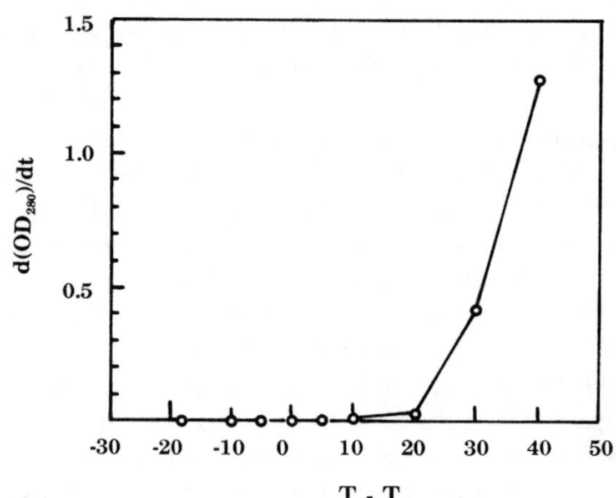

Figure 26. Rate of nonenzymatic browning reaction in a system of lactose:carboxymethylcellulose:xylose:lysine as a function of $(T - T_g)$. Reproduced with permission from Karel *et al.* (1993). Copyright 1993.

lidone (PVP) in order to address the question of which is more important, the glass transition temperature or residual water activity. The solid-state degradation of aspartame via rearrangement to a diketopiperazine was measured at 25°C after equilibrating the model formulations containing different molecular weights of PVP at different relative humidities. It was found that reaction rates at constant water activity, but different T_g values, were not significantly different. However, rates at similar $(T - T_g)$ values, but different water activities, were significantly different. Therefore, in this system, water activity is more important than $(T - T_g)$. This same approach has been applied in the authors' laboratory, where the model proteins LDH and β-galactosidase were freeze-dried in matrices consisting of different molecular weights of maltodextrin. The freeze-dried cakes were equilibrated for 2 days in desiccators containing saturated salt solutions at a range of relative humidities, and recovery of enzyme activity was measured at 40°C. The results are shown for β-galactosidase and LDH in Fig. 27 and Fig. 28, respectively, where the rate constant (first order for β-galactosidase, second order for LDH) is plotted against $(T - T_g)$ for different water activities. For β-galactosidase, the contrast of the rate versus $(T - T_g)$ plot to Fig. 26 is striking. The relatively high rates of loss of activity recovery at large values of $(T - T_g)$ are probably attributable to overdrying. There is no sharp increase in degradation rate above T_g, although it was not feasible to examine large values of $(T - T_g)$ and still maintain moderate accelerated stability temperatures for this system.

For LDH, the rate constant versus $(T - T_g)$ plot more closely resembles that of Fig. 26, but significant differences remain. First, the rate of loss of activity is easily measurable far below T_g. Second, in looking at the rate versus $(T - T_g)$ pattern for different water activities, it can be seen that, similar to the data reported by Bell and Hageman (1994), there is not a significant difference in rate constant at 40°C for the 11% RH group, despite a difference in $(T - T_g)$ of more than 50°C. This same pattern holds for the samples at higher water activity, despite some scatter in the data. The data above T_g are very limited for the reason mentioned above, but only maltose exhibits a sharp increase in rate constant above T_g. Based on width of the glass transition region, maltose appears to be the most fragile of the excipients studied, and would be expected to show the sharpest increase above T_g, assuming that the concept of strong and fragile glasses applies quantitatively to stability assessment of freeze-dried protein formulations. The maltodextrin system seems to offer a convenient model system for further investigation. Nonetheless, data from both the β-galactosidase and the LDH model systems support the conclusion that water activity plays a more important role in the stability than glass transition-associated mobility.

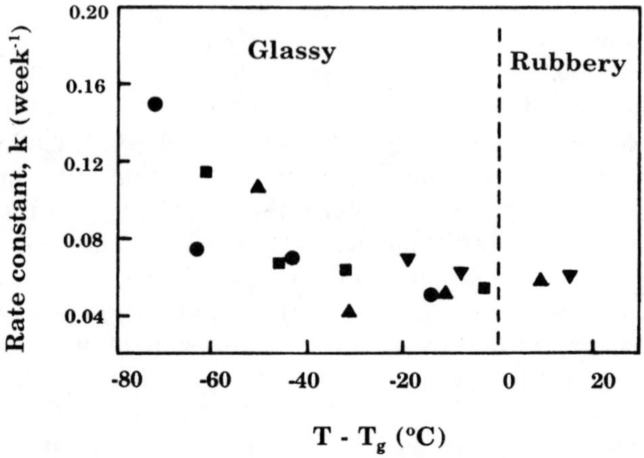

Figure 27. Solid-state stability of freeze-dried β-galactosidase versus $(T - T_g)$. (●) Solids equilibrated at 11% RH; (■) solids equilibrated at 23% RH; (▲) solids equilibrated at 33% RH; (▼) solids equilibrated at 43% RH.

6.2. Cake Shrinkage and Crystallization of Excipients during Storage

The glass transition temperature is certainly relevant to physical stability of the lyophilized solid, particularly during accelerated stability testing, because glass transition-associated mobility can cause shrinkage of the cake as the storage temperature approaches the glass transition temperature. This is particularly evident for predominantly amorphous solids, as illustrated in

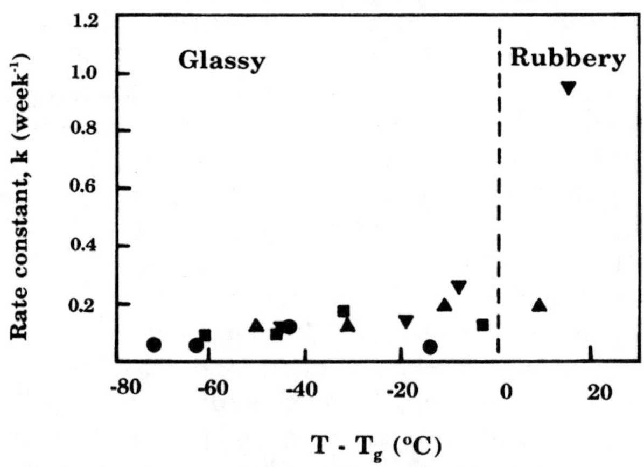

Figure 28. Solid-state stability of freeze-dried lactate dehydrogenase versus $(T - T_g)$. (●) Solids equilibrated at 11% RH; (■) solids equilibrated at 23% RH; (▲) solids equilibrated at 33% RH; (▼) solids equilibrated at 43% RH.

Fig. 29 for a series of sugars having a range of glass transition temperatures. If the freeze-dried solid contains a substantial crystalline fraction, this cake shrinkage may not be observed, because the collapsed amorphous phase is supported by the crystalline component, and little if any cake collapse is observed.

Crystallization of stabilizing excipients, usually sugars and polyols, has been shown to result in decreased stabilization of model proteins. Izutsu *et al.* (1993) used β-galactosidase in mannitol, inositol, and glucose to study the effects of excipient crystallinity on the short-term stability of the model protein. As the relative amount of mannitol and inositol in the formulation increases, the stabilizer crystallizes more readily, with a corresponding decrease in recovery of protein activity after freeze-drying. Annealing during the process was also shown to promote crystallization of the excipient and to decrease recovery of protein activity. This work was extended to the enzymes lactate dehydrogenase and L-asparaginase in a later study by the same authors (Izutsu *et al.*, 1994).

Sarciaux and Hageman (1997) examined the effect of relative concentration of recombinant bovine somatotropin (rbSt) and sucrose on the tendency of sucrose to crystallize. Sucrose crystallization was measured by MTDSC (see discussion above), using the temperature of the crystallization exotherm for sucrose. Addition of up to 20% rbSt to sucrose did not measurably change the T_g of sucrose; however, the crystallization temperature of sucrose increased from 110°C for sucrose alone to 140°C when 20% sucrose was added. Sucrose crystallization could be monitored gravimetrically by use of a controlled atmosphere microbalance at room temperature and elevated relative humidity. Induction time for sucrose crystallization at room temperature and 45% RH increased almost 10-fold with addition of 10% rbSt to sucrose. The authors conclude that using high protein concentrations results in a more robust formulation with respect to physical stability.

No reports have been published on the effect of excipient crystallization on long-term stability of a protein formulation. Herman *et al.* (1994) examined the long-term stability of a steroid, methylprednisolone sodium succinate, in the presence of either mannitol or lactose. The mannitol formulation, initially amorphous by X-ray powder diffraction, was shown to crystallize during storage at 40°C after a few weeks, and the steroid hydrolyzed at a relatively rapid rate as a freeze-dried solid. When mannitol was replaced with lactose, the formulation remained amorphous over the duration of the study and the drug was stable. This effect was attributed in part to redistribution of water in the freeze-dried solid as a result of mannitol crystallization. The increased water activity in the microenvironment of the drug is a likely cause of accelerated reactions in the solid state.

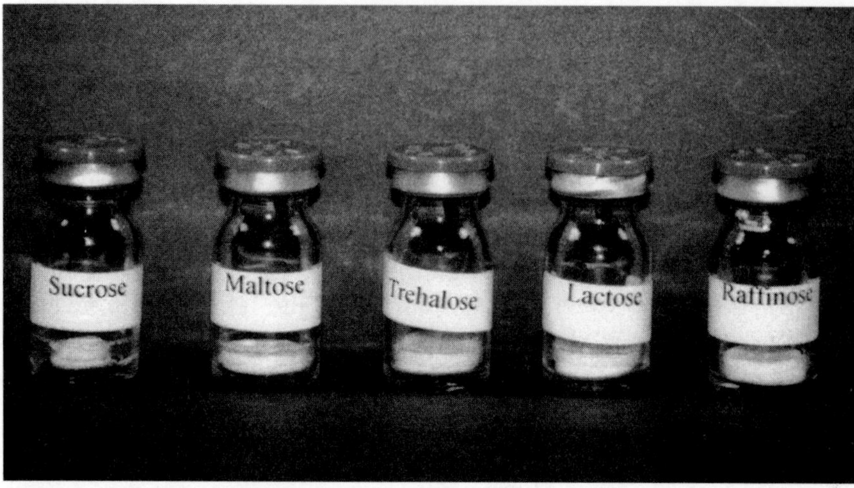

Figure 29. Freeze-dried disaccharides, showing collapse during storage at 50°C for 24 hr. As storage temperature increases, the solids collapse in order of increasing glass transition temperature.

It is good practice to include periodic examination of the physical state of freeze-dried protein formulations by X-ray powder diffraction during stability studies. Designing formulations with high T_g, although no assurance against chemical instability, is a rational method for minimizing physical instability such as collapse or crystallization of a critical component of the formulation.

6.3. Phase Separation during Freezing: Protein Stability Implications

As discussed in the section on freezing behavior, liquid–liquid phase separation can take place when enthalpically unfavorable interactions between solutes outweigh the favorable entropy of a homogeneous solution. Examples are polymer–polymer phase separation and polymer–phosphate buffer phase separation. This phase separation can also occur between different proteins and between proteins and polysaccharides. Heller *et al.* (1996) used recombinant human hemoglobin as a model protein and combinations of various concentrations of PEG and dextran to study the effects of phase separation on the secondary structure of hemoglobin after both freeze-thawing and freeze-drying. A solution of 7% PEG and 7% dextran separates into PEG-rich and dextran-rich phases at 0°C, and the concentration of hemoglobin in the dextran-rich phase is approximately three times that in the PEG-rich phase. The α-helical content, as measured by FTIR at 1656 cm^{-1}, was used to estimate the effects of freezing and freeze-drying on both phases. Both polymers have about the same protective effect during freezing, but these effects differ significantly during drying, with the PEG-rich phase showing the greater protective effect.

In an extension of this work, Heller *et al.* (1997) used 4% PEG and 4% dextran in a phosphate-buffered saline solution containing either NaCl or KCl. This combination of polymer concentrations is below the two-phase region at 0°C, and the concentrations of salt used were shown to have little effect on the phase diagram at room temperature. However, freeze-concentration would be expected to force this system into the two-phase region. Evidence of phase separation was observed by SEM for the formulation containing sodium chloride, and this phase separation was enhanced by annealing at −7°C. Loss of secondary structure was shown to correlate with the extent of phase separation during freezing. This effect was particularly striking for annealed versus unannealed samples in the sodium chloride-containing system. Some protection of native structure was observed for unannealed samples; however, the amount of α-helical structure remaining after freeze-drying decreases with annealing time up to about 18 hr. The decrease in α-helix content with annealing does not occur for the system containing potassium chloride, however. The authors attribute this observation to the formation of a stable glass in the potassium chloride system, which kinetically prevents phase separation. Regardless of the mechanism, preventing phase separation prevents loss of the native structure of hemoglobin.

7. SUMMARY

Given the increasing importance of reducing development time for new pharmaceutical products, formulation and process development scientists must continually look for ways to "work smarter, not harder." Within the product development arena, this means reducing the amount of trial and error empiricism in arriving at a formulation and identification of processing conditions which will result in a quality final dosage form. Characterization of the freezing behavior of the intended formulation is necessary for developing processing conditions which will result in the shortest drying time while maintaining all critical quality attibutes of the freeze-dried product. Analysis of frozen systems was discussed in detail, particularly with respect to the glass transition as the physical event underlying collapse during freeze-drying, eutectic mixture formation, and crystallization events upon warming of frozen systems. Experiments to determine how freezing and freeze-drying behavior is affected by changes in the composition of the formulation are often useful in establishing the "robustness" of a formulation. It is not uncommon for seemingly subtle changes in composition of the formulation, such as a change in formulation pH, buffer salt, drug concentration, or an additional excipient, to result in striking differences in freezing and freeze-drying behavior.

With regard to selecting a formulation, it is wise to keep the formulation as simple as possible. If a buffer is needed, a minimum concentration should be used. The same principle applies to added salts: If used at all, the concentration should be kept to a minimum. For many proteins a combination of an amorphous excipient, such as a disaccharide, and a crystallizing excipient, such as glycine, will result in a suitable combination of chemical stability and physical stability of the freeze-dried solid.

Concepts of heat and mass transfer are valuable in rational design of processing conditions. Heat transfer by conduction—the dominant mechanism of heat transfer in freeze-drying—is inefficient at the pressures used in freeze-drying. Steps should be taken to improve the thermal contact between the product and the shelf of the freeze dryer, such as eliminating metal trays from the drying process. Quantitation of the heat transfer coefficient for the geometry used is a useful way of assessing the impact of changes in the system such as elimination of product trays and changes in the vial. Because heat transfer by conduction through the vapor increases with increasing pressure, the commonly held point of view that "the lower the pressure, the better" is not true with respect to process efficiency. The optimum pressure for a given product is a function of the temperature at which freeze-drying is carried out, and lower pressures are needed at low product temperatures. The controlling resistance to mass transfer is almost always the resistance of the partially dried solids above the submination interface. This resistance can be minimized by avoiding fill volumes of more than about half the volume of the container. The development scientist should also recognize that very high concentrations of solute may not be appropriate for optimum freeze-drying, particularly if the resistance of the dried product layer increases sharply with concentration.

Although the last 10 years has seen the publication of a significant body of literature of great value in allowing development scientists and engineers to "work smarter," there is still much work needed in both the science and the technology of freeze-drying. Scientific development is needed for improving analytical methodology for characterization of frozen systems and freeze-dried solids. A better understanding of the relationship between molecular mobility and reactivity is needed to allow accurate prediction of product stability at the intended storage temperature based on accelerated stability at higher temperatures. This requires that the temperature dependence of glass transition-associated mobility, particularly at temperatures below the glass transition, be studied in greater depth. The relevance of the concept of strong and fragile glasses to frozen systems and freeze-dried solids has only begun to be explored. The list of pharmaceutically acceptable protective solutes is very short, and more imagination—and work—is needed in order to develop pharmaceutically acceptable alternative stabilizers.

There is a need for technology development in process monitoring, particularly in developing a way to measure the status of the product during freezing and freeze-drying without placing temperature measurement probes in individual vials of product. The current practice of placing thermocouples in vials is uncertain with respect to reliability of the data, inconsistent with elimination of personnel in close proximity to open vials of product in an aseptic environment, and incompatible with technology for automatic material handling in freeze-drying. In addition, a method for controlling the degree of supercooling during freezing would allow better control of freezing rate and would, in many cases, result in more consistent product quality.

REFERENCES

Ablett, S., Izzard, M. J., and Lillford, P. J., 1992a, Differential scanning calorimetric study of frozen sucrose and glycerol solutions, *J. Chem. Soc. Faraday Trans.* **88:**789–794.

Ablett, S., Clark, A. H., Izzard, M. J., and Lillford, P. J., 1992b, Modelling of heat capacity–temperature data for sucrose–water systems, *J. Chem. Soc. Faraday Trans.* **88:**795–802.

Anchordoquy, T. J., and Carpenter, J. F., 1996, Polymers protect lactate dehydrogenase during freeze–drying by inhibiting dissociation in the frozen state, *Arch. Biochem. Biophys.* **332:**231–238.

Angell, C. A., 1995, Formation of glasses from liquids and biopolymers, *Science* **267:**1924–1935.

Arakawa, T., Prestrelski, S. J., Kinney, W., and Carpenter, J. F., 1993. Factors affecting short-term and long-term stability of proteins, *Adv. Drug Deliv. Rev.* **10:**1–28.

Bell, L. N., and Hageman, M. J., 1994, Differentiating between the effects of water activity and glass transition dependent mobility on a solid state chemical reaction: Aspartame degradation, *J. Agric. Food Chem.* **42:**2398–2401.

Blond, G., Simatos, D., and Catte, M., 1997, Modeling of the water–sucrose state diagram below 0°C, *Carbohydrate Research* **298:**139–145.

Bogardus, J. B., 1982, Phase equilibria of nafcillin sodium–water, *J. Pharm. Sci.* **71:**105–109.

Carpenter, J. F., Prestrelski, S. J., Anchordoguy, T. J., and Arakawa, T., 1994, Interactions of stabilizers with proteins during freezing and drying, in: *Formulation and Delivery of Peptides and Proteins* (J. L. Cleland and R. Langer, eds.), American Chemical Society, Washington, DC, pp. 134–147.

Cavatur, R. K., and Suryanarayanan, R., 1998, Characterization of frozen aqueous solutions by low temperature X–ray powder diffractometry, *Pharm. Res.* **15:**194–199.

Chang, B. S., Kendrick, B. S., and Carpenter, J. F., 1996, Surface–induced denaturation of proteins during freezing and its inhibition by surfactants, *J. Pharm. Sci.* **85:**1325–1330.

Chongprasert, S., Griesser, U. J., Bottorff, A. T., Byrn, S. R., and Nail, S. L., 1997, Effects of process conditions on crystallization of pentamidine isethionate during freeze drying, *J. Pharm. Sci.* **87:**1155–1160.

Chongprasert, S., Knopp, S. A., and Nail, S. L., 2001, Characterization of frozen solutions of glycine, *J. Pharm. Sci.* **90:**1720–1728.

Coleman, N. J., and Craig, D. Q. M., 1996, Modulated temperature differential scanning calorimetry: A novel approach to pharmaceutical thermal analysis, *Int. J. Pharm.* **135:**13–29.

Connelly, J. P., and Welch, J. V., 1993, Monitor lyophilization with mass spectrometer gas analysis, *J. Parenteral Sci. Technol.* **47:**70–75.

Costantino, H. R., Nguyen, T. H., and Hsu, C. C., 1996a, Fourier–transform infrared spectroscopy demonstrates that lyophilization alters the secondary structure of recombinant human growth hormone, *Pharm. Sci.* **2:**229–232.

Costantino, H. R., Schwendeman, S. P., Griebenow, K., Klibanov, A. M., and Langer, R., 1996b, The secondary structure and aggregation of lyophilized tetanus toxoid, *J. Pharm. Sci.* **85:**1290–1293.

Costantino, H. R., Andya, J. D., Shire, S. J., and Hsu, C. C., 1997a, Fourier–transform infrared spectroscopic analysis of the secondary structure of recombinant humanized immunoglobulin G, *Pharma. Sci.* **3:**121–128.

Costantino, H. R., Griebenow, K., Langer, R., and Klibanov, A. M., 1997b, On the pH memory of lyophilized compounds containing protein functional groups, *Biotechnol. Bioeng.* **53:**345–348.

Dong, A., Prestrelski, S. J., Allison, S. D., and Carpenter, J. F., 1995, Infrared spectroscopic studies of lyophilization and temperature–induced protein aggregation, *J. Pharm. Sci.* **84:**415–424.

Duddu, S. P., and Dal Monte, P. R., 1997, Effect of glass transition temperature on the stability of lyophilized formulations containing a chimeric therapeutic monoclonal antibody, *Pharm. Res.* **14:**591–595.

Duddu, S. P., and Weller, K., 1996, Importance of glass transition temperature in accelerated stability testing of amorphous solids: Case study using a lyophilized aspirin formulation, *J. Pharm. Sci.* **85:**345–347.

Duddu, S. P., Zhang, G., and Dal Monte, P. R., 1997, The relationship between protein aggregation and molecular mobility below the glass transition temperature of lyophilized formulations containing a monoclonal antibody, *Pharm. Res.* **14:**596–600.

Dushman, S., and Lafferty, J. M., 1962, *Scientific Foundations of Vacuum Technique*, Wiley, New York, Chapter 1.

Eckhardt, B. M., Oeswein, J. Q., and Bewley, T. A., 1991, Effect of freezing on aggregation of human grown hormone, *Pharma. Res.* **11:**1360–1364.

Elliot, A., and Ambrose, E. J., 1950, Structure of synthetic polypeptides, *Nature* **165:**921–922.

Evans, S. A., Morris, K. R., MacKenzie, A. P., and Lordi, N. G., 1995, Dielectric characterization of thermodynamic first order events in model frozen systems intended for lyophilization, *J. Parenteral. Sci. Technol.* **49:**2–8.

Flink, J. M., and Gejl-Hansen, F., 1978, Two simple freeze drying microscope stages, *Rev. Sci. Instrum.* **49:**269–271.

Flink, J. M., Gejl-Hansen, F., and Karel, M., 1973, Microscopic observations of the freeze drying of volatile–containing model food compounds, *J. Food Sci.* **38:**1174–1178.

Franks, F., 1990, Freeze–drying: From empiricism to predictability, *Cryo–Letters* **11:**93–110.

Freedman, M., Whittum, J. H., and Rosano, H. L., 1972, Temperature gradient freeze drying microscope stage, *J. Food Sci.* **37:**492–493.

Gatlin, L. A., 1991, Kinetics of a phase transition in a frozen solution, *Dev. Biol Stand.* **74:**93–104.

Gomez, G., Pikal, M. J., and Rodriguez–Hornedo, N., 2001, Effect of initial buffer composition on pH changes during far from equilibrium freezing of sodium phosphate buffer solutions, *Pharm. Res.* **18:**90–97.

Hancock, B. C., and Zografi, G., 1997, Characteristics and significance of the amorphous state in pharmaceutical systems, *J. Pharm. Sci.* **86:**1–11.

Hancock, B. C., Shamblin, S. L., and Zografi, G., 1995, Molecular mobility of amorphous pharmaceutical solids below their glass transition temperatures, *Pharm. Res.* **12:**799–806.

Hancock, B. C., Dalton, C. R., Pikal, M. J., and Shamblin, S. L., 1998, A pragmatic test of a simple calorimetric method for determining the fragility of some amorphous pharmaceutical materials, *Pharm. Res.* **15:**762–67.

Harris, R. K., 1986, *Nuclear Magnetic Resonance Spectroscopy. A Physicochemical View*, Wiley, New York, Chapter 3.

Hatley, R. H. M., Franks, F., Day, H., and Byth, B., 1986, Subzero temperature preservation of reactive fluids in the undercooled state. I. The reduction of potassium ferrricyanide by potassium cyanide, *Biophys. Chem.* **24:**41–46.

Hatley, R. H. M., Franks, F., and Day, H., 1986, Subzero temperature preservation of reactive fluids in the undercooled state. II. The effect on the oxidation of ascorbic acid of freeze concentration and undercooling, *Biophys. Chem.* **24:**187–192.

Heller, M. C., Carpenter, J. F., and Randolph, T. W., 1996, Effects of phase separating systems on lyophilized hemoglobin, *J. Pharm. Sci.* **85:**1358–1362.

Heller, M. C., Carpenter, J. F., and Randolph, T. W., 1997, Manipulation of lyophilization–induced phase separation: Implications for pharmaceutical proteins, *Biotechnol. Prog.* **13:** 590–596.

Hemminga, M. A., Roozen, M., and Walstra, P., 1993, Molecular motions and the glassy state, in: *The Glassy State in Foods* (J. M. V. Blanshard and P. J. Lillford, eds.), Nottingham University Press, Loghborough, Leicestershire, UK, pp. 157–171.

Her, L. M., and Nail, S. L., 1994, Measurement of glass transition temperatures of freeze–concentrated solutes by differential scanning calorimetry, *Pharm. Res.* **11:**54–59.

Her, L. M., Deras, M., and Nail, S. L., 1995, Electrolyte–induced changes in glass transition temperatures of freeze–concentrated solutes, *Pharm. Res.* **11:**768–772.

Herman, B. D., Sinclair, B. D., Milton, N., and Nail, S. L., 1994, The effect of bulking agent on the solid–state stability of freeze–dried methylprednisolone sodium succinate, *Pharm. Res.* **11:**1467–1473.

Hsu, C. C., Ward, C. A., Pearlman, R., Nguyen, H. M., Yeung, D. A., and Curley, J. G., 1992, Determining the optimum residual moisture in lyophilized protein pharmaceuticals, *Dev. Biol. Stand.* **74:**255–271.

Hsu, C. C., Nguyen, H. M., Yeung, D. A., Brooks, D. A., Koe, G. S., Bewley, T. A., and Pearlman, R., 1995, Surface denaturation at solid–void interface—A possible pathway by which opalescent particulates form during the storage of lyophilized tissue-type plasminogen activator at high temperature, *Pharm. Res.* **12:**69–77.

Izutsu, K., Yoshioka, S., and Terao, T., 1993, Decreased protein-stabilizing effects of cryoprotectants due to crystallization, *Pharm. Res.* **10:**1232–1237.

Izutsu, K., Yoshioka, S., and Terao, T., 1994, Effect of mannitol crystallinity on the stabilization of enzymes during freeze-drying, *Chem. Pharm. Bull.* **42:**5–8.

Izutsu, K., Yoshioka, S., and Kojima, S., 1995, Effect of cyroprotectants on the eutectic crystallization of NaCl in frozen solutions studied by differential scanning calorimetry and broad-line pulsed NMR, *Chem. Pharm. Bull.* **43:**1804–1806.

Izutsu, K., Yoshioka, S., Kojima, S., Randolph, T. W., and Carpenter, J. F., 1996, Effects of sugars and polymers on crystallization of poly(ethylene glycol) in frozen solutions: Phase separation between incompatible polymers, *Pharm. Res.* **13:**1393–1400.

Jackson, M., and Mantsch, H. H., 1995, The use and misuse of FTIR spectroscopy in the determination of protein structure, *Crit. Rev. Biochem. Mol. Biol.* **30:**95–120.

Jiang, S., and Nail, S. L., 1998, Effect of process conditions on recovery of protein activity after freezing and freeze-drying, *Eur. J. Pharm. Biopharm.* **45:**249–257.

Johari, G. P., Hallbrucker, A., and Mayer, E., 1987, The glass–liquid transition of hyperquenched water, *Nature* **330:**552–553.

Karel, M., Buera, M. P., and Roos, Y., 1993, Effects of glass transitions on processing and storage, in: *The Glassy State in Foods* (J. M. V. Blanchard and P. J. Lillford, eds.), Nottingham University Press, Nottingham, UK, pp. 13–34.

Kim, A. I., Akers, M. J., and Nail, S. L., 1998, The physical state of mannitol after freeze-drying: Effects of mannitol concentration, freezing rate, and a non-crystallizing cosolute, *J. Pharm. Sci.* **87**:931–935.

Klug, H. P., and Alexander, L. E., 1976, *X-Ray Diffraction Procedures for Polycrystalline and Amorphous Materials*, Wiley, New York.

Knopp, S. A., Chongprasert, S., and Nail, S. L., 1998, The relationship between the MDSC thermogram of frozen sucrose solutions and collapse during freeze-drying, *J. Thermal Anal.* **54**:659–672.

Kochs, M., Schwindke, P., and Korber, C., 1989, A microscope stage for the dynamic observation of freezing and freeze-drying in solutions and cell suspensions, *Cryo-Letters* **10**:401–420.

Lam, X. M., Costantino, H. R., Overcashier, D. E., Nguyen, T. H., and Hsu, C. C., 1996, Replacing succinate with glycolate buffer improves the stability of lyophilized interferon-γ, *Int. J. Pharm.* **142:**85–95.

Larsen, S. S., 1973, Studies on stability of drugs in frozen systems. VI. The effect of freezing upon pH for buffered aqueous solutions, *Arch. Pharm. Chem. Sci. Ed.* **1**:433–445.

MacKenzie, A. P., 1964, Apparatus for microscopic observations during freeze-drying, *Biodynamica* **9**:213–222.

MacKenzie, A. P., 1977a, The physico-chemical basis for the freeze drying process, *Dev. Biol. Stand.* **36:**51–67.

MacKenzie, A. P., Derbyshire, W., and Reid, D. S., 1977, Non-equilibrium freezing behavior of aqueous systems, *Phil. Trans. R. Soc. Lond. B* **278:**167–176.

Mansfield, M. L., 1993, An overview of theories of the glass transition, in: *The Glassy State in Foods* (J. M. V. Blanshard and P. J. Lillford, eds.), Nottingham University Press, Loghborough, Leicestershire, UK, pp. 103–122.

Millqvist-Fureby, A., Malmsten, M., and Bergenstahl, B., 1999, Surface characterisation of freeze-dried protein/carbohydrate mixtures, *Int. J. Pharm.* **191:**103–114.

Milton, N., and Nail, S. L., 1996, The physical state of nafcillin sodium in frozen aqueous solutions and freeze-dried powders, *Pharm. Dev. Technol.* **1:**269–277.

Milton, N., Pikal, M. J., Roy, M. L., and Nail, S. L., 1997. Evaluation of manometric temperature measurement as a method of monitoring product temperature during lyophilization, *PDA J. Pharm. Sci. Technol.* **51:**7–16.

Morris, K. R., Evans, S. A., MacKenzie, A. P., Scheule, D., and Lordi, N. G., 1994, Prediction of lyophile collapse temperature by dielectric analysis, *J. Parenteral Sci. Techol.* **48:**318–329.

Murase, N., and Franks, F., 1989, Salt precipitation during the freeze-concentration of phosphate buffer solutions, *Biophys. Chem.* **34:**293–300.

Nail, S. L., 1980, The effect of chamber pressure on heat transfer in the freeze-drying of parenteral solutions, *J. Parenteral Drug Assoc.* **34:**358–368.

Nail, S. L., and Johnson, J. W., 1992, Methodology for in-process determination of residual water in freeze dried products, *Dev. Biol. Std.* **74**:137–151.

Nail, S. L., Her, L. M., Proffitt, C. P. B., and Nail, L. L., 1994, An improved microscope stage for direct observation of freezing and freeze-drying, *Pharm. Res.* **11:**1098–1100.

Oguchi, T., Okada, M., Yonemochi, E., Yamamoto, K., and Nakai, Y., 1990, Freeze-drying of drug-additive binary systems III. Crystallization of α-cyclodextrin inclusion complex in freezing process, *Int. J. Pharm.* **61:**27–34.

Orii, Y., and Morita, M., 1977, Measurement of the pH of frozen buffer solutions by using pH indicators, *J. Biochem.* **81:**163–168.

Osterberg, T., and Wadsten, T., 1999, Physical state of L-histidine after freeze-drying and long-term storage, *Eur. J. Pharm. Sci.* **8:**301–308.

Pikal, M. J., 1985, Use of laboratory data in freeze-dry process design: Heat and mass transfer coefficients and computer simulation of freeze-drying, *J. Parenteral Sci. Tech.* **39:**115–138.

Pikal, M. J., and Shah, S., 1997, Intravial distribution of moisture during the secondary drying stage of freeze-drying, *PDA J. Pharm. Sci. Technol.* **51:**17–24.

Pikal, M. J., Lukes, A. L., Lang, J. E., and Gaines, K., 1978, Quantitative crystallinity determinations for β-lactam antibiotics by solution calorimetry: Correlations with stability, *J. Pharm. Sci.* **67:**767–772.

Pikal, M. J., Shah, S., Senior, D., and Lang, J. E., 1983a, Physical chemistry of freeze-drying: Measurement of sublimation rates for frozen aqueous solutions by a microbalance technique, *J. Pharm. Sci.* **72:**635–650.

Pikal, M. J., Shah, S., and Roy, M. L., 1984, Mass and heat transfer in vial freeze-drying of pharmaceuticals: Role of the vial, *J. Pharm. Sci.* **73:**1224-1237.

Pikal, M. J., Shah, S., Roy, M. L., and Putman, R., 1990, The secondary drying stage of freeze-drying: Drying kinetics as a function of temperature and chamber pressure, *Int. J. Pharm.* **60:**203–217.

Pincock, R. E., and Kiovsky, T. E., 1966, Kinetics of reactions in frozen solutions, *J. Chem. Ed.* **43:**358–360.

Powell, M. F., Fleitman, J., Sanders, L. M., and Si, V. C., 1994, Peptide liquid crystals: Inverse correlation of kinetic formation and thermodynamic stability in aqueous solution, *Pharm. Res.* **11:**1352–1354.

Prestrelski, S. J., Pikal, K. A., and Arakawa, T., 1995, Optimization of lyophilization conditions for recombinant human interleukin-2 by dried-state conformational analysis using Fourier-transform infrared spectroscopy, *Pharm. Res.* **12:** 1250–1259.

Prestrelski, S. J., Arakawa, T., and Carpenter, J. F., 1994, Structure of proteins in lyophilized formulations using Fourier transform infrared spectroscopy, in: *Formulation and Delivery of Peptides and Proteins* (J. L. Cleland and R. Langer, eds.), American Chemical Society, Washington, DC, pp. 148–170.

Randolph, T. W., 1997, Phase separation of excipients during lyophilization: Effects on protein stability, *J. Pharm. Sci.* **86:**1198–1203.

Reading, M., Luget, A., and Wilson, R., 1994, Modulated differential scanning calorimetry, *Thermochim. Acta* **238:**295–307.

Remmele, R. L., Stoshnoff, C., and Carpenter, J. F., 1997, Real-time *in-situ* monitoring of lysozyme during lyophilization using infrared spectroscopy: Dehydration stress in the presence of sucrose, *Pharm. Res.* **14:**1548–1555.

Rey, L. R., 1957, Dispositif pour l'examen microscopique aux basses température, *Experientia* **13:**201–202.

Roy, M. L., and Pikal, M. J., 1989, Process control in freeze-drying: Determination of the end point of sublimation drying by an electronic moisture sensor, *J. Parenteral Sci. Technol.* **43:**60–66.

Roy, M. L., Pikal, M. J., Rickard, E. C., and Maloney, A. M., 1991, The effects of formulation and moisture on the stability of a freeze-dried monoclonal antibody–vinca conjugate: A test of the WLF glass transition theory, *Dev. Biol. Stand.* **74:**323–340.

Sarciaux, J. M. E., and Hageman, M. J., 1997, Effects of bovine somatotropin (rbSt) concentration at different moisture levels on the physical stability of sucrose in freeze-dried rbSt/sucrose mixtures, *J. Pharm. Sci.* **86:**365–371.

Sarciaux, J. M., Mansour, S., Hageman, M. J., and Nail, S. L., 1999, Effects of buffer composition and processing conditions on aggregation of bovine IgG during freeze-drying, *J. Pharm. Sci.* **88:**1354–1361.

Schuster, M., Aaviksaar, A., Haga, M., Ullmann, U., and Jakubke, H. D., 1991, Protease-catalyzed peptide synthesis in frozen aqueous systems: The "freeze concentration" model, *Biomed. Biochim. Acta* **50:**S84–S89.

Searles, J. A., Carpenter, J. F., and Randolph, T. W., 2001a, Ice nucleation temperature determines the primary drying rate of lyophilization for samples frozen on a temperature-controlled shelf, *J. Pharm. Sci.* **90:**860–871.

Searles, J. A., Carpenter, J. F., and Randolph, T. W., 2001b, Annealing to optimize the primary drying rate, reduce freezing-induced drying rate heterogeneity, and determine T_g' in pharmaceutical lyophilization, *J. Pharm. Sci.* **90:**872–887.

Shalaev, E. Y., and Franks, F., 1995, Structural glass transitions and thermophysical processes in amorphous carbohydrates and their supersaturated solutions, *J. Chem. Soc. Faraday Trans.* **91:**1511–1517.

Smith, G., Duffy, A. P., Shen, J., and Olliff, C. J., 1995, Dielectric relaxation spectroscopy and some applications in the pharmaceutical sciences, *J. Pharm. Sci.* **84:**1029–1044.

Spieles, G., Marx, T., Heschel, I., and Rau, G., 1995, Analysis of desorption and diffusion during secondary drying in vacuum freeze-drying of hydroxyethyl starch, *Chem. Eng. Process* **34:**351–357.

Strambini, G. B., and Gabellieri, E., 1996, Proteins in frozen solutions: Evidence of ice-induced partial unfolding, *Biophys. J.* **70:**971–976.

Surewicz, W. K., Mantsch, H. H., and Chapman, D., 1993, Determination of protein secondary structure by Fourier-transform infrared spectroscopy: A critical assessment, *Biochemistry* **32:**389–394.

Tarelli, E., and Wheeler, S. F., 1994, Drying from phosphate-buffered solutions can result in the phosporylation of primary and secondary alcohol groups of saccharides, hydroxylated amino acids, proteins, and glycoproteins, *Anal. Biochem.* **222:**196–201.

Townsend, M. W., Byron, P. R., and DeLuca, P. P., 1990, The effects of formulation additives on the degradation of freeze-dried ribonuclease A, *Pharm. Res.* **7:**1086–1091.

Vadas, E. B., Toma, P., and Zografi, G., 1991, Solid-state phase transitions initiated by water vapor sorption of crystalline L-660,711, a leukotriene D_4 receptor antagonist, *Pharm. Res.* **8:**148–155.

van den Berg, L., 1959, The effect of addition of sodium and potassium chloride to the reciprocal system: KH_2PO_4–Na_2HPO_4–H_2O on pH and composition during freezing, *Arch. Biochem. Biophys.* **84:**305–315.

van den Berg, L., 1960, pH measurement at low temperatures using modified glass and calomel electrodes, *Anal. Chem.* **32:**628–631.

van den Berg, L., 1966, pH changes in buffers and foods during freezing and subsequent storage, *Cryobiology* **3:**236–242.

van den Berg, L., and Rose, D., 1959, Effect of freezing on the pH and composition of sodium and potassium phosphate solutions: The reciprocal system KH_2PO_4–Na_2HPO_4–H_2O, *Arch. Biochem. Biophys.* **81:**319–329.

Walter, H., Johansson, G., and Brooks, D. E., 1991. Partitioning in two-phase systems: Recent results, *Anal. Biochem.* **197:**1–18.

Williams-Smith, D. L., Bray, R. C., Barber, M. J., Tsopanakis, A. D., and Vincent, S. P., 1977, Changes in apparent pH on freezing aqueous buffer solutions and their relevance to biochemical electron-paramagnetic-resonance spectroscopy, *Biochem. J.* **167:**593–600.

Yoshioka, S., Aso, Y., and Kojima, S., 1996, Determination of molecular mobility of lyophilized bovine serum albumin and γ-globulin by solid-state ^1H NMR and relation to aggregation susceptibility, *Pharm. Res.* **13:**926–930.

Yoshioka, S., Aso, Y., and Kojima, S., 1997a, Softening temperature of lyophilized bovine serum albumin and γ-globulin as measured by spin–spin relaxation time of protein protons, *J. Pharm. Sci.* **86:**470–474.

Yoshioka, S., Aso, Y., and Kojima, S., 1997b, Dependence of the molecular mobility and protein stability of freeze-dried γ-globulin formulations on the molecular weight of dextran. *Pharm. Res.* **14:**736–741.

Yoshioka, S., Aso, Y., Nakai, Y., and Kojima, S., 1998, Effect of high molecular mobility of poly(vinyl alcohol) on protein stability of lyophilized γ-globulin formulations, *J. Pharm. Sci.* **87:**147–151.

Zaslavsky, B. Y., *et al.*, 1989, Structure of water as a key factor of phase separation in aqueous mixtures of two nonionic polymers, *Polymer* **30:**2104–2110.

7

Quality Assurance and Quality Control for Biopharmaceutical Products

John Geigert

1. INTRODUCTION

Quality assurance (QA) and quality control (QC) play important roles in both assuring compliance to current good manufacturing practices (cGMP) and ensuring the consistency, quality, and safety of the manufactured drug product. Fortunately, many QA and QC systems are common to different manufacturing processes, such as chemical synthesis, isolation of biologics from natural sources, and production of biopharmaceuticals through recombinant DNA technology or monoclonal antibody hybridoma processes: All involve raw material and component control, facility support, batch record control, process control, product testing, and auditing (see Table I). Application of these quality systems throughout the pharmaceutical industry is described in numerous reference books (Bryant, 1984; Peine, 1994; Sharp, 1991; Willig and Stoker, 1992).

However, there are special QA and QC concerns associated with biopharmaceutical manufacturing and the quality of biopharmaceutical drug products (see Table II). The main component of a production process is generally a living organism which has been genetically altered. Hence, a complete characterization of this organism, as well as an evaluation of the genetic stability of the system, are primary concerns. The manufacturing processes used to produce biopharmaceuticals can be fairly compli-

---

John Geigert • Biopharmaceutical Quality Solutions, Carlsbad, California 92009.

Development and Manufacture of Protein Pharmaceuticals, edited by Nail and Akers. Kluwer Academic/Plenum Publishers, New York, 2002.

Table I
Quality Systems Common to Pharmaceuticals and Biopharmaceuticals

Raw material/component control	Process control
Acceptance criteria	Control limits
Test method standard operating procedures	Process step disposition
Batch testing and release	Yield calculations
Label control	Time limits
Facility support	Product testing
Utility system performance	Test method standard operating procedures
Water quality monitoring	Critical reagent supply
Air quality monitoring	Assay controls
Action level closeout	Qualified reference standards
Equipment validation	Methods development
Product changeover	Methods qualification
Room/line clearance	Methods validation
Facility release	Specifications and stability
Batch record control	Auditing
Preparation	External vendors
Approval and issuance	Internal groups
Review and closeout	Contract manufacturers

cated and multistepped, with extended production times of weeks or months. Many and complex raw materials are used in the manufacture of these products. To further complicate matters, it is also not unusual to combine several fermentation runs and/or purification runs to produce a single lot of product. Due to these complexities, an extremely rigorous approach is needed to protect both the process and the product from contaminating viruses and other adventitious agents. Adequate in-process

Table II
Special QA and QC Concerns Associated with the Manufacturing Control and Quality of Biopharmaceuticals

Correct coding sequence for the product incorporated into the host cell, and maintenance of the integrity of the expression construct during culture to the end of production
Prevention of viruses and other adventitious agents from entering the process and contaminating the product
Ever-increasing regulatory expectation for "well characterization" of biopharmaceuticals
Difficulty in predicting real-time stability profiles for large, complex molecules

testing, thorough process validation, and a comprehensive assessment of final product quality using a complete battery of analytical methodologies are needed to demonstrate lot-to-lot consistency. The analysis of these products involves the use of different analytical methods, which are strategically designed to be complementary with respect to specificity, because no single test method is sufficient for the evaluation of these large molecules. The virtual explosion in the availability of sophisticated and reliable analytical instrumentation, as well as extensive and sensitive chromatographic, immunological, and spectrometric methods of product analysis, continues to push the envelope of the definition of "well-characterized" for biopharmaceuticals. In addition, evaluation of the multi-path degradation profiles possible with these large molecules cannot be dependent on any single analytical method. In particular, at best, elevated-temperature stability stress studies are mildly predictive of real-time shelf lives.

The focus of this chapter will be on the special QA and QC concerns associated with the manufacturing control and quality of all biopharmaceutical products, whether protein-based (e.g., recombinant proteins, monoclonal antibodies) or nucleic acid-based (e.g., plasmid DNA vaccines, gene therapy products). It will be noticed that this chapter does not contain a single chromatogram, mass spectrum, or other analytical graph. An overabundance of these can be found in numerous reference books (Pearlman and Wang, 1996; Wang and Pearlman, 1993).

2. DEFINING QUALITY ASSURANCE AND QUALITY CONTROL IN BIOPHARMACEUTICAL COMPANIES

Defining the separate roles and responsibilities of quality assurance and quality control has never been easy, even in traditional pharmaceutical companies. Some companies have separate QA and QC departments, others have either only a QA department or only a QC department, whereas others avoid wrestling with the definition altogether by just having a "Quality Department" or a "Quality Management Department" or a "Quality Systems Department."

To further complicate the issue, there is also no regulatory agreement on even the definition of QC and QA (Chapman *et al.*, 1996). In the U.S. *Code of Federal Regulations* (CFR) on GMPs (U.S. Food and Drug Administration [FDA], 2001), QA is not even mentioned, and the term "quality control unit" is used; however, in the corresponding European Union (EU) document on

GMPs (European Community, 2001), QC is defined as a department, whereas QA is defined as a system.

During the public comment period for the FDA's recent proposed revisions to the GMPs, the FDA received comments from the pharmaceutical industry asking for clarification and change of the term "quality control unit." Although no change in terminology resulted, the FDA did attempt to clarify its position on the definition of QC and QA (Federal Register, 1996a):

> The agency understands that some manufacturers would prefer that the term "quality control" be replaced with "quality assurance," that the functions of quality control and quality assurance be somehow differentiated, or that a number of other terms be incorporated into the regulation to reflect the distribution of quality oversight responsibilities in various manufacturing settings.
>
> FDA does not believe that such changes in terminology would be useful. The difference between "quality assurance" and "quality control" is recognized to be operational. The quality control unit is usually responsible for performing the testing to assure that proper specifications and limits are adhered to, while the quality assurance unit is responsible for auditing methods, results, systems, and processes, and for performing trend analyses.

For the biopharmaceutical companies, many of which have not been in existence long enough to have established traditions and for which job titles are generously handed out, the distinction between QC and QA is even more diffuse. For example, in a bench-marking study in which the author participated involving several biopharmaceutical companies with FDA-licensed products, the responsibility for validation of the release test methods used by QC was placed in different departments carrying the title of either QC, QA, Technical Services, Analytical Resources, or Pharmaceutical Technology, depending upon the biopharmaceutical company involved.

As biopharmaceutical companies evolve from startup to drawing close to having a drug product approved for market release, two things generally change in the quality area. First, the roles and responsibilities of QC and QA are generally brought together under the same group, with the title of "Quality Department" being the most popular today. Second, the head of the quality program is generally assigned to a person with an advanced degree in a scientific field (frequently analytical or organic chemistry), reflecting the complex decision making that occurs in the assessment of biopharmaceutical manufacturing control and the quality of the biopharmaceutical products themselves.

3. INVOLVEMENT OF QUALITY IN BIOPHARMACEUTICAL MANUFACTURING CONTROL

"Genetically altered living organisms," "complicated, multistepped, extended production cycles," and "extensive in-process testing and interpretation" are phrases frequently used to describe a biopharmaceutical manufacturing process. Therefore, it is not surprising that the quality program must be extensively involved in all decisions about manufacturing control that may impact the quality of the biopharmaceutical products. For biopharmaceuticals, the following major quality issues need to be adequately addressed:

- Expression construct studies: thorough characterization and genetic stability
- Cell substrate studies: thorough characterization
- Raw material selection: prevention of virus and other adventitious agent introduction
- Process validation studies: impurity removal and viral clearance safety factor
- Facility control: environmental monitoring control and multiuse utilization

3.1. Expression Construct Studies

The expression construct is the expression vector (e.g., the plasmid) which contains the coding sequence of the recombinant protein (the product gene) and the elements necessary for its expression (e.g., the promoter, enhancer, and splicing sequences that are adjacent to the 5′ and the 3′ ends of the product gene). A thorough characterization study of the expression construct is necessary to establish that the correct coding sequence of the product has been incorporated into the production cell bank, and a genetic stability study of the expression construct is necessary to establish that it is maintained with integrity during culture to the end of production. These expression construct studies, as part of an overall quality evaluation, ensure that the recombinant proteins which are produced in living cells do not undergo mutations that could alter the properties of a protein with potential adverse consequences to patients, such as alterations in pharmacodynamics or pharmacokinetics or induction of antibody responses.

The analysis of the expression construct is accomplished by both nucleic acid techniques (e.g., DNA sequencing) and protein structural testing (e.g., peptide mapping). Both approaches are necessary, because no single QC approach can detect all possible modifications to a protein product. Nucleic acid analysis is used to verify the coding sequence and the physical state of the expression construct. The nucleic acid analysis is performed to ensure that the expressed protein will have the correct amino acid sequence, but is not intended to detect low levels of variant sequences. Protein structural testing is performed to assess the many structural features of the expressed protein as well as changes due to posttranslational modification systems such as those affecting protein proteolytic processing, glycosylation, phosphorylation, and acetylation and other alterations to the protein, such as deamidation and oxidation. However, analysis of the final purified protein may not detect every change in a protein's structure that could result from a mutation in the protein coding sequence. For example, conservative amino acid substitutions may be missed by peptide mapping techniques and yet could cause structural changes in the protein. Such changes might not affect *in vitro* activity, but could lead to significant *in vivo* effects. It is for these reasons that the regulatory agencies strongly suggest that both nucleic acid and protein testing approaches be considered in the analysis of the expression construct.

The analysis of the expression construct is necessary through the entire cell bank system from the cell clone (also referred to as the parent seed stock) to the master cell bank (MCB; an aliquot of a single pool of cells which generally has been prepared from the selected cell clone under defined conditions, dispensed into multiple containers, and stored under defined conditions) to the working cell bank (WCB; prepared from aliquots of a homogeneous suspension of cells obtained from culturing the MCB under defined culture conditions). The analysis of the expression construct is also necessary at the end of the production cell culture process, referred to as the *in vitro* cell age limit.

Specific requirements and details about the analyses for each aspect of the cell bank system and at the end of the production cell culture process are provided by the International Congress on Harmonization (ICH, 2001a) and the U.S. FDA (1992).

3.2. Cell Substrate Studies

The cell substrate used in production could be a microbial cell culture (bacteria, fungi, yeast) or a metazoan cell culture (animal or human cells). Historically, quality concerns for cell-derived biological products have originated from the introduction of adventitious contaminants or from the properties of the cells used to prepare the product. Examples include yellow fever vaccine, which was contaminated by avian leukosis virus by virtue of its production in naturally infected hens eggs, and simian virus 40

(SV40), which was a contaminant of poliovirus and adenovirus vaccines prepared in the 1950s on primary cultures of kidney cells obtained from Rhesus monkeys naturally harboring a clinically inapparent infection with SV40.

A thorough characterization study of the cell substrate used in the manufacture of a biopharmaceutical needs to include the source and history of the cell substrate, controls during the cell bank laydown, and appropriate QC release testing of the prepared banks. It is important to document the source and history of the cell substrate that is used in the manufacture of a biopharmaceutical product, as well as any parental cell line from which it was totally or partially derived. Events during the research and development phases of the cell substrate could contribute significantly to assessment of the risks associated with the use of that particular cell substrate for production. Therefore, careful records of the manipulation of the cell substrate should be maintained throughout its development. During the generation of the cell substrate, one or more specific procedures may be utilized in the ultimate development of the desired characteristics. These may include, for example, cell fusion, transfection, selection, colony isolation, cloning, gene amplification, and adaption to specific culture conditions or media. Information regarding the methodologies utilized in developing the cell substrate can help to provide a clear understanding of the history of the cell substrate.

Once a cell line is chosen as the biological source of a product, a cell bank system is generated to assure that an adequate supply of equivalent cells exists for use over the entire life span of the product. In addition to providing a constant supply of starting material, the advantages of a cell bank system include allowing for a detailed characterization of the cell line and decreasing the likelihood and increasing the detection of both cell line cross-contamination and adventitious agent contamination. As mentioned previously, the cell bank system generally consists of two tiers: the MCB and WCB. A single-tiered banking system consisting only of the MCB, but no WCBs can be used, for example, if relatively few containers were needed each year to produce the desired product.

It is important to prevent a contaminated cell bank from being used in production and to avoid a loss of product availability or development time resulting from the need to recreate a cell bank found to be unusable due to contamination. In this regard, it is recognized that no cell bank testing regimen is able to detect all potential contaminants; therefore, the use of preventive control principles (i.e., enforcement of GMP practices) during cell banking is important for providing reasonable assurance of the absence of contamination and for providing a reliable source of the cell substrate. For example, the cell banks should be prepared under batch record control and in a manufacturing suite that preferably contains no other cell lines present at the same time that open cell banking activities are carried out. Process-specific (also referred to as dynamic) environmental monitoring should be performed and evaluated on all open steps (usually the culture inoculation and ampouling steps, which generally occur in a laminar flow bench).

It is also important to properly test the prepared cell banks for QC release. The objective of this testing is to confirm the identity, purity, and stability of the cell substrate for manufacturing use. Cell substrate release testing, however, does not come cheap. For mammalian cell banks, primarily due to the extensive virus testing required, QC release test costs can readily exceed $70,000 per cell bank. Thus it is worth the effort to evaluate which tests are absolutely necessary. Factors such as choice of host (bacterial vs. mammalian), knowledge about host (established hosts such as *Escherichia coli* vs. a host not previously used with marketed biopharmaceuticals), location of product gene in host (extrachromosomal plasmid vs. chromosomal integration), status of product development (Phase 1 clinical vs. moving to market), and anticipated area of marketing (United States only vs. worldwide) all need to be considered in the selection of QC release tests.

Tests for identity confirm that the cell bank is what it is represented to be. For microbial cells, analysis of growth on selective media, with and without antibiotic present, is generally used to confirm host cell identity at the species level for these cell lines. For metazoan cells, isozyme analysis can be sufficient to confirm the species of origin for these cell lines. Expression of the desired product can also be used to confirm the identity of the expression system in the transformed cell line.

Tests for biological purity confirm that the cell bank is free from microbial, mycoplasmal, and adventitious agents. Purity testing needs to be designed based on the cultivation history of the cell substrate and the ability of the cell line to propagate specific adventitious agents such as viruses. Examples of viral adventitious testing for cell lines of human or animal origin are presented in Table III.

Evidence for banked cell stability under defined storage conditions is usually generated during production of clinical trial material from the banked cells. Data from the determination of cell viability when the preserved cells are reconstituted or thawed for production of clinical trial supplies can verify that the revived cells have survived the preservation process. Data from the preparation of clinical materials can demonstrate that the revived cells can be used to prepare the desired product.

Specific requirements and details about the analyses for each aspect of the cell substrate studies are provided by the ICH (2001b) and the U.S. FDA (1993).

Table III
Virus Tests for Cell Lines of Human or Animal Origin[a]

	Master cell bank	Working cell bank	Cells at the limit
Tests for retroviruses			
Infectivity	+	–	+
Electron microscopy	+	–	+
Reverse transcriptase	+	–	+
Tests for endogenous or adventitious viruses			
In vitro assays	+	–	+
In vivo assays	+	–	+
Antibody production tests (e.g., MAP, RAP, HAP)[b]	+	–	–
Other virus-specific tests (e.g., human viruses)	As appropriate	–	As appropriate

[a] ICH (2001b).
[b] MAP, Mouse antibody production; RAP, Rat antibody production; HAP, Hamster antibody production.

3.3. Raw Material Selection

Raw materials used to prepare the media for the cell banks and cell culture must be carefully selected to provide the proper rate of growth and the essential nutrients for the organisms producing the desired product. Raw materials used to prepare the buffers for purification and components such as column resins must also be carefully selected to ensure that they do not contain any undesirable and toxic components that may be carried through the purification process to the finished product. Historically, quality concerns have originated from the choice of raw materials used for manufacture, specifically the potential presence of adventitious agents. Perhaps the most notable was yellow fever vaccine contaminated with hepatitis B virus present in the human serum used as a stabilizer in the 1940s. More recently, adventitious agent contamination from raw materials has been implicated in contamination of biopharmaceutical manufacturing, for example, the contamination of Chinese hamster ovary (CHO) cell culture medium by minute virus of mice, MVM (Darling and Spaltro, 1996).

Therefore, the selection and control of raw materials derived from animal or human sources remains a critical function of the quality program. These animal- and human-derived raw materials can be present throughout the entire manufacturing process, as shown in Table IV. Control over the source of these raw materials is most important for minimizing the risk of contamination, whether it be from mycoplasma, viruses, or transmissible

Table IV
Animal- and Human-Derived Raw Materials Possible in a Biopharmaceutical Manufacturing Process

Process stage	Possible animal-/human-derived raw materials
Cell bank	Fetal bovine serum
	Porcine trypsin
Cell culture	Human transferrin
	Bovine fetuin
	Animal protein hydrolyzates
Purification	Protein A (affinity column material isolated with human immunoglobulin G)
Formulation	Polysorbate (oleate derived from beef tallow) human serum albumin

spongiform encephalopathies (TSEs). Vendor testing for the absence of mycoplasma and viruses, provided in a certificate of analysis, is standard practice; however, it is apparent that vendor testing is not always a guarantee of the absence of contamination. For example, bovine viruses have been found in certified fetal bovine serum lots sent to manufacturers for use in cell culture processes (Kappeler *et al.*, 1996). Unfortunately, there are no rapid tests for TSEs. TSEs include scrapie in sheep and goats, chronic wasting disease in mule deer and elk, bovine spongiform encephalopathy (BSE) in cattle, as well as Kuru and Creutzfeldt–Jakob disease (CJD) in humans. Agents causing these diseases replicate in infected individuals generally without evidence of infection detectable by available diagnostic tests. After incubation periods of up to several years the agents cause disease and finally lead to death. Therefore, in lieu of testing, source control procedures, such as country of origin and herd selection, as well as strict processing conditions, such as treatment of raw materials with heat or solvents, are used to minimize risk. The European Union has issued guidelines for risk control of TSE from raw materials of animal origin (Committee for Proprietary Medical Products [CPMP], 2001).

It is becoming more apparent that the regulatory agencies want manufacturers to get the message that animal-derived and human-derived raw materials need to be removed from biopharmaceutical processes. Several withdrawals and recalls in 1996 and 1997 of human albumin, human immunoglobulin and human transferrin lots due to blood donors subsequently being diagnosed with CJD triggered a letter from the Center for Biologics Evaluation and Research (CBER) to biological product manufactures requesting detailed information on the processing steps using blood-derived materials as well as procedures for dealing with product made using such materials if a lot of the material is withdrawn in the future.

CBER further requested information concerning plans for replacement of blood-derived materials and an anticipated time frame for doing so.

3.4. Process Validation Studies

Although the regulatory agencies have been moving away from having the biopharmaceutical product defined by its manufacturing process, it is important to point out that a well-characterized process is necessary for a well-characterized product.

Validation of the cell culture process is centered around controlling the growth parameters and ensuring the absence of contaminating microorganisms and adventitious agents. Table V presents a list of what can be included in cell culture process validation studies. Even in a validated cell culture process, it should be noted that the established operating parameters may still allow product quality impact. For example, cell culture media, if deficient in certain amino acids, may cause substitutions in the produced protein product. This is readily noted when *E. coli* is starved of methionine while growing. The organism will synthesize norleucine and incorporate it in the positions normally occupied by methionine, yielding an analogue of the wild-type protein (Randhawa *et al.*, 1994). From a quality perspective, this can be a major challenge to control, especially if the QC analytical release tests do not readily detect the analogue in the presence

Table V
Process Validation Studies Performed at the Cell Culture Stage

Cell growth parameter ranges
pH
Temperature
Dissolved O_2
Viable cell density
Cell productivity
Run time
Waste byproduct levels
Control over microbes and adventitious agents
Cleaning-in-place of process equipment and tanks
Sterilizing-in-place of process equipment and tanks
End-of-production adventitious agent testing
Justification of harvest hold
Time period study
Temperature study

of the native protein product. Another example is the impact that growth conditions can have on the biosynthesis of glycoforms. Because glycoform synthesis is a multienzyme process, glucose starvation can shut down glycoform synthesis, whereas ammonium ion buildup can disrupt sialylation capping. Dissolved oxygen levels, hormone, growth rate, cell density, and age of the cells can also affect the composition of the protein-linked carbohydrate chains (Damm, 1995). Thus, in addition to the concern about the complexity of the protein due to the composition and sequence of its amino acids, QC must also address the additional complexity due to the composition and sequence of the attached monosaccharides on the synthesized protein product.

Validation of the purification process is centered around controlling the extent of purification required for the product and providing an acceptable viral clearance safety factor. The extent of purification of the product needs to be consistent with its intended use. Products which are to be administered repeatedly or at high doses need to be adequately pure to prevent the development of undesired immune or toxic reactions due to contaminants. Although biopharmaceutical products may be demonstrated to be 99% pure by physiochemical characterization, special attention needs to be directed toward the removal of certain contaminants which may be present in small amounts, specifically detectable viruses, microbial and nucleic acid contamination, and undesirable antigenic or pyrogenic materials. Table VI presents a list of what can be included in purification process validation studies.

Removal of process-related impurities is commonly determined either by direct measurement of the purification process streams or with spiking studies on the small scale. For front-end purification steps, the reduction factor can be determined by measuring the component levels before and after the particular purification step. For the back-end purification process, because the component levels may be reduced below the detectable concentration in various process sample matrices, spiking experiments can evaluate the efficiency of removal of these purification steps. An example of a process-related impurities validation summary for a biopharmaceutical is presented in Table VII.

Removal of pyrogenic lipopolysaccharides (i.e., endotoxins) are essential during the purification process of a biopharmaceutical. Endotoxins can be introduced into the process either from contaminated raw materials or from the cell substrate itself (e.g., *E. coli* or other Gram-negative bacteria). Because of the wide size distribution of endotoxins, purification methods that rely on size will be unreliable unless all of the factors that influence size distribution are well controlled. Use of a specific purification column just to remove endotoxins (e.g., immobilized polymixin B) is generally unwarranted, because

Table VI
Process Validation Studies Performed at the Purification Stage

Removal of host-related impurities
Host cellular DNA
Host cellular protein
Removal of cell culture media impurities
Antibiotics and hormones
Bovine IgG
Removal of purification process impurities
Endotoxin
Detergents
Protein A leachates
Control of chromatographic columns
Cleaning
Inactivation of viruses and bioburden
Use life
Justification of holding time periods
Material at each purification step
Viral clearance safety evaluation

traditional methods such as ion exchange have been shown to be extremely effective, either by binding the negatively charged endotoxins to an anion exchanger or by binding the product to a cation exchanger. Other purification techniques, such as hydrophobic interaction and affinity chromatography,

Table VII
Example of a Process-Related Impurities Validation Summary[a]

Process-related impurity	Residual per dose	Log reduction factor
Bovine serum albumin	< 4.3 ng	>7.0
Bovine IgG	60 ng	3.4
IgG1	<250 ng	NA
Host cell DNA	<10 pg	>8.1
Gentamicin	<0.5 μg	2.6
Papain	0.5 ng	4.2
Iodoacetamide	NA	5.8
Guanidine	<1.3 μg	5.2
Protein A	<0.6 ng	0.8

[a] Imciromab pentetate reported in the FDA's Summary Basis for Approval, July 3, 1996 on the FDA website (www.fda.gov/cber).

that isolate a product from feedstream impurities can be useful in the removal of endotoxins (Sofer, 1994).

Historically, the removal or inactivation of viruses during biological product isolation and purification has been a major quality concern. Viruses present in human plasma [e.g., human immunodeficiency virus (HIV)] have contaminated Factor VIII preparations, and human growth hormone extracted from the pituitaries of human cadavers has been implicated in the transmission of the TSE agent responsible for CJD. For biopharmaceutical products, the evaluation and characterization of the virus removal and/or inactivation procedures during purification play important roles in establishing viral safety. The objective of these viral clearance studies is to assess process step(s) that can be considered to be effective in inactivation/removing viruses and to estimate quantitatively the overall level of virus reduction obtained by the process. This is achieved by deliberately adding (spiking) significant amounts of a virus to the crude material and/or to different fractions obtained during the various purification process steps and demonstrating its removal or inactivation during the subsequent steps. Some expected ranges of retrovirus removal using various robust inactivation/removal purification steps are listed in Table VIII.

Specific requirements for each aspect of these viral clearance studies, including the choice of the spiking viruses, design of the small-scale purification model, details of the analyses, and calculation of reduction factors, are provided by the ICH (2001e). It is very important to stress that the small-scale model design must accurately reflect the normal purification process scale operations in order to obtain a valid result (CPMP, 1996; Sofer, 1996).

Table VIII

Expected Range of Retrovirus Clearance during Various Purification Process Steps[a]

Step	Expected log virus clearance
Inactivation step	
Low pH (3.9 or lower)	3–4
Heat	4
Solvent/detergent	5
Filtration (15–40 nm)	4–8
Removal steps	
Protein A affinity chromatography	3–6
Ion-exchange chromatography	2–6
Size-exclusion chromatography	1–5

[a] Data from Darling and Spaltro (1996); Revie and Vincent (1997); U.S. FDA (1997).

3.5. Facility Control

Ten years ago environmental monitoring (EM) during drug substance manufacturing of a biopharmaceutical was not a major concern, with little monitoring actually taking place during this stage of manufacturing. This has changed rapidly, and all cell culture and purification processing now includes a rather extensive EM program, due to the ever-increasing regulatory expectations. Some insight into the current FDA expectation for EM at the cell culture and purification stages of biopharmaceutical manufacturing can be obtained form the following statements (Roscioli *et al.*, 1996):

3.5.1. CELL CULTURE STAGE

> We generally recommend that scale-up, fermentation, and harvest procedures be performed in Class 100,000 areas. For manufacturing areas containing validated closed systems, the limits may exceed those recommended for Class 100,000 environments, but these areas should continue to be environmentally monitored. Additionally, action limits should be established based on historical data and the nature of the operations that occur in the area. We recommend Class 100 conditions for any inoculations, additions, or sampling procedures for which the connections cannot be appropriately sterilized prior to transfer. Some manufacturers have achieved Class 100 conditions for these operations through the use of portable laminar flow hoods.

3.5.2. PURIFICATION STAGE

> Purification or other processing steps that have minimal or no potential for environmental contact should be validated as closed systems. When a purification system is shown to be closed to environmental factors, operations performed in environmentally controlled areas of at least Class 100,000 are acceptable. When closed systems are successfully validated, alternatives to these recommendations may be considered, as described above. Air in purification areas should be monitored for viable and nonviable particulate quality during dynamic conditions. Such monitoring should occur during various stages of purification for each batch because of the variability of the environmental quality introduced by personnel practices.
>
> Purification or other processing steps that expose product to environmental conditions are considered open systems and should be performed in controlled environments, meeting at least Class 100,000 conditions. As stated above, controlled environments should be

> monitored for viable and nonviable particulates at least weekly during dynamic conditions. However, where possible, it is recommended that air be monitored for viable and nonviable particulate quality during each processing step. When a greater potential for product contamination exists, more stringent limits and/or controls should be established.

Biopharmaceutical manufacturing facilities generally are not operated as single-product-dedicated facilities due to the financial burden this would place on a company. To contain costs, many biopharmaceutical companies are now electing to develop multiuse manufacturing facilities where two or more products are processed within the same, or shared, manufacturing space. In addition to the engineering design and proper operation of the facility under GMP, the implementation of a validated cleaning/changeover program by the quality unit is imperative to prevent cross-contamination and ensure product safety (Williams, 1997).

To illustrate the quality considerations that enter into a cleaning validation study, an example of how this can be done for a large-scale bioreactor by addressing five key questions is provided:

1. What components need to be cleaned from the bioreactor?

The first concern, of course, is the removal of residues having pharmacological activity from the previous biopharmaceutical product produced in the bioreactor. However, concern also has to be given to removal of residues from the host cell components (e.g., cellular proteins, genomic DNA), the media components used to support the cell culture (e.g., glucose, complex nitrogen nutrient feeds), and the possible cleaning agent to be used (e.g., sodium hydroxide, detergent).

2. What is a scientifically sound and appropriate sampling plan for the bioreactor?

For a bioreactor, sampling the liquid from the rinse cycles of the equipment and direct surface sampling (e.g., swabs) are common approaches to collecting samples. As with any sample collection procedure, there are limitations. If the components of concern are insoluble in water (e.g., recombinant protein in the form of cytoplasmic inclusion bodies), the approach of collecting rinse samples may give a false-negative result. Certain recombinant proteins may be difficult to clean out of the ancillary lines of the hard-piped system of a bioreactor. Although for a bioreactor, the areas generally hardest to clean may also be the areas which are not readily accessible for direct surface sampling, it is still necessary to swab the accessible surface sites that have a probability for residue retention (e.g., around valves).

3. How will the test samples from the bioreactor be handled prior to analysis?

Components to be measured may be rendered unstable by the handling conditions chosen (e.g., hydrophobic proteins at dilute concentrations either may stick to the plastic surfaces of the collection containers, leading to loss of mass, or self-associate, leading to loss of activity). Change may occur upon storage of the samples prior to analysis.

4. Are the test methods chosen scientifically sound and appropriate for the bioreactor samples?

The ideal analytical method is one that is capable of measuring a very broad spectrum of potential contaminants, including the active protein product, at the appropriate sensitivity, in minimal time period, and for a reasonable cost. While no ideal assay exists, an abundance of test methods are available to choose from for the detection of contaminants at the ppm level. Total organic carbon (TOC) analysis is a nonspecific method capable of detecting all carbon-containing contaminants, whereas enzyme-linked immunosorbent assay (ELISA), *Limulus* amebocyte lysate (LAL), and bioassay methods are contaminant-specific methods. Bioassays are the only methods guaranteed to measure the active protein product if it is present and if it was not inactivated due to the cleaning agents.

5. What is a "clean" bioreactor?

Setting cleaning limits is one of the most important steps in the cleaning validation. In U.S. FDA (1991), some guidance is provided:

> Part of the answer to the question 'how clean is clean?' is, "how good is your analytical system?" The sensitivity of modern analytical apparatus has lowered some detection thresholds below parts per million (ppm), down to parts per billion (ppb). The residue limits established for each piece of apparatus should be practical, achievable, and verifiable.

However "clean" is defined, each manufacturer has to be able to document, by means of data, that the residual level permitted has a scientifically sound basis.

The response to these five key questions has changed over time due to increased regulatory expectations. Five years ago, the expectation would be that the active substance primarily would need to be demonstrated to be removed from the bioreactor, that swabbing is less important compared to rinse water sampling in the cleaning validation program, that biological or product-specific ELISAs would be performed, and that "clean" would be defined at the level set by the individual company (Geigert *et al.*, 1994). Currently, the expectation would be that all components would need to be demonstrated to be removed form the bioreactor, that swabbing is a critical component of the cleaning validation program, that TOC analysis would be performed, and that "clean" would be defined as the level at or near the limit of quantitation of the test methods employed (Sherwood *et al.*, 1996; Voss *et al.*, 1996).

Along with a cleaning validation program, establishment of a product changeover program by the quality unit is another key to the proper use of the manufacturing facility (Bader *et al.*, 1992). When bringing a new biopharmaceutical product into a multiuse facility, the quality program should evaluate the potential impact of that product on all operations of the facility. This requires that the quality program establish appropriate procedures for evaluating the new process and implement appropriate controls, as necessary, prior to new product entry. Evaluation should include, at a minimum, potential areas of cross-contamination, available cleaning validations, biocontainment, and presence of infectious or adventitious agents. In addition, the host system, biological effects, and toxicity of products manufactured in the facility should be considered.

The operation of a biopharmaceutical manufacturing facility in a multiuse mode can be economically attractive for the company, but the quality challenges associated with cleaning validation and product changeover programs should not be underestimated.

4. OVERALL QUALITY CONTROL STRATEGY FOR BIOPHARMACEUTICAL PRODUCTS

The overall quality control strategy is to assure that the biopharmaceutical product is consistently produced with an acceptable degree of quality and that it retains such through its intended shelf life. When discussing control of a product to ensure its identity, purity, potency, quantity, and safety, one usually focuses on the product specifications. However, product specifications are only one aspect of the overall strategy. For biopharmaceutical drug products, all of the following elements are part of the overall quality control strategy:

- Product characterization: "well-characterized"
- Process validation versus release testing: elimination of release tests
- QC release assays: identity, purity, potency, quantity, and safety
- Specifications: appropriate and scientifically based
- Product stability: quality through the intended shelf life

4.1. Product Characterization

Historically, biological pharmaceutical products have been derived from natural sources and frequently the therapeutic agent has been difficult to characterize in detail with available analytical methods. However, in the past decade there has been a significant advance in the purification capabilities of biopharmaceuticals and a virtual explosion in the availability of sophisticated and reliable analytical instrumentation, as well as extensive and sensitive chromatographic, immunological, and spectrometric methods of product analysis. These advances continue to push the envelope of the definition of "well-characterized" for biopharmaceuticals.

Product characterization not only provides a comprehensive understanding of the properties of the molecule, but also guides the quality program in selecting the appropriate test methods to be used for product release and stability monitoring. Clearly, those properties of the molecule that lead to heterogeneity, loss of potency, or change over time need to be evaluated in the release of the product lots. Specifications should be chosen to confirm, rather than establish, molecular structure and should focus on those product characteristics found to be useful in ensuring the safety and clinical utility of the biopharmaceutical product. A list of what is generally included in characterization of a biopharmaceutical is presented in Table IX.

4.2. Process Validation versus Release Testing

Process validation is documented evidence that the manufacturing process is effective and reproducible and yields a product that consistently meets predetermined quality attributes. Process validation data should result in the elimination of release specifications. Based on validated impurity reduction factors, it should be possible to predict and guarantee the residual level of the impurity in the final product. Thus, such impurities would not need to be tested for release of each lot of product. This "validation approach" to impurities is becoming more acceptable to the regulatory agencies. This approach has been used for a number of years in assuring effective and acceptable removal of antibiotics present in cell culture medium (e.g., tetracycline from bacterial processes and methotrexate from mammalian processes) and of removal of putative viral contaminants through the purification process. The FDA has also accepted this approach for assuring effective and acceptable removal of host cell contaminants such as DNA and proteins. However, the European counterpart, the European

Table IX
Typical Product Characterization Studies for a Biopharmaceutical

Definitive determination of composition: amino acid analysis/sequence, carbohydrate composition/sequence
Evaluation of higher order structures: disulfide bonds, circular dichroisim
Physical properties: molecular weight, extinction coefficient
Biological properties: potency, specific activity
Evaluation of posttranslational modifications: microheterogeneity
Evaluation of impurity profile
Determination of quantity
Evaluation of stability under stressed conditions to define degradation pathways

Medicines Evaluation Agency (EMEA, Website www.emea.eu.int) is currently more conservative in accepting the "validation approach" for these host cell contaminants (CPMP, 1997).

4.3. Quality Control Release Assays

The role of QC release testing is to ensure product identity, purity, potency, quantity, and safety, whether the biopharmaceutical drug product is recombinant human insulin, which has only 51 amino acids and no attached carbohydrate, or recombinant human Factor VIII which has over 2000 amino acids and a significant percentage of its molecular weight deriving from attached carbohydrate. Extracting useful information from the analysis of these biopharmaceutical products is a challenge for QC. With an abundance of test methods available for use with biopharmaceuticals (see Tables X and XI), QC must wisely select the appropriate test methodologies for a product's release. Performing every available test method on a biopharmaceutical product does not guarantee an improvement in product quality. With increasing corporate pressure for cost-of-goods considerations, the selection of the necessary release assays thus becomes a very important challenge for QC.

For biopharmaceutical drug products, the following QC release assays will be further discussed:

- Identity: peptide mapping and capillary isoelectric focusing (IEF)
- Purity: high-performance liquid chromatography (HPLC)
- Potency: bioassays
- Quantity: active component and excipient testing
- Safety: CFR requirements for biologics

Table X
Analytical Test Methods Applicable to Protein-Based Biopharmaceutical Products

Analytical methods	Common applications
Amino acid analysis	Quantity, identity
Bioassay	Potency, identity
C-Terminal sequencing	Microheterogeneity
Carbohydrate composition analysis	Identity, microheterogeneity
ELISA/RIA	Identity, quantity
IEF, cIEF	Identity, charge
IEX-HPLC	Purity, charge
Mass spectrometry	Identity
N-Terminal sequencing	Identity, microheterogeneity
Oligosaccharide analysis	Microheterogeneity
Peptide map	Identity, microheterogeneity
RP-HPLC	Identity, purity, impurities, stability
SDS–PAGE	Purity, molecular weight determination
SEC-HPLC	Identity, purity, aggregates
UV_{280}	Quantity

Peptide mapping is routinely used as an identity test for biopharmaceutical products. At the drug substance stage, peptide mapping can uniquely identify the molecular structure of the active drug component. At the drug product stage, peptide mapping can distinguish the drug product form other drug products manufactured at the same facility to prevent mixups and/or mislabeling. Peptide mapping involves either the chemical or enzymatic treatment of a protein resulting in the formation of peptide components. The peptide mixture is then subjected to HPLC separation, resulting in a "peptide fingerprint." Peptide mapping is a comparative procedure, because the information obtained, compared to a reference standard similarly treated, confirms the primary structure of the protein (USP, 2002c). Illustration of the

Table XI
Analytical Test Methods Applicable to Nucleic Acid-Based Biopharmaceutical Products[a]

Analytical method	Common applications
Agarose gel electrophoresis	Purity
Bioassay	Potency, identity
DNA sequencing	Identity
Restriction enzyme analysis	Identity
$UV_{260/280}$	Quantity

[a] U.S. FDA (1996a,b).

use of peptide mapping as a QC identity release assay can be found in the USP monograph (USP, 2002a), for alteplase (recombinant human tissue plasminogen activator, TPA) as follows:

> Chromatographic system (*Chromatography* <621>): The liquid chromatograph is equipped with a 214-nm detector and a 4.6 mm × 10 cm column that contains packing L1. The flow rate is about 1 ml/min. The system is programmed to provide a mobile phase consisting of variable mixtures of mobile phase A (6.9 ng/ml monobasic sodium phosphate, pH 2.85) and mobile phase B (acetonitrile). The system is equilibrated with 100% *mobile phase A*. After injection of the solution under test, the proportion of *mobile phase B* is increased linearly from 0% to 30% at a rate of 0.33%/min. The proportion of *mobile phase B* is then increased linearly at a rate of 1.0%/min until the proportion of *mobile phase B* is 60%, and is held at that composition for 10 min.
>
> Procedure: Separately inject equal volumes (about 100 μl) of the standard preparation, the test preparation, and a mixture of the standard preparation and the test preparation, (1:1) into the chromatograph, record the chromatograms, and measure the responses for not less than 20 major peaks as defined in the USP Alteplase RS Data Sheet: The retention times of corresponding peaks from the standard preparation and the test preparation do not differ by more than 0.4 min, and the peak area ratios relative to peak 19 (as shown on the USP Alteplase RS Data Sheet) do not differ by more than 20%. No additional significant peaks or shoulders are found, a significant peak or shoulder being defined as one having a peak area response of not less than 5% of peak 19.

It should be noted that peptide mapping can also provide much useful information to QC about the active drug ingredient itself. Truncated forms of the protein due to cellular peptidases that catalyze the removal of N-terminal or C-terminal amino acids can be detected by changes noted in the peptides containing these termini. Chemical modification of amino acids (e.g., oxidation of methionine to its sulfoxide derivative, evidence of a disulfide bridge) can also be detected. As an example, the detection of a trisulfide structure (cysteine$_2$ S_3) in recombinant methionyl human growth hormone in the Cys^{53}–Cys^{165} bridge was made possible by peptide mapping coupled with mass spectrometry (Canova-Davis *et al.*, 1996).

The increasing interest by QC in the use of capillary isoelectric focusing (cIEF) over the past few years is due to its value as a fast, accurate, precise, and linear means for the determination of identity and charge distribution of a biopharmaceutical. This, as well as the other, more obvious advantages of decreased analysis times, automation, and quantitative capabilities, make cIEF an extremely attractive methodology for routine use in QC. In cIEF, the product to be identified is applied to the capillary together with the ampholytes. Typically, the entire capillary is filled with the ampholyte/sample mixture and cIEF is accomplished through two consecutive events: focusing and mobilization. The ampholytes create a pH gradient within a short time upon application of an electric field. Under the influence of this electric field, charged solutes migrate through the medium until the local pH is equal to their isoelectric points (pI), whereby the solutes become neutral and remain focused in narrow zones. The pattern obtained during this first step is characteristic of the sample. The focused zones will then be mobilized past the detector using chemical, pressure, or electrosmotic flow mobilization so that a tracing of the separation can be obtained.

"Purity," which is clearly dependent upon the test method used, means the freedom form extraneous matter in the product, whether or not harmful to the recipient or deleterious to the product. For a biopharmaceutical, the types of possible product-related impurities are numerous (see Table XII). These product-related impurities can arise from infidelity of the genetic transcription and translation steps during protein synthesis, posttranslational modification of the produced protein product, contamination with host cell proteins which copurify with the product, interaction and inactivation of the protein product with chemicals used during the purification process, or instability of the drug product during storage. Liquid chromatographic methods, usually HPLC, are often the most useful techniques for biopharmaceutical quality control. These methods are rapid, accurate, sensitive, and readily validated, because they can be automated. The most commonly used HPLC methods are reversed-phase (RP), ion-exchange (IEX), and size-exclusion (SEC) or molecular sieving methods. RP-HPLC is a high-resolution technique, which separates protein species on the basis of their relative hydrophobicity. In this technique, a hydrophobic stationary phase is equilibrated in a hydrophilic solvent. The protein solution is injected, interacts with the stationary phase, and is eluted by increasing the mobile-phase hydrophobicity. IEX-HPLC is similarly useful, separating species on the basis of charge instead of hydrophobicity. SEC-HPLC is a rapid method for determining the size distribution of proteins in a solution. It has low resolving capability in comparison to RP-HPLC or electrophoretic methods, but is an especially useful technique for detecting protein aggregation and fragmentation. Illustration of the use of HPLC as both a QC purity and impurity release assay can be found in the proposed USP monograph (USP, 1997) for somatropin (recombinant human growth hormone), as follows:

> Chromatographic system (*Chromatography* <621>): The liquid chromatograph is equipped with a 220-nm detector and a 4.6 mm × 25 cm column that contains packing L7 and is maintained at 45°C. The flow

Table XII
Potential Product-Related Impurities with Biopharmaceuticals[a]

Aggregates
Deamidation products
Denatured products
Dimers and higher oligomers
Disulfide scrambling
Formyl methionine
Mistranslation products
Oxidized methionine
Proteolytic cleavage
Truncated forms

[a] USP (2002b).

rate is about 0.5 ml/min of 0.05 *M* Tris, pH 7.5, and *n*-propyl alcohol (71:29).

Procedure: Separately inject equal volumes of standard solution and the test solution into the chromatograph, record the chromatograms, and measure the responses for the major peaks. The retention time of the somatropin peak in the chromatogram of test solution corresponds to that of the somatropin peak in the chromatogram of the standard solution. Calculate the percentage of impurities in the portion of somatropin taken by the formula $100(r_i/r_s)$, in which r_i is the total of all extraneous peak responses of impurities and r_s is the sum of the responses of all peaks: not more than 6.0% of total impurities is found. Determine the area of the main peak and the peak areas of any peaks eluting earlier than the main peak in the chromatogram of the test solution. Calculate the percentage of high molecular weight proteins in the portion of somatropin taken by the formula $100A_H/(A_H + A_M)$, in which A_H is the area of the high molecular weight proteins in the chromatogram of the test solution, and A_M is the monomer peak in the test solution: not more than 4% is found.

One of the possible posttransitional modifications that occur on proteins is the covalent attachment of oligosaccharide chains. Glycosylation is a characteristic of recombinant proteins that are expressed from eukaryotic cell lines. Although the polypeptide chain of a glycoprotein is synthesized under the direct control of the genetic code, oligosaccharides are not primary gene products, but are synthesized by enzymes known as glycosyltransferases. This synthesis results in microheterogeneity of the carbohydrate chains. In addition, glycosylation is cell-line-dependent, so glycoproteins with identical polypeptide chains made in different cell lines may have considerably different carbohydrate structures. The sugars commonly found in glycoproteins include neutral sugars (D-galactose, D-mannose, and L-fucose), amino sugars (N-acetylglucosamine and N-acetylgalactosamine), and the acidic sugar sialic acid. Two main approaches can be taken to identify the sugars covalently attached to the glycoprotein. Both are based on the understanding that microheterogeneity is a common phenomenon among glycoproteins, and that the information represents either average composition or representative structures. The first approach is the determination of the composition of sugars in a glycoprotein, which can be performed after digestion of the protein by several methods. Neutral sugars and sialic acid can be determined by simple colorimetric tests. The second approach for determining the carbohydrate composition is to release and separate individual oligosaccharide structures covalently attached to the glycoprotein. This requires an understanding of the types of structures attached. The attachment of sugars to proteins can occur in two major ways: through an O-glycosidic bond involving the hydroxyl group of serine, threonine, or modified amino acids such as hydroxylysine or hydroxyproline, or through the N-glycosidic bond of asparagine. O-Linked oligosaccharides can be released from the protein following β-elimination under alkaline conditions and reduction of the reducing end sugar with sodium borohydride. N-Linked oligosaccharides can be released chemically by hydrazinolysis or enzymatically by one of a variety of specific glycosidases, such as endo H, endo F, or peptido-*N*-glycanase. The oligosaccharides can then be separated by anion exchange HPLC at high pH and quantitated with pulsed amperometric detection (PAD). This results in an oligosaccharide or carbohydrate map analogous to the peptide map for the protein (Spellman, 1990).

The biological activity of a biopharmaceutical drug product must be measured and quantified. Bioassays are utilized by QC to determine this biological potency. Bioassays are analytical procedures that utilize a biological reporter system (resulting in a biological/functional response), the purpose of which is to measure the amount of analyte or effective constituent in a biologic. This contrasts with binding assays, such as ELISAs and radioimmunoassays (RIAs) and receptor–ligand methods, that do not measure the ability of the biopharmaceutical to induce a biological response. The most appropriate method for assessing potency is to compare the biological activity of a sample to that of a well-characterized potency reference standard. However, such internationally recognized potency reference standards are available for only a few biopharmaceutical products (Federici *et al.*, 1991), and the process of reaching international recognition (i.e., assignment of International Units, IU) can take up to 5 years, as illustrated in Table XIII.

Where possible, it is preferable to use a bioassay that reflects the biological role of the material, although it is not necessary to attempt to

Table XIII
Five-Year World Health Organization Process for Establishing an International Standard: Example of the Granulocyte-Macrophage Colony-Stimulating Factor (GM-CSF) International Standard for Potency Assignment[a]

1988	Expert Committee on Biological Standardization (ECBS) approval to proceed on a standard for GM-CSF; National Institute for Biological Standards and Control (NIBSC) prepares GM-CSF Reference Standard ampoules
1989	NIBSC carries out stability study
1990	International collaborative laboratory study initiated
1991	NIBSC compiles/analyzes data from 29 participating labs
1992	ECBS accepts NIBSC's recommendation.
1993	First International Standard for GM-CSF potency assignment available to industry

[a] Data from Geigert and Ghrist (1996).

mimic therapeutic activity for a QC test that is used to assess quality or efficacy. Bioassays can be carried out *in vivo* or *in vitro*. The choice of bioassay depends upon the nature of biological material available for *in vitro* tests and whether the biological activity of the biologic is only measurable in whole animals. It is desirable to develop *in vitro* biological tests where possible, because they offer distinct economic, automation, and precision advantages over using live animals. A list of *in vitro* bioassays used for biopharmaceuticals is presented in Table XIV.

For biopharmaceutical products, the quantity of both the active compound and the excipients present must be measurable. The protein content of the drug product can be determined by a variety of methods, ranging from colorimetric methods, which complex specific dyes to the

Table XIV
In vitro Bioassays Used for Biopharmaceutical Drug Products[a]

Biopharmaceutical[b]	*In vitro* assay	Biological response quantified
GM-CSF	Stimulation of cell lines (TF-1)	Cellular proliferation
HGH	Stimulation of cell lines (3T3)	Cellular differentiation
IFN-α	Antiviral activity on cell lines (WISH)	Cell survival
TNF	Cytotoxicity on cell lines (L929)	Cell survival

[a] Data from Mire-Slius *et al.* (1996).

[b] GM-CSF, Granulocyte-macrophage colony-stimulating factor; hGH, human growth hormone; IFN-α, interferon-α; TNF, tumor necrosis factor.

protein, to quantitative amino acid analysis, to chromatographic methods. A most useful approach which has been employed for the determination of protein content for purified proteins involves determining the molar extinction coefficient of the protein, thus allowing protein content to be determined directly from ultraviolet absorbance (USP, 2002c).

It should be noted that QC needs to also measure the concentration of the excipients used to compound the biologic. The presence of the recombinant protein as well as the mixture of excipients used in the formulation frequently makes these measurements difficult. An indication of the diversity of excipients used with biopharmaceuticals is presented in Table XV.

Safety testing for QC release of biopharmaceutical products is described in the CFR on General Biological Product Standards (U.S. FDA, 2001, Part 610). The sterility of the final product is determined using the specified methods of either membrane filtration or direct inoculation. Pyrogenic, LAL endotoxin, measurement must not exceed 5.0 EU/kg of patient weight (U.S. FDA, 1987b). The general safety of the final product is determined using the specified guinea pigs and mice. However, it should be noted that the requirement for general safety testing has been waived for specified biologics (Federal Register, 1996b).

Biopharmaceutical drug products can be either protein-based (recombinant protein or monoclonal antibody) or nucleic acid–based (plasmid DNA or gene therapy). Because of this diversity and complexity of

Table XV
Indication of the Diversity of Excipients Used in the Formulation of Biopharmaceuticals[a]

Leukine®, sargramostim, recombinant human granulocyte-macrophage colony-stimulating factor
Mannitol
Sucrose
Tromethamine (Tris)
Benzyl alcohol
Epogen®, epoetin alpha, recombinant human erythropoietin
Sodium chloride
Sodium citrate
Albumin, human
Nutropin AQ™, somatropin, recombinant human growth hormone
Sodium chloride
Sodium citrate
Polysorbate 20
Phenol

[a] Data from PDR (1999).

biopharmaceutical drug products, the regulatory agencies have accepted a flexible approach to QC release testing. This has permitted tailoring of the release testing program based on the specific biologic, the intended use of the drug product, and the introduction of new analytical technologies. It should also be pointed out that the QC test methods can only do what they are designed to do. These release test methods cannot cover every conceivable eventuality or detect every possible impurity or product change. The application of multiple QC release tests is the only way to continue to ensure that a biopharmaceutical drug product is appropriately meeting regulatory requirements. General guidance on QC release testing strategies for biopharmaceutical products can be found in published sources (see Table XVI) and in several of the FDA's Points to Consider documents (see Table XVII).

Regulatory oversight of lot release testing has undergone significant changes in recent years. With the introduction of "specified biologics" by the FDA comes the advantage of not having to submit a Certificate of Analysis to CBER for each lot of product for their preapproval prior to market release (Federal Register, 1996b). Of course, to take advantage of this, one's biopharmaceutical must fall into one of the following four categories:

- Therapeutic recombinant DNA-derived products
- Monoclonal antibodies for *in vivo* use

Table XVI
QC Release Testing Strategies used for Recombinant Human Erythropoietin[a]

Strategy for drug substance	Strategy for drug product
Western blot	Appearance
IEF	Activity, RIA
SEC-HPLC	Protein content
Peptide mapping	Sterility
N-Terminal sequencing	Safety
Activity, RIA	Pyrogenicity
Activity, *in vivo*	Endotoxin
HPLC, purity	Excipient content
DNA	
Host cell proteins	
Mycoplasma	
Sterility	
Protein content	
Endotoxin	

[a] Data from Federici (1994).

Table XVII
General Guidance on QC Lot Release Testing Strategies from the FDA[a]

Quality parameter	Testing strategy: Monoclonal antibodies	Plasmid DNA vaccines
Identity	Ig classification	DNA sequence restriction enzyme
Purity	SDS–PAGE	Agarose gel
	IEF	
	Molecular integrity	Contaminants: host DNA (Southern blot), host RNA (Northern blot), host protein (SDS–PAGE)
	Contaminants: animal components, process contaminants	
Potency	Close possible relationship to the putative pharmacological activity	Ability of vaccine to effect an immune response
Quantity	Protein content (A_{280})	DNA content (A_{260}/A_{280})
Safety	Sterility	Sterility
	Endotoxin (LAL)	LAL

[a] U.S. FDA (1996b, 1997b).

- Therapeutic DNA plasmid products
- Therapeutic synthetic peptide products of 40 or fewer amino acids

On the other hand, the requirement by the EMEA for full QC rerelease testing for marketed products imported into Europe is a major disadvantage for U.S. manufacturer's. This duplicative testing by a competent quality laboratory in Europe adds significantly to the cost of the products, with little quality value being added.

4.4. Specifications

A specification is defined by a test method and appropriate acceptance criteria, which are generally numerical limits or ranges. It establishes the set of criteria to which a drug substance or drug product should conform to be considered acceptable for its intended use. Specifications are binding quality standards that are agreed to between a regulatory agency and the biopharmaceutical manufacturer. Setting appropriate specifications is both a very important and a most difficult process for a biopharmaceutical company. It is very important, because the respect for specifications ensures that, regardless of the batch, the patient will receive a strictly equivalent product, despite the complexity of the production process of the biological product. It is a most difficult process, in that should the wrong specifications be set, the end result may be product recalls or the inability to release a product, jeopardizing the future of the biopharmaceutical company.

Regulatory guidance on how to set appropriate and scientific-based specifications for biopharmaceuticals is now available. To assist the quality program in setting specifications at the start of human clinical trials, referred to as Phase 1 studies, the U.S. FDA (1995) published an appropriate Guidance for Industry document. The following are the stated acceptable limits that can be used at this stage of the drug development cycle:

4.4.1. DRUG SUBSTANCE

> A brief description of the test methods used should be submitted. Proposed acceptable limits supported by simple analytical data, (e.g., IR spectrum to prove the identity, and HPLC chromatograms to support the purity level and impurities profile) of the clinical trials material should be provided. Submission of a copy of the certificate of analysis for the drug substance used in the clinical batch is also suggested. The specific methods will depend on the source and type of drug substance (e.g., animal source, plant extract, radiopharmaceutic, other biotechnology-derived products). Validation data and established specifications ordinarily need not be submitted at the initial stage of drug development. However, for some well-characterized, therapeutic biotechnology-derived products, preliminary specifications and additional validation data may be needed in certain circumstance to ensure safety in Phase 1.

4.4.2. DRUG PRODUCT

> A brief description of the proposed acceptable limits and the test methods used should be submitted. Tests that should be submitted will vary according to the dosage form. For example, for sterile products, sterility and pyrogenicity tests (or endotoxin assays, as appropriate) should be submitted. Submission of a copy of the certificate of analysis of the clinical batch is also suggested. Validation data and established specifications need not be submitted at the initial stage of drug development. For well-characterized, therapeutic, biotechnology-derived products, adequate assessment of bioactivity and preliminary specifications should be available.

"Preliminary specifications" and "assessment of bioactivity" are therefore acceptable at this early stage of clinical development, and this FDA approach makes good common sense. Because companies today are expediting the entry of their products into clinical programs, they are now finding that at the time of filing the investigational new drug (IND) application they have less experience, both with the control of the manufacturing process and with the release testing, than previously. Typically, today, companies may file their IND with three or fewer lots total. Therefore, this minimal approach at the IND filing with further evaluation and analysis over the clinical development program, eventually leading to established specifications by the time of filing the biologies license application (BLA), works to everyone's advantage. An illustration of the evolution of a potency specification starting from an "adequate assessment of bioactivity" and ending with a "quantitative potency assignment" over the course of the clinical development is presented in Table XVIII.

The FDA has issued a Guidance for Industry document on setting specifications for later stage clinical development products, that is, Phases 2 and 3 (U.S. FDA, 1999).

An ICH guidance document (Q6B) on specifications for marketed biopharmaceutical drug substances and drug products is now available (ICH, 2001c). A general understanding of the specification requirements for a marketed biopharmaceutical drug product can be obtained from FDA-released documents (see Tables XIX and XX).

A final point on specifications is that regulatory acceptable limits do occasionally "loosen." The most notable example is acceptable residual levels of cellular DNA in biopharmaceuticals. Because of possible product contamination by potentially oncogenic DNA from mammalian cell lines or from a transforming vector, both European and U.S. regulatory agencies

Table XVIII

Evolution of a Potency Specification over the Course of the Clinical Development Program[a]

		Changes in potency specification	
Qualitative	Phase 1 IND	Report units per mass	U/mg result
↓	↓	Lower limit assigned	1 U/mg
		Reasonable range assigned	1–8 U/mg
Quantitative	BLA filing	Justified range, expressed in International Units	3.3–5.5 IU/mg

[a] Data from Geigert (1997a).

Table XIX

Summary of Tests and Specifications for Release of the Recombinant Human Platelet-Derived Growth Factor Drug Substance Regranex[a]

Test	Specification
Appearance	Clear/colorless to straw-colored
pH	5.0–7.0
Protein content by A_{280}	8–12 mg/ml
Potency by mitogenic assay	70–150% of reference material
Identity by Lys-C peptide map	Comparable to reference material; co-mix reveals no new peaks
Purity by SDS–PAGE	Comparable to reference material
Purity by cIEF	
Purity by HPSEC	
Purity by RP-HPLC	
Total DNA	100 pg/mg
Host cell protein (yeast, ELISA)	50 ng/mg
Monosaccharide content	4 mol mannose/mol PDGF
Endotoxin (LAL)	< 2 EU/mg
Bioburden	1 CFU/ml

[a] Reported in the FDA's Summary Basis of Approval, December 16, 1997, on the FDA website (www.fda.gov/cber).

Table XX

Summary of Tests and Specifications for Release of the Murine Monoclonal Antibody Drug Product Imciromab Pentetate[a]

Test[a]	Specification
Color	Colorless
Appearance	Clear liquid
PH	6.4–6.6
Particles	Free of foreign particles; essentially free of proteinaceous particles
Fill volume	range 1.05–1.20 ml; mean 1.10 ml
Protein concentration by OD_{280}	0.45–0.55 mg/ml
Purity and identity by GF-HPLC	98% as main component
Purity of Fab-DTPA by SDS–PAGE	97% purity
Charge distribution by IEF	Conforms to reference standard
Endotoxin	20 EU/mg
Sterility	No growth
General safety	Nontoxic

[a] Reported in the FDA's Summary Basis for Approval, July 3, 1996, on the FDA website (www.fda.gov/cber).

[b] GF-HPLC, Gel filtration HPLC; DTPA, diethylenetriaminepentaacetate.

have historically required manufacturers to ensure DNA removal. The regulatory agencies initially set very low safety limits for DNA, regardless of the specific DNA origin. Today, these safety limits are much higher (see Table XXI).

4.5. Product Stability

The stability monitoring of biopharmaceutical drug products requires a carefully thought-out and well-developed quality program, which is based on the knowledge of both basic molecular chemistry and regulatory requirements. Biopharmaceutical products are distinguished, because the maintenance of their molecular conformation and hence of biological activity is dependent on noncovalent as well as covalent forces. These products can be particularly sensitive to environmental factors such as temperature changes, oxidation, light, ionic content, and shear. In order to ensure maintenance of biological activity and to avoid degradation, stringent conditions for their storage are usually necessary.

Similar to the logic that is used by QC in the design of release testing strategies, the analysis of the degradation profiles for biopharmaceuticals does not rest on one particular test method or approach. A multiplicity of degradation pathways exists for compounds composed of long chains of the 20 different amino acids. The amide bonds which link the amino acids together can undergo hydrolysis cleaving the protein into fragments. Various amino acids in the protein chain are susceptible to chemical modification. Asparagine and glutamine can lose α-amino groups, and the resulting deamidation results in aspartic acid and glutamic acid residues, respectively. Methionine can be oxidized to its sulfoxide derivative. Cystine

Table XXI
Chronology of Regulatory Safety Limits for DNA Contamination in Biopharmaceuticals

Year	Safety limit for DNA contamination (pg/dose)	Agency	Reference
1985	<10	FDA	FDA (1985)
1987	<10	FDA	FDA (1987c)
1987	<100	WHO	WHO (1987)
1993	<100	FDA	FDA (1993)
1997	<100	FDA	FDA (1997c)
1997	<10,000	WHO	Annecy proposal

can undergo either an intramolecular rearrangement (disulfide scrambling) or an intermolecular rearrangement (oligomer formation). Hydrophobic interaction between protein molecules can lead to noncovalent bonding (aggregation) and covalent bonding (crosslinking). Perturbation of the threedimensional structure of the protein leads to denaturation, which may or may not be reversible.

In order to detect the many pathways of degradation for these products, QC must employ a full range of test methods. No single test method can be guaranteed to be stability-indicating for all proteins. In fact, each test method has to be evaluated for its stability-indicating capability on a biopharmaceutical-by-biopharmaceutical basis. For example, a liquid formulation of recombinant human tumor necrosis factor (TNF) was studied under refrigeration conditions (Geigert *et al.*, 1988). By the sodium dodecyl sulfate–polyacrylamide gel electrophoresis (SDS–PAGE) test method, the TNF product appeared to be stable for a period of 1 year; however, by the IEF test method, the TNF product was undergoing significant deamidation during storage. In another example, a lyophilized formulation of recombinant human interleukin-2 (IL-2) was studied under refrigeration conditions (Geigert *et al.*, 1993). By both the SDS–PAGE and cell line-derived bioassay test methods, the IL-2 product appeared to be stable for a period of 2 years; however, by the RP-HPLC test method, the IL-2 product was decreasing in purity due to the formation of noncovalent aggregates. With different test methods yielding different stability results, QC assigns an expiration date for each biopharmaceutical product using the most relevant stability-indicating parameter.

Regulatory guidance for stability requirements of biopharmaceuticals during the clinical development stages has been provided by the FDA. According to the U.S. FDA (1987b) Guideline, as the clinical studies advance, the understanding of the biopharmaceutical's stability profile is expected to increase:

4.5.1. IND PHASES 1 AND 2

> The objectives of stability testing during Phases 1 and 2 are (a) to evaluate the stability of the investigational formulation used in the clinical trials and (b) to obtain the additional information needed to develop a final formulation (e.g., compatibility studies of potential interactive effects between the drug substance and other components of the system). Required stability information is limited to that needed to demonstrate that the clinical product is stable for the duration of the investigation.

4.5.2. IND PHASE 3

> The emphasis in stability testing during Phase 3 is on final formulations in their probable market packaging, on expiration dating, and on the study of degradation products when encountered. Studies to support the proposed expiration dating period should be completed, where possible, during Phase 3 for inclusion in the license submission

Regulatory guidance for stability requirements of biopharmaceuticals for market authorization is provided in ICH (2001d). For regulatory approval, stability data from at least three batches of drug substance and three batches of drug product, validated stability-indicating test methods, and precisely defined storage temperatures and storage conditions all must be provided to the regulatory agency. When designed properly, the ongoing QC stability program can yield an acceptable shelflife for the biopharmaceutical drug product when released to the market (see Table XXII).

Many biopharmaceuticals also require some degree of handling prior to administration to the patient. Intravenous infusions require the product to be diluted. Hospital policy may allow a pharmacist to prepare the drug solution ahead of time and then have it administered to the patient by a nurse on a later shift. The physical stability of biopharmaceuticals in these delivery environments is a critical determinant of biological potency and treatment efficacy and yet is often taken for granted. Interactions between formulated biopharmaceutical drugs and the surfaces of a delivery device may induce conformational

Table XXII

Approved Shelf Lives for Biopharmaceutical Drug Products on the Market[a]

Biopharmaceutical	Manufacturer	Expiry dating period
Avonex IFN - β-1a	Biogen	18 months at 2–25°C
Neumaga IL-11	Genetics Institute	24 months at 2–8°C
Rapilysin TPA	Boehringer Mannheim	24 months at 2–25°C
Rituxan Chimeric monoclonal antibody	IDEC Pharmaceuticals/Genentech	24 months at 2–8°C
Zenapax Humanized monoclonal antibody	Hoffmann-La Roche	12 months at 2–8°C

[a] Information obtained from the FDA website (www.fda.gov/CBER) for biologic licenses approved in 1997 and from the EMEA website (www.eudra.org/frame/frametest3.html) for European Public Assessment Reports (EPAR) issued in 1996 and 1997.

changes and aggregation that will compromise the intended therapeutic benefit (Geigert, 1989). To ensure that the handling of the drug product does not affect its safe administration, extensive stability and compatibility tests on various treatments of the product need to be performed.

5. CLOSING COMMENTS ON QUALITY IN BIOPHARMACEUTICAL COMPANIES

5.1. Meeting Increasing Compliance Expectations

Overall, the track record for biopharmaceutical companies meeting the regulatory agency compliance requirements is quite good. Even though the expectations from these agencies continue to increase with time, quality programs keep stepping up to the challenge. The major rewrite of the CFR cGMPs (Federal Register, 1996a), expected to be finalized shortly, increases the quality program's responsibility by formalizing its role in such areas as process validation, test method validation, and out-of-specification investigations. The FDA's expectation from quality programs is clearly stated in the *Federal Register* notice:

> Proposed 211.22(a) would require that firms be accountable with respect to validation provisions and would give quality control units the additional responsibility of reviewing and approving validation protocols to assess their adequacy. Quality control units would also be responsible for reviewing product, process, equipment, or other changes to determine if and when revalidation is warranted. This change is intended to make the quality control unit responsible for keeping validation current and is a logical extension of the quality control unit's role in ensuring product quality. The agency believes that, by making clear such accountability, compliance with the validation provisions will be more consistent and reliable.
>
> FDA believes, however, that CGMP requires written procedures to be in place to determine the cause of any apparent failure, discrepancy, or out-of-specification result. If the out-of-specification result cannot be clearly attributed to laboratory error, then the quality control unit should ensure that a thorough investigation is conducted and supported by a written record.

For biopharmaceuticals, there will always be more that the quality program can do to assist its company's level of compliance. As Table XXIII illustrates, biopharmaceutical companies' adherence to compliance is not always perfect. However, the product recall record for biopharmaceuticals is

Table XXIII
Quality Issues Raised by the FDA for Licensed Biopharmaceuticals[a]

Improve production oversight by increasing the number of personnel in manufacturing and quality areas
Revise production batch records and conduct assessments to minimize variances
Continue to monitor trends in changes of the relative levels of amino-terminal variants and the correlation of these changes with all measured fermentation parameters
Reevaluate the hold times for in-process intermediates, based on revised drug substance specification, and submit the study report
Provide a validation report for, and stability data from, the revised method for the reversed-phase HPLC assay
Repeat the photostability study on lyophilized drug product and on the final packaging configuration as well as on the immediate container/closure, in accordance with the ICH Photostability Guidance

[a] Information obtained on the FDA website (www.fda.gov/cber), from biologic licenses approved in 1997.

quite good, with only a few recalls having occurred in the history of these products (see Table XXIV). This reflects the overall commitment to quality by the biopharmaceutical industry.

5.2. Quality Cost-Effectiveness

Cost-effective quality is emerging at the forefront of biopharmaceutical company concern. Gone are the days of limitless spending by the quality groups. Two financial cost centers in quality programs are undergoing constant challenge: (1) the increasing sophistication (hence, cost) of the QC tests and (2) the outsourcing of the quality program's workload.

Biopharmaceutical products have provided a rich opportunity for analytical method development. "State-of-the-art" analytical methods of

Table XXIV
Product Recalls Involving Biopharmaceutical Drug Products

Manufacturer (year)	Product	Reason for recall
Amgen (1989)	Recombinant human erythropoetin	Unstoppered vials distributed
Ortho Biotech (1989)	OKT3 monoclonal antibody	Stability not maintained
Nova Nordisk (1993)	Recombinant human insulin	Adverse storage conditions at distributor
Baxter (1997)	Recombinant Factor VIII	Potential sterility compromise

yesterday, such as amino acid analysis and IEF, are being supplemented or replaced by new state-of-the art test methods, such as mass spectrometry and capillary electrophoresis. The ability to analyze for protein and carbohydrate microheterogenity has never been greater. The challenge for quality programs is to balance the very increasing capability of analytical method development against the cost-effective and relevant needs of QC testing. Regulations require the QC testing to be appropriate and scientifically sound, but they do not require the testing to be state of the art.

Biopharmaceutical companies are actively redefining themselves, their goals, and their priorities as a result of regulatory reform, limited capital financing, mergers, and acquisitions and other market changes. One way for these companies to maximize their investment is to outsource their workload. Although this has been widely practiced in several areas, such as clinical research development and fill/finish manufacturing steps, outsourcing is now also affecting quality groups. For the small, financially limited biopharmaceutical companies (commonly referred to as virtual or loosely structured), the dream of becoming a fully integrated pharmaceutical company (commonly referred to as FIPCO) is out of reach at least until sufficient product revenue can be materialized. The quality program workload must still get done, and it is in general the same amount of work whether virtual (i.e., contracted out) or FIPCO, as shown in Fig. 1. Rather than invest finances in QC labs and quality program staff, biopharmaceutical companies are now more open to outsourcing this workload.

However, outsourcing a quality program is not without its challenges. Successful contract quality groups must meet the basic elements of capability, proven performance, ethics, and willingness (Geigert, 1997b). Not every

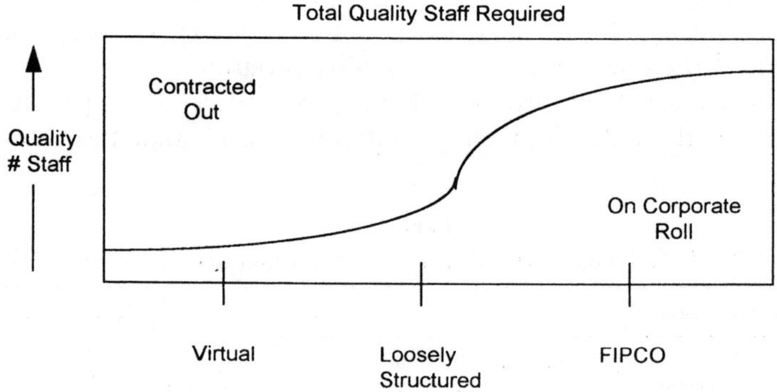

Figure 1. Total quality staff workload to support a biopharmaceutical product remains the same, whether the work is contracted out or performed within a fully integrated pharmaceutical company (FIPCO).

quality service in support of a biopharmaceutical product can be penned into the contract, for example, the timeliness of the contract quality group in performing investigations when out-of-specification results are obtained. In addition, of course, the key questions of whether the contract quality group will even be in existence 2–3 years down the road when seeking marketing approval for the product can be difficult to answer. For the virtual or loosely structured biopharmaceutical company, their choice of quality group partner can mean the difference between hitting its milestones and not.

5.3. Change of Pace

When product approvals took up to 2 years from the date of filing a marketing authorization request, a quality program had more than adequate time to accomplish its goals for the filing and to prepare for an inspection. Now, with the ever-decreasing time between filing a marketing application and receiving approval to market from a regulatory agency, possibly as soon as within 6 months of filing, the challenge for a quality program is now shifting to planning early enough for the information required for the filing and preparing early enough for the preapproval inspection (PAI).

For a regulatory filing of a biopharmaceutical, major chemistry and manufacturing control (CMC) sections must be prepared, such as specified biological product characterization, QC test method validation packages, process validation studies, and stability assessment. At the time of filing they must be complete, because there may be little time to prepare additional material before the regulatory agency must issue its decision. Ensuring that the filing will occur on schedule requires, the quality program to begin active planning 1–2 years ahead of time. If planning is not started early enough, the company can be faced with loss of flexibility for its manufacturing and quality operations (see Table XXV). It should also be noted that due to this ever-decreasing time between filing and receiving approval, it is most important that the quality program has the manufacturing facility, as well as all manufacturing and quality staff, "inspection-ready" at the time of filing. A PAI can now occur as soon as 1–2 months after the filing.

5.4. World Wide Web Access to Information

Never before has it been so easy to access information relevant to quality considerations. Because of the Internet, information is instantly available to everybody, anywhere. Two of the more important websites of

Table XXV
Possible Trade-offs between Filing on Time versus Giving up Flexibility in Manufacturing and Quality Operations

BLA filing requirement	Loss of flexibility trade-off
In vitro cell age limit justification	Limited cell age in cell culture operations (consume more cell bank ampoules over time)
Column use life	Limited runs on column (discard expensive resins more often)
Real-time stability	Limited expiry date (strain on distribution systems and rejection of outdated product)

interest to biopharmaceutical quality groups are those linked to regulatory agencies (the FDA for the United States and the EMEA for Europe). The FDA website (www.fda.gov) gives links to the Center for Biologics Evaluation and Research (CBER) and the Center for Drugs Evaluation and Research (CDER). At the CBER link (www.fda.gov/CBER), organization information, product information (recall/withdrawal/safety issues, BLA product approvals), FDA documents (Points to Consider, Guidelines; ICH documents), and CBER public conferences (upcoming meetings, meeting slide presentations) are available for reading and downloading. At the CDER link (www.fda.gov/CDER), organizational information, product information (safety issues, NDA product approvals), FDA documents (Guidelines, guidance documents, ICH documents), and CDER public conferences are available for reading and downloading. At the EMEA website (www.emea.eu.int), product information (European Public Assessment Reports) and the Committee for Proprietary Medicinal Products (CPMP) documents (position papers, Notes for Guidance) are available for reading and downloading. Another important website is the one for the National Archives and Records Administration (www.access.gpo.gov/nara/index.html), which provides access to the *Code of Federal Regulations* and the *Federal Register*.

REFERENCES

Bader, F. G., Blum, A., Garfinkle, B. D., MacFarlane, D., Massa, T., and Copmann, T. L., 1992, Multiuse manufacturing facilities for biologicals, *BioPharm* **5**:34–42.

Bryant, R., 1984, *The Pharmaceutical Quality Control Handbook*, Aster, Eugene, OR.

Canova-Davis, E., Baldonado, I. P., Chloupek, R. C., Ling, V. T., Gehant, R., Olson, K., and Gillece-Castro, B. L., 1996, Confirmation by mass spectrometry of a trisulfide variant in methionyl human growth hormone biosynthesized in *Escherichia coli*, *Anal. Chem.* **68:**4044–4051.

Chapman, K. G., Fields, T. J., and Smith, B. C., 1996, The case of the Q. C. unit, *Pharm. Technol.* **20:**74–79.

Committee for Proprietary Medicinal Products, 1996, *Note for Guidance on Virus Validation Studies: The Design, Contribution and Interpretation of Studies Validating the Inactivation and Removal of Viruses*, CPMP/BWP/268/95.

Committee for Proprietary Medical Products, 2001, *Note for Guidance on Minimizing the Risks of Transmitting Animal Spongiform Encephalopathy Agents Via Medicinal Products*, EMEA/410/01.

Committee for Proprietary Medicinal Products, 1997, *Position Statement on DNA and Host Cell Proteins (HCP) Impurities, Routine Testing Versus Validation Studies*, CPMP/BWP/382/97.

Damm, J. B. L., 1995, Application of glycobiology in the biotechnology production of pharmaceuticals, *BioPharm* **8:**43–48.

Darling, A. J., and Spaltro, J. J., 1996, Process validation for virus removal: Considerations for design of process studies and viral assays, *BioPharm* **9:**42–50.

European Community, 2001, *Good Manufacturing Practice for Medicinal Products in the European Community*, Vol. 4, Commission of the European Communities.

Federal Register, 1996a, 21 CFR Parts 210 and 211; Current good manufacturing practice: Amendment of certain requirements for finished pharmaceuticals; Proposed Rule, *Fed. Rep.* **61:**20104–20115.

Federal Register, 1996b, Elimination of establishment license application for specified biotechnology and specified synthetic biological products, *Fed. Reg.* **61:**24227–24233.

Federici, M. M., 1994, The quality control of biotechnology products, *Biologicals* **22:**151–159.

Federici, M. M., Garnick, R. L., and Geigert, J., 1991, A perspective on reference standard and reference material requirements in biotechnology-derived pharmaceutical protein products, *Pharmacopeial Forum* **17:**2683–2687.

Geigert, J., 1989, Overview of the stability and handling of recombinant protein drugs, *J. Parenteral Sci. Technol.* **43:**220–224.

Geigert, J., 1997a, Appropriate specifications at the IND stage, *PDA J. Pharma. Sci. Technol.* **51:**78–80.

Geigert, J., 1997b, The challenge of managing quality control/quality assurance for loosely structured and virtual biopharmaceutical companies, *Drug Information J.* **31:**97–100.

Geigert, J., and Ghrist, B. F. D., 1996, Development and shelf-life determination of recombinant human granulocyte-macrophage colony-stimulating factor (LEUKINE®, GM-CSF), in: *Formulation Characterization and Stability of Protein Drugs: Case Histories* (R. Pearlman, and Y. J. Wang, eds.), Plenum Press, New York, Chapter 6.

Geigert, J., Panschar, B. M., Taforo, C., Paolo, J., Fong, S., Huston, H. N., Wong, D. E., and Wong, D. Y., 1988, Parameters for the evaluation of long-term stability of tumour necrosis factor preparations, *Dev. Biol. Stand.* **69:**129–138.

Geigert, J., Solli, N., Woehleke, P., and Vermuri, S., 1993, Development and shelf-life determination of recombinant interleukin-2 (proleukin), in: *Stability and Characterization of Protein and Peptide Drugs: Case Histories* (Y. J. Wang, and R. Pearlman, eds.), Plenum Press, New York, Chapter 8.

Geigert, J., Klinke, R., Carter, K., and Vahratian, A., 1994, Role of quality control in validation of biopharmaceutical processes: Case example of clean-in-place (CIP) procedure for a bioreactor, *PDA J. Pharm. Sci. Technol.* **48:**236–240.

International Congress on Harmonization, 2001a, *Analysis of the Expression Construct in Cells Used for Production of r-DNA Derived Protein Products*, Q5B, ICH, Geneva.

International Congress on Harmonization, 2001b, *Derivation and Characterization of Cell Substrates Used for Production of Biotechnological/Biological Products*, Q5D, ICH, Geneva.

International Congress on Harmonization, 2001c, *Specifications: Test Procedures and Acceptance Criteria for Biotechnological/Biological Products*, Q6B, ICH, Geneva.

International Congress on Harmonization, 2001d, *Stability Testing of Biotechnological/Biological Products*, Q5C, ICH, Geneva.

International Congress on Harmonization, 2001e, *Viral Safety Evaluation of Biotechnology Products Derived for Cell Lines of Human or Animal Origin*, Q5A, ICH, Geneva.

Kappeler, A., Lutz-Wallace, C., Sapp, T., Sidhu, M., 1996, Detection of bovine polyomavirus contamination in fetal bovine sera and modified live viral vaccines using polymerase chain reaction, *Biologicals* **24:**131–135.

Mire-Slius, A. R., Gerrard, T., Das, R. G., Padilla, A., and Thorpe, R., 1996, Biological assays: Their role in the development and quality control of recombinant biological medicinal products, *Biologicals* **24:**351–362.

Pearlman, R., and Wang, Y. J., 1996, *Formulation, Characterization, and Stability of Protein Drugs: Case Histories*, Plenum Press, New York.

PDR, 1999, *Physicians' Desk Reference*, 53rd ed., Medical Economics Company, Montvale, NJ.

Peine, I. C., 1994, *Quality Assurance Compliance: Procedures for Pharmaceutical and Biotechnology Manufacturers*, Interpharm Press, Buffalo Grove, IL.

Randhawa, Z. I., Witkowska, H. E., Cone, J., Wilkins, J. A., Hughes, P., Yamanishi, K., Yasuda, S., Masui, Y., Arthur, P., Kletke, C., Bitsch, F., and Shackleton, C. H. L., 1994, Incorporation of norleucine at methionine positions in recombinant human macrophage colony stimulating factor (M-CSF, 4–153) expressed in *Escherichia coli*: Structural analysis, *Biochemistry* **33:**4352–4362.

Revie, D. R., and Vincent, D. W., 1997, Manufacturing concerns regarding the viral safety of biological products, *J. cGMP Compliance* **2:**78–85.

Roscioli, N. A., Renshaw, C. A., Gilbert, A. A., Kerry, C. F., and Probst, P. G., 1996, Environmental monitoring considerations for biological manufacturing, *BioPharm* **9:**32–40.

Sharp, J., 1991, *Good Manufacturing Practice: Philosophy and Applications*, Interpharm Press, Buffalo Grove, IL.

Sherwood, D., Fisher, D., Clifford, J., and Slade, S., 1996, Experiences with clean-in-place validation in a multiproduct biopharmaceutical manufacturing facility *Eur. J. Parenteral Sci.* **1**:35–41.

Sofer, G., 1994, Clearance of impurities from protein and peptide biotherapeutics, *BioPharm* **7**:46–50.

Sofer, G., 1996, Validation: Ensuring the accuracy of scaled-down chromatography models, *BioPharm* **9**:51–54.

Spellman, M. W., 1990, Carbohydrate characterization of recombinant glycoproteins of pharmaceutical interest, *Anal. Chem.* **62**:1714–1722.

U.S. Food and Drug Administration, 1985, *Points to Consider in the Production and Testing of New Drugs and Biological Produced by Recombinant DNA Technology*, U.S. FDA, Rockville, MD.

U.S. Food and Drug Administration, 1987a, *Guideline for Submitting Documentation for the Stability of Human Drugs and Biologics*, U.S. FDA, Rockville, MD.

U.S. Food and Drug Administration, 1987b, *Guideline on Validation of the Limulus Amebocyte Lysate Test as an End-Product Endotoxin Test for Human and Animal Parenteral Drugs, Biological Products, and Medical Devices*, U.S. FDA, Rockville, MD.

U.S. Food and Drug Administration, 1987c, *Points to Consider in the Characterization of Cell Lines Used to Produce Biologicals*, U.S. FDA, Rockville, MD.

U.S. Food and Drug Administration, 1991, *Biotechnology Inspection Guide: Reference Materials and Training Aids*, U.S. FDA, Rockville, MD.

U.S. Food and Drug Administration, 1992, *Supplement to the Points to Consider in the Production and Testing of New Drugs and Biologicals Produced by Recombinant DNA Technology: Nucleic Acid Characterization and Genetic Stability*, U.S. FDA, Rockville, MD.

U.S. Food and Drug Administration, 1993, *Points to Consider in the Characterization of Cell Lines Used to Produce Biologicals*, U.S. FDA, Rockville, MD.

U.S. Food and Drug Administration, 1995, *Guidance for Industry on the Content and Format for Investigational New Drug Applications (INDs) for Phase 1 Studies of Drugs, Including Well-Characterized, Therapeutic, Biotechnology-Derived Products*, U.S. FDA, Rockville, MD.

U.S. Food and Drug Administration, 1996a, *Points to Consider (Addendum) in Human Somatic Cell and Gene Therapy*, U.S. FDA, Rockville, MD.

U.S. Food and Drug Administration, 1996b, *Points to Consider on Plasmid DNA Vaccines for Preventative Infectious Disease Indications*, U.S. FDA, Rockville, MD.

U.S. Food and Drug Administration, 1997a, *December 11th Letter to Biologic Product Manufacturers Concerning Use of Human Blood-Derived Raw Materials*, U.S. FDA, Rockville, MD.

U.S. Food and Drug Administration, 1997b, *Points to Consider in the Manufacture and Testing of Monoclonal Antibody Products for Human Use*, U.S. FDA, Rockville, MD.

U.S. Food and Drug Administration, 1999, *Guidance for Industry on the CMC Content and Format of INDs for Phases 2 and 3 Studies of Drugs, Including*

Specified Therapeutic Biotechnology-Derived Products, U.S. FDA, Rockville, MD.

U.S. Food and Drug Administration, 2001, Title 21, Part 211, *Current good manufacturing practice for finished pharmaceuticals*, Code of Federal Regulations, U.S. Government Printing Office, Washington, DC.

USP, 1997, Somatropin Monograph (proposed), *Pharmacopeial Forum* **23**:4662–4673.

USP, 2002a, United States Pharmacopeia 25, *Alteplase Monograph*, United States Pharmacopeial Convention Inc., Rockville, MD.

USP, 2002b, United States Pharmacopeia 25, Article 1045, *Biotechnology-Derived Articles*, United States Pharmacopeial Convention Inc., Rockville, MD.

USP, 2002c, United States Pharmacopeia 25, Article 1047, *Biotechnology-Derived Articles—Tests*, United States Pharmacopeial Convention Inc., Rockville, MD.

U.S. Code of Federal Regulations, Title 21, Part 610, *General Biological Products Standards*.

Voss, J., Brunkow, R., Delucia, D., Haft, S., Hyde, J., Lindsay, J., McEntire, J., Murphy, R., Myers, J., Nichols, K., Terranova, B., and White, E., 1996, *Cleaning and Cleaning Validation: A Biotechnology Perspective*, Parenteral Drug Association, Bethesda, MD.

Wang, Y. J., and Pearlman, R., eds., 1993, *Stability and Characterization of Protein and Peptide Drugs: Case Histories*, Plenum Press, New York.

Williams, R. L., 1997, Big biotech: Designing the plant of the future, *BioPharm* **10**:44–52.

Willig, S. H., and Stoker, J. R., 1992, *Good Manufacturing Practices for Pharmaceuticals: A Plan for Total Quality Control*, 3rd ed., Marcel Dekker, New York.

World Health Organization, 1987, *Acceptability of Cell Substrates for Production of Biologicals*, Technical Report Series No. 747, WHO, Geneva.

8

Regulatory Considerations in the Development of Protein Pharmaceuticals

Michael G. Beatrice

1. INTRODUCTION

In 1980, the first investigational new drug (IND) application for a biological product made by utilizing biotechnology was submitted to the U.S. Food and Drug Administration (FDA). It could be argued that biotechnology, as a discipline, was not new. Depending on the definition one selected, biotechnology could be considered as old as traditional beer- or bread-making, as indicated by one of the early definitions: "Biotechnology involves the application of biological systems to technical and industrial processes" (Bonk, 1987). However, as applied to production of pharmaceutical products, it was a new technology for products intended to diagnose and treat diseases or conditions in humans.

After this early seed was planted, the number of biotechnology-related INDs grew at an exponential rate from 1980 through 1991, when a slow, but noticeable decline became evident. This decline lasted for almost 5 years due in large part to the fact that an old regulatory paradigm, which was based on the regulation of traditional biological products, continued to be applied to a product development technology that had matured to the point where it resembled the methodology for conventional pharmaceutical products.

Michael G. Beatrice • Corporate Regulatory and Quality Science, Abbott Laboratories, Abbott Park, Illinois 60064-6091.

Development and Manufacture of Protein Pharmaceuticals, edited by Nail and Akers. Kluwer Academic/Plenum Publishers, New York, 2002.

This maturation, however, was preceded by a number of product development problems, which were identified by the Center for Biologics Evaluation and Research (CBER) of the FDA. The following problems were identified with the first submissions for biotechnology-derived products, as regulatory/scientific concerns that complicated the review of these new technology products resulting from application of the same regulatory paradigm as utilized for traditional biological products:

1. Conventional drug product development methodology principles were not applied to biotechnology products. For example, human albumin was the stabilizer of first choice in product development, even though there are potential viral contamination issues with human source materials.
2. Specific and sensitive analytical methods were not available resulting in the requirement that a battery of tests be applied during and after production.
3. Interlaboratory test variability was difficult to control.
4. Stability was difficult to maintain and the mechanisms for degradation or loss of conformational integrity were not understood.
5. Product equivalence after routine process changes was not easily documented.
6. There were problems unique to large molecules that complicated production, for example, determination of the optimal structure or the effects of glycosylation on activity.
7. Process-related product failures were common and production variables which contributed to such failures were difficult to identify.
8. There was a fear in the biotechnology industry to make product changes due to the regulatory requirements, which mandated review and approval of all changes before implementation. For example, process scale-up frequently results in a change in product yield and characteristics such as ratio of high and low molecular weight species or glycosylation patterns.
9. Points to Consider documents were evolving at the FDA as the technology matured, but were considered to be academic documents, which were not easily amenable to clinical and commercial manufacturing methods.
10. It was difficult to demonstrate a correlation between a product's observed biological activity and expected clinical activity.

These issues became symbolic of the problems which forced a new scientific technology to be fit to an old regulatory system, which had been engineered to provide adequate assurance of safety, purity, and potency for traditional biological products that could not be completely characterized in intermediate or final dosage form. Application of this 50-year-old regulatory paradigm limited the biotechnology industry's flexibility to experiment with novel solutions to obvious development problems. For example, the traditional biological product regulatory system required the applicant seeking approval to conduct all manufacturing steps, from propagation of the source material (growth and expansion of the cell line) to final formulation, filling, labeling, and packaging, in facilities that were under their direct supervision and control, which translated into a requirement that the applicant actually own all manufacturing facilities [Title 21, U.S. *Code of Federal Regulations* (21 CFR), Parts, 600.3(t), 600.3(u), and 601.2]. The challenge from the receipt of the first application for a biotechnology-derived product was to create a new regulatory paradigm based on sound scientific principles, which would expand the field of biotechnology product development and permit regulatory flexibility to optimize the product and manufacturing process.

It became obvious that, in order to change the old regulatory paradigm, it would be necessary to substantially revise or eliminate many of the existing restrictive regulations in 21 CFR, Parts 600–680. This was attempted by this author, twice between 1980 and 1991 but without success due in part to the political climate, which made the promulgation of new regulations and even the revision of existing regulations difficult to impossible. This is exemplified by an early attempt to change a regulation which required a manufacturer's logo or trademark to be less prominent than the proper name of the product on the label, a requirement which was not imposed on manufacturers of traditional drug products approved under the New Drug Application (NDA) System. It took 18 months to publish a one-page proposed change in the regulations. Acceptance of this proposed regulatory change would have permitted manufacturers to utilize any size trademark or logo in any position on the label of a biological product (as is permitted for drug products). However, within 1 month of publication of the proposed change in the *Federal Register*, the proposal was rescinded due to pressure from one large corporation, not engaged in pharmaceutical production, that objected to the FDA changing any existing standards related to trademarks or logos.

1.1. Early Regulation of Biotechnology

As soon as the commercial potential of biotechnology processes began to be recognized, the federal government saw the need to examine the adequacy its existing regulatory framework, including its compatibility or lack thereof between the various regulatory agencies. The White House Office of Science and Technology Policy was commissioned to study this issue in the early 1980s in cooperation with the FDA, Environmental Protection Agency (EPA), U.S. Department of Agriculture (USDA), and the National Institutes of Health (NIH). Following an exhaustive examination of the existing regulatory framework by representatives of the aforementioned regulatory agencies, including this author, a *Federal Register* Notice entitled "Coordinated framework for the regulation of biotechnology," was published (Federal Register, 1984). This Notice was updated in June 1986 and served as the basis for defining which agencies would regulate specific products in the field of biotechnology (Federal Register 1985, 1986).

The deliberations which resulted in the publication of the coordinated framework documents also concluded that laws in existence at the time were sufficient for the regulation of genetic engineering and that new legislation should not be proposed. This decision was applauded by many in industry, who feared the imposition of a complex new regulatory scheme, simply because of political concerns at that time over possible environmental consequences and the ethical issues to which the new technology had given rise. Experience with this new technology over the next 10 years would completely change the attitude of the public and industry. This change of attitude on the part of the industry was based on the potential use of biotechnology processes in the manufacture of a broad range of commercial products, including drugs, biologics, devices, and foods, combined with the experience in the safe and consistent production of genetically engineered products.

To examine the magnitude of the paradigm shift that was necessary, it is important to understand the system for regulation of traditional biological products (the old regulatory paradigm), which was in place at the time and was initially identified in the coordinated framework document as appropriate for regulation of these new technology products.

1.2. Traditional Biological Product Regulation

The decade of the 1990s saw the most dramatic change in the regulation of biological products since the consolidation of all health-related laws into the *United States Public Health Service Act* in 1944. The previous, highly complex regulatory paradigm was based on the premise that biological products were complex mixtures produced by living organisms, including vaccines, coagulation factors, and other blood products and a variety of extracted materials, and biologically derived therapeutic agents. In many cases (e.g., Coagulation factor VIII, a product derived from human blood) active biological substances were obtained from pools of source materials derived from thousands of different humans or animals. Thus, such products were considered impossible to define in precise molecular terms. This is in sharp contrast to the situation for conventional drugs, for which molecular structures can be completely defined. Alternative methods have been developed for the assessment of safety, efficacy, and consistency of biological products. These have incorporated a careful evaluation of the manufacturing process, safety, and potency testing in models appropriate for each product, and development of analytical procedures for characterization of the product at various stages of production (Chiu *et al.*, 1987; Garnick *et al.*, 1988; Marino, 1989; Marshak and Liu, 1988; Werner *et al.*, 1988).

Traditional biological product regulations in 21 CFR, Part 600 required issuance of an establishment license to an applicant that demonstrated the capability to manufacture a safe, pure, and potent biological product within a facility under their supervision and control. No establishment license could be issued unless a product license was issued simultaneously with the establishment license and such product was available for examination at the facility intended for licensure [21 CFR, 601.10(b)]. Demonstration of adequate control over the manufacturing of a biological product from start to finish was documented by the manufacturer (i.e., legal entity engaged in actual manufacture of the product) performing *all* steps in the production of a product within facilities under its direct supervision and control.

Even though the mechanism for regulation of biological products differed from that of other drug and device products, the basic philosophy of providing safe and effective products manufactured in establishments which meet current good manufacturing practice regulations (21 CFR, Part 211) is all-encompassing. In the early part of this century biological product development was regulated mainly by the requirement that manufacturing could only be performed in licensed and inspected laboratories under the strict control of the legal entity (licensee). This type of control seemed to contribute to a dramatic decrease in the death rates for pertussis, measles, and diphtheria, because the phenomenal growth in the biologics industry, as evidenced by the exponential increase in licensed establishments, appeared to be correlated with the establishment of standards for manufacturing laboratories.

From their historical inception, biological regulations have defined "manufacture" to mean all steps in the preparation of the product, including, but not limited to, propagation, filling, testing, labeling, packaging, and storage by the manufacturer [21 CFR, Part 600.3(u)]. Further, "manufacturer" has, until recently, been defined as the *legal entity or person* engaged in the manufacture of a licensable biologic [21 CFR, Part 600.3(t)]. These regulatory definitions have formed the basis of the policy decision that required the legal entity (the legal preparer of the product which is intended to be licensed) to perform all steps in the "manufacture" of the biologic in a licensed facility (facilities) under their direct supervision and control.

The above licensing policy is one example of a major difference that was in place between regulation of drugs and biologics under the NDA and licensing systems, respectively. In order to receive a product and establishment license under the previous interpretation of licensing requirements, the sponsor of a biological product application must also have established ownership or supervision and control over all facilities where manufacturing and testing operations for the biological product were performed. Some exceptions were permitted for use of contract facilities to perform certain *limited* operations occurring usually after the preparation of a final purified concentrate (e.g., aseptic filling, lyophilization, labeling, *or* limited product testing), provided that the licensee demonstrate adequate supervision and control over the contract operation.

In addition, regulatory policy interpretation required submission of additional product and establishment license applications for a legal entity that moved manufacturing operations to another country or location where the local law required that a separate legal entity be established. For example, if a large multinational company based in the United States held a currently approved license for a biological product and establishment and desired to move manufacturing operations to a foreign country, a complete new filing for an establishment and product license would have been requested, because the government of the importing country would require that a distinct legal entity recognized by that country be established. As a part of the new filing, the company would have been asked to perform limited clinical testing in addition to complete product characterization in order to demonstrate similar biological and clinical activity of products manufactured in different facilities.

The rationale behind this restrictive interpretation of the regulations can best be illustrated by pinpointing a major difference in the analytical character of drugs and traditional biologics. Drugs other than biological products can for the most part be purified, characterized, and identified by accurate and *reproducible* methodology. Traditional biologics, such as viral and bacterial vaccines, blood fractionation products, and allergenic extracts, are not uniform mixtures of easily characterized materials. The regulations, as originally interpreted, required strict control by the legal entity over all facilities and processes utilized for manufacturing and testing the biological product at all stages in development. The process defined the product. From this point of view, it is easy to understand why the random "farming out" of manufacturing steps in the preparation of biological products was not acceptable under U.S. licensure.

This legal policy interpretation was predicated upon the fact that traditional biological products are not easily characterized. However, recent advances in biotechnology have permitted more flexibility in interpretation of the licensing regulations as the product in question becomes more like a traditional drug in its indications for use as well as its ability to be easily characterized using reproducible methodology. It should be noted that this flexibility may not be appropriate for products which best fit the traditional definition of a biologic or which exhibit difficulty or inconsistency in propagation, manufacturing, or analysis.

1.3. Advances in Genetic Engineering

Though genetic engineering has allowed much greater control over certain aspects of manufacture, recombinant products are produced by living organisms and so are unavoidably heterogeneous to some degree. Despite the enormous technological advances in their manufacture, the products of biotechnology thus remain strongly analogous to earlier biological products derived from natural sources. A distinctive characteristic of biological products is that every product is unique and must be evaluated accordingly. To some extent, every individual batch of a biological product has the potential for having different characteristics. In addition, changes in the manufacturing process, equipment, and environment, which are an integral part of product development and scaleup, may also adversely affect the final product. In some instances different characteristics will have little or no effect on the biological and clinical activity of the molecule. The concern arises in instances when either the effect of such differences is known to change the product's safety or effectiveness or the effect of such a change is unknown.

As an example of how the product development process proceeds from pre-IND stage through Phases I, II, and III of investigational studies and ultimate commercialization, the development of a recombinant DNA (rDNA)-derived product will be described in detail. A recombinant DNA product is manufactured by a complex process which is difficult to control due to many variables. The process begins with genetic engineering steps and ends with preparation of a finished dosage form. Unlike conventional pharmaceuticals, an identifiable and stabilized bulk drug substance is not ordinarily isolated during the entire process and each of the products manufactured at intermediate production stages (intermediate products) has its own set of specific testing and storage requirements, which must meet established specifications to continue in the manufacturing process. These specifications are established empirically and must also comply with an overarching set of regulatory requirements.

Until the mid 1990s, the definition of "control" was differently applied to drug and biological products, even though the operating principle for regulation of drugs and biological products has always been based on the concept of process control. Biological products were thought to always be defined by the process used to produce them, regardless of product type. Therefore, control of the manufacture of biological products must be documented and validated beginning with propagation of the source material and extending to all manufacturing steps. Forces such as heat, shear, and environmental conditions may significantly change the biological and/or clinical activity of the product. In legal terms, appropriate control over product manufacture must be demonstrated by a sponsor in order to receive approval and place their name on the product label for either a drug or biologic [21 CFR, Parts 201.1 and 600.3(r)].

Strict confidentiality regulations prohibit the disclosure of details related to specific product failures or successes. Development of a rational tool to document the product attributes necessary to support production of an optimized formulation at all stages of development and to justify process changes was sought by this author as an alternative to disclosing confidential information or relying only on regulatory guidance documents. This product development tool came to be known as the Biotechnology Product Development Profile (PDP).

Thus the original hypothesis for a biotechnology product development research project was that such a development tool could be designed and constructed to fit a broad range of biotechnology products. Integration of the results of analytical product characterization studies, combined with the momentum to retrospectively review existing outdated regulations, which had grown to disproportionate size and complexity (i.e., regulatory creep), helped establish the concept of the existence of distinct classes of products which could be sufficiently characterized so that they could be treated more like conventional drug products.

Development of a rational PDP as the first phase of this regulatory science evolution could only have been accomplished after extensive examination of the successes and failures of biotechnology products submitted since the early 1980s. Most applications submitted by large and small manufacturing establishments to FDA demonstrated a wide range of problems encountered during product development. In some instances, these problems were resolvable; in others, they resulted in fatal flaws, which rendered an application not approvable. Confidentiality and trade secret regulations permit only a discussion of failures which are publicized in the literature or otherwise by product sponsors. For example, one product that encountered a major problem which resulted in a significant delay in approval was tissue plasminogen activator (TPA), a thrombolytic agent produced in mammalian cell culture. It is an 80-kDa protein containing a number of disulfide bonds and four N-linked glycosylation sites. The protein can also readily undergo proteolytic processing to a two-chain form. TPA produced using roller bottles was made up of two polypeptide chains and was found to have an optimal therapeutic dose of 80 mg/kg. When a seemingly simple manufacturing change was initiated to increase product yield, switching from roller bottles to a suspension culture technique for propagation of the host cells, a mixture of the one- and two-chain forms was observed that had an optimal dose of 120 mg/kg. A change in the manufacturing process was thus clearly associated with a significant change in clinically important characteristics of the biological product.

Three reasons may be cited as complicating factors in the evaluation of product failures: (1) lack of a comprehensive database due to the confidentiality of commercial and trade secret information, (2) the complex nature and minimal information concerning the physical and chemical attributes of biological products, and (3) the lack of clear guidance in the product development process.

The complex nature of protein chemistry can be elucidated by the fact that the actual enzymology or binding site activity is usually based on the activity of three or four amino acids. The role of the numerous other amino acids is to fold the protein into a very specific structure, which configures the side chains involved in the actual chemistry into the proper spatial and temporal orientation (McPhie, 1995).

2. BIOTECHNOLOGY PRODUCT DEVELOPMENT

Biotechnology product development has become both a science and an art. Frequently the developmental scientist is presented with several candidate products and asked to develop a stable, biologically active formulation for use in the clinic in the absence of information necessary to appropriately characterize the product. Product heterogeneity, combined with the lack of knowledge of the physical chemistry of proteins and the scarcity of literature on formulation of biotechnology products, adds to the challenge.

Product failures during late development stages have been observed to be process-related or formulation-related. It is often difficult to isolate the cause of such failure, because the two can be interrelated. Formulation failures can be overcome by modifications, which tend to be trial and error exercises unless the molecule has been characterized to the extent that compatible excipients, stabilizers, and manufacturing conditions can be easily selected. However, extensive characterization is generally not performed until a candidate product demonstrates desired biological activity and lack of measurable toxicity in preclinical and early clinical studies. Another reason that products utilized in early animal or clinical studies (i.e., Preclinical or IND Phases I and II) are less well characterized is that their *in vivo* activity (or lack thereof) is difficult or impossible to predict from mechanistic or *in vitro* models and must be elucidated in the clinic. To further complicate matters, the FDA and other regulatory authorities do not permit a poorly characterized product to be introduced into commercial manufacturing facilities due to the potential for cross-contamination with the product or adventitious agents. The economics of such situations therefore dictates that investigators utilize many crude formulations made on less than a pilot scale (i.e., bench-scale production) during these early clinical trials.

Process-related failures can be remedied by changing the process until optimal results are obtained. The problem that arises when process changes are made during the clinical development phase is the manufacturer's lack of flexibility to initiate such changes in the product and/or process during development without significant documentation. In addition, the traditional biological product regulatory paradigm required such documentation to be submitted for review and approval before such changes could be implemented.

Changes in process and/or formulation are necessary as more information on biological and clinical activity becomes available. Until recently, the regulatory documentation requirements needed to support such changes were not based on rational scientific principles and the rationale for requiring such documentation was not clearly defined.

A product that demonstrates the desired activity in early clinical trials is thus analyzed for the purpose of providing regulatory authorities the characterization information that is essential for progressing from bench to pilot scale followed by intermediate- and large-scale production, if applicable.* This translates into a number of desirable process or product changes, which may occur after the preclinical phase or between any of the phases of investigational study. Regulators consider the impact of such

*The FDA's publication of the guidance document concerning use of pilot manufacturing facilities for the development and manufacture of biological products (Federal Register, 1995a) signaled a major change in regulatory policy interpretation by permitting use of product manufactured in pilot facilities for clinical trials as well as for commercialization.

changes on the data generated in each of these investigational studies prior to the pivotal clinical study as well as the comparability of products manufactured utilizing different methodology. If the process or product changes are such that the product intended for use in large-scale studies or for commercialization is distinctly different in its biological or clinical properties from that used in early clinical trials, the data from the early studies may be considered invalid.

Therefore, product sponsors had a distinct fear of initiating any change in product or process, the consequences of which are ordinarily unknown until the product is tested *in vivo*. Full product characterization at early developmental stages is one way to address this concern. However, as previously described, the time and expense involved in full characterization of all candidate products for use in pre-pivotal and early investigational studies limits the number of potential candidates and may result in early elimination of potentially viable products. A simple and reliable algorithm for identifying and providing a reasonable amount of documentation for production and process changes would therefore have a major beneficial impact on the biologics and biotechnology industry. Such an algorithm, in order to be efficient, must also be capable of defining the battery of tests necessary for complete product characterization at all phases of study and supporting the comparability of products manufactured by different procedures. Defining the appropriate battery of tests also means providing a rational explanation of why certain test methods were not selected.

Once a product candidate has demonstrated the desired activity in preclinical studies, the product sponsor is faced with the challenge of developing a prototype formulation for initiation of clinical studies in humans. This early formulation development is considered a vital part of the overall formulation optimization process and must be sufficiently documented to regulatory authorities.

Formulation optimization cannot be economically achieved without initially considering all of the potential process variables inherent in biological manufacturing processes. Developers have sought guidance from the literature and FDA on parameters for adequate characterization and on outcome measures to gauge the effect of product and process changes on intermediate and final product characteristics. FDA has responded to this need with a number of Points to Consider (PTC) documents (Table I). These documents address certain product and process concerns, such as the requirements for characterization of cell lines used to produce recombinant DNA products. However, the PTC documents have been criticized as being academic documents, which are too specific and not easily amenable to integration into clinical and commercial manufacturing processes.

Table I
Points to Consider (PTC) Documents Issued by the U.S. Food and Drug Administration

Title	Date	Revised
In Vitro Monoclonal Antibody Products Which Are Identified as Biological Products	June 20, 1983	April 10, 1985, March 1992, and August 2, 1994
PTC in the Production and Testing of New Drugs and Biologicals Produced by Recombinant DNA Technology	April 10, 1985	April 6, 1992
PTC in the Collection, Processing, and Testing of *ex-Vivo* Activated Mononuclear Leukocytes for Administration to Humans	August 22, 1989	—
Draft PTC in the Manufacture and Clinical Evaluation of *in Vitro* Tests to Detect Antibodies to Human Immunodeficiency Virus, Type 1 (1989)	August 8, 1989	—
PTC: Computer Assisted Submissions for License Applications	July 1990	March 1996 (CAPLAR[a] Manual)
PTC in the Safety Evaluation of Hemoglobin-Based Oxygen Carriers	August 21, 1990	—
Cytokine and Growth Factor Pre-Pivotal Trial Information Package	April 2, 1990	Undated, Interferon Test Procedures
PTC in Human Somatic Cell Therapy and Gene Therapy	August 1991	January 1996
PTC in the Manufacture of *in Vitro* Monoclonal Antibody Products for Further Manufacturing into Blood Grouping Reagent and Anti-Human Globulin	March 1992	—
PTC in the Design and Implementation of Field Trials for Blood Grouping Reagents and Anti-Human Globulin	1992	—
Supplement to the PTC in the Production and Testing of New Drugs and Biologicals Produced by Recombinant DNA Technology: Nucleic Acid Characterization and Genetic Stability	April 6, 1992	—
Draft PTC in the Characterization of Cell Lines Used to Produce Biologicals (1993)	July 12, 1993	—
PTC in the Manufacture and Testing of Monoclonal Antibody Products for Human Use (1994)	August 2, 1994	—
Interferon Test Procedures: PTC in the Production and Testing of Interferon Intended for Investigational Use in Humans	Undated	—
PTC in the Manufacture and Testing of Therapeutic Products for Human Use Derived from Transgenic Animals	August 22, 1995	—

Table I
(*continued*)

Title	Date	Revised
PTC for the Evaluation of Combination Vaccines: Production, Testing, and Clinical Study (1995)	1995	—
Addendum to the PTC in Human Somatic Cell and Gene Therapy (1991)	January 1996	—

[a]CAPLAR, computer assisted product license application review.

2.1. Selection of a Model Biotechnology Product

An effective tool for documentation of all phases of product development therefore is needed to (1) define analytical test requirements and justify exclusion of unnecessary tests, (2) provide a rationale for making product and process changes, and (3) assist in demonstrating product comparability after such changes are made. However, experience has demonstrated that, in addition to having a thorough understanding of the product attributes, the manufacturing processes and production variables that need to be controlled or have the potential to adversely effect the product must be considered. These elements need to be integrated in the design of an efficient tool for documenting product development. Therefore, it was decided that studying one of the first biotechnology-derived products to be approved for widespread clinical and commercial use, interferon α, would address the major regulatory issues, specifically in relation to early interferon products:

1. Conventional product development methodology was not applied to early production. The initial development approach was similar to traditional biological product development.
2. Specific and sensitive analytical methods were not available, especially after microgram quantities of the active cytokine were formulated with massive volumes of excipients in the final vial formulation.
3. Interlaboratory testing was highly variable, as evidenced by the fact that the FDA field biotechnology laboratory could not duplicate test results from the manufacturer or the laboratories of the CBER FDA headquarters.

4. Stability was not completely defined, as evidenced by the fact that one experimental formulation for an early biotechnology product lost conformational integrity due to minor agitation in transport.
5. Demonstration of product equivalence was not attempted. In fact, manufacturers were careful not to make significant process or product changes for fear that equivalence could not easily be demonstrated.
6. Several problems were encountered with one biotechnology product formulation, even though it was not the largest or most complicated molecule to be studied at the time.
7. Process-related product failures occurred with early biotechnology products, causing delays in proceeding from the clinic to the marketplace.
8. Significant product changes were not attempted, for reasons explained in item 5 above.
9. The original interferon products were under development concurrently with the development of early Points to Consider documents for characterization of cell lines used to produce biologicals, for production and testing of new drugs and biologicals produced by recombinant DNA technology, and for production and testing of interferon.
10. There was initially no direct correlation between biological and clinical activity. Results of various biological assay methods were therefore associated with activity demonstrated in the clinic.

2.2. The Biotechnology Product Development Profile as a Component in Support of Product Characterization

The biotechnology PDP was the first important product development tool to emerge from formulation research and it complemented efforts to define well-characterized products and develop an international regulatory standard for evaluation of such products. In parallel efforts that were supported by the FDA, the European Community developed draft CPMP guidelines on production and quality control (EC Committee for Proprietary Medicinal Products, 1994). These working guidelines incorporated principles outlined in several of the FDA's Points to Consider documents.

The result of all of these domestic and international efforts to provide guidance on aspects of drug development has been to describe specific required submission elements in greater detail rather than to provide a roadmap for drug development. These international guidance documents assist both regulators and corporate regulatory officials in the assembly and review of investigational and marketing applications, but they do not provide guidance for optimizing the full spectrum of drug development issues due in part to the fact that no single pathway exists for describing the development of all possible drugs and biological products.

The PDP was therefore developed in a distinct, but parallel effort to serve as a rational tool for product development scientists to map a rational development pathway. An effective product development algorithm should accomplish the following objectives;

1. Define the candidate product and relate its activity either to a natural product or to a standard with known pharmacological activity
2. Describe a method for assessing critical biochemical, analytical, and structural features of the product to facilitate rational product characterization
3. Examine available manufacturing process pathways and describe interactions with the process or equipment which may adversely influence product recovery
4. Document product attributes necessary to develop an optimally active formulation at all stages of investigation
5. Establish rational in-process, release, and end-of-stability specifications

2.2.1. THE CONCEPT OF PRODUCT DEVELOPMENT

The term product development has been applied to preclinical and clinical research as well as product manufacturing. The PDP which emerged from collaborative laboratory research efforts primarily focuses on the product from propagation of the active biological drug substance to preparation of a final biological drug product. Concepts were derived from principles applied to the development of traditional drug products, which were further modified to describe the complexities associated with biological product development. Many of these concepts were incorporated into Points to Consider and other guidance documents, including the 1996 Guideline for Describing Chemistry Manufacturing and controls in an application for a well-characterized biotechnology product.

Biological and biotechnology drug development matured in the 1990s, with emphasis on product characterization. The required extent of product characterization should be assessed at each stage of development, from preclinical through clinical research phases. Further, characterization should be based on identified attributes specific to the product under development.

Specific characterization requirements necessary to implement for each phase of a clinical research program are defined by product attributes, which are influenced by factors deriving from conditions encountered during propagation and production. Some attributes are predictable and require specific, defined manufacturing and testing alternatives, such as designing steps to achieve viral inactivation for products derived from human blood.

2.2.1.1. The Biological Drug Substance

Biological drug substance attributes can be influenced by engineering the source and method of propagation to minimize or avoid toxic impurities, contaminants, and denaturing conditions. Theoretical optimization at this early stage should be verified by the first level of product characterization, basic biological drug substance identification. This includes measurements of structure, conformation, and biological activity.

Biological activity can be determined by comparison of the candidate molecule with a naturally derived analogue or independent biological activity measurements optimally demonstrated in living systems. Animal models are predominantly utilized as predictors of activity in living systems. These models should mimic the expected activity in humans. If the product or the model requires extensive reengineering to confer activity, such as artificially inducing a Parkinsonian condition in mice, the utility of the model comes into question. The FDA has observed numerous examples of activity in animal model systems which cannot be correlated with biological or clinical activity in humans.

In vitro models usually follow initial animal model studies, because experience suggests that activity observed in such systems is not always predictive of *in vivo* potency. The word potency is interpreted to mean the specific ability or capacity of the product, as indicated by appropriate laboratory tests or by adequately controlled clinical data obtained through the administration of the product in the manner intended, to effect a given result [21 CFR, Part 600.3(s)]. This definition is the basis for the difference in the requirement that drugs have two adequate and well-controlled studies, whereas biological products have been approved on the basis of a single controlled study and/or laboratory *in vivo* potency determinations.

It should be noted that product development does not follow strict rules of order. For example, a biological drug substance with well-documented *in vitro* activity may be developed into a formulation appropriate for preclinical studies without preliminary testing in early animal models. These attempts to define and quantitate activity and correlate *in vitro* and *in vivo* potency constitute early characterization of biological activity. Once this level of biological activity has been demonstrated, an effective preclinical program can be designed to assess the pharmacological and toxicological effects as well as the pharmacokinetics and biological disposition of the drug in animals.

2.2.1.2. Traditional Drug Development versus Biotechnology Product Development

Traditional drug development follows principles which have been established over decades of trial and error. The steps include the following:

1. Drug candidate synthesis, screening, and selection
2. Activity/potency assessment (may include structure/activity relationships or creation of "designer drugs")
3. *In vitro/in vivo* potency assessment
4. Development of an initial formulation for human use
5. Formulation and yield optimization
6. Scaleup studies
7. To-be-marketed formulation development
8. Drug stability studies for the drug substance and final container product
9. Drug product and process optimization
10. Selection of optimized drug product (final specifications and manufacturing process)

Efficient drug development requires coordination of the steps described above with clinical research and development. In addition, all steps should be conducted in facilities appropriate to the stage of production utilizing equipment and processes which have been validated sufficiently to demonstrate reproducibility. The product development chart in Fig. 1 correlates the stages of drug development with clinical research, characterization, and facility and process documentation and validation.

Development and characterization of a recombinant protein such as interferon α exemplifies a method of construction of a PDP for a product of biotechnology. The types of information that FDA utilizes in its evaluation of a final recombinant product are not all related to the final product itself. It is important to document a consistent source of propagation without contamination by unwanted materials.

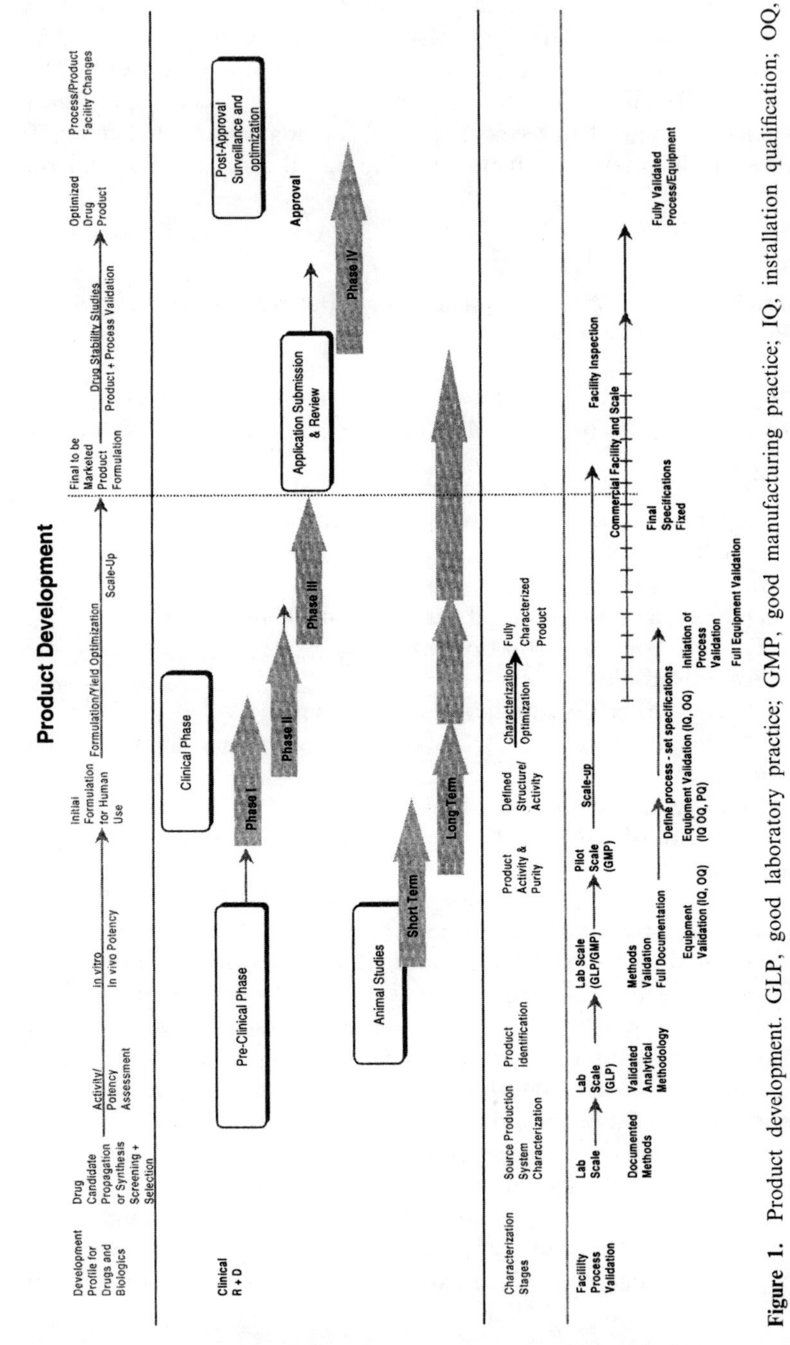

Figure 1. Product development. GLP, good laboratory practice; GMP, good manufacturing practice; IQ, installation qualification; OQ, operational qualification; PQ, performance qualification.

Recombinant DNA technology involves the systematic arrangement and manipulation of specific segments of nucleic acid for construction of composite molecules which, when placed into an appropriate host environment, will yield a desired product (U.S. FDA, 1985). Several methods for obtaining the optimal coding segment may be utilized. Prior to selection of a method for preparing the segment coding for the desired product, verification of the integrity of the source material and characterization of production cells should be performed following appropriate guidance, including Points to Consider documents for characterization of Cell Lines, recombinant DNA production, and nucleic acid characterization and genetic stability.

Characterization of the nucleic acid segment is the first step in a series of characterization events which continues until all components from the source of production to the final product have been adequately described. This is followed by characterization of the expression constructs, which includes the initial cell clone and cell expression clone selected for production of the desired recombinant protein. A major concern at this stage of development involves the stability of the coding sequence of the transferred segment into the host expression system. In the numerous generations of growth from starter culture to final growth in a large reaction vessel, it has been observed that a formerly stable plasmid may develop subpopulations of molecules which are altered in structure and activity.

Genetic stability through the generational growth steps necessary for production of a final recombinant product is usually achieved by engineering the plasmid to contain additional segments necessary for growth of the desired cell types. These engineered segments usually confer drug resistance or complement an auxotropic mutation in the host cell. Verification that the desired host cells predominate in culture can be achieved by examination of a batch for phenotypic traits and other host cell markers.

Following successful engineering of the host cell and expression vector system, a master cell bank (MCB) and working cell bank (WCB) are established. The MCB is a collection of cells of uniform composition stored in aliquots under defined conditions, from which subsequent cell banks are made (U.S. FDA, 1992). The identity and purity of the cells in a MCB must be established and potential adventitious agents identified. Requirements for maintenance of the MCB have been described in the *Points to Consider for Nucleic Acid Characterization and Genetic Stability* and include periodic verification of the integrity of the expression construct.

A manufacturers working cell bank (MWCB) can subsequently be derived by expansion of one or more ampoules of the MCB. The series of steps described above leading to the development of a MWCB of uniform composition which results in the production of a stable biological drug candidate places biotechnology drug development at the level of traditional drug development, where drug design and/or screening studies have identified a candidate with desired activity and the preferred route of synthesis has been defined (including levels of impurities). FDA has described requirements for documenting the synthesis of drug products at this stage of development in a guidance document (U.S. FDA, 1991). Requirements for the MWCB have also been described by FDA and include documentation of conditions which, if validated to ensure specificity of testing methodology, ensure that a product of defined identity can be consistently manufactured.

2.2.1.3. Correlation of Biological and Clinical Activity

Experience derived from years of reviewing applications at FDA has demonstrated that the next phase of biotechnology drug development is frequently either not considered or inadequately assessed. This involves establishing a correlation between the biological and the clinical activity of the candidate product.

Clinical activity can only be completely determined following a clinical development program in humans, which usually begin with studies in normal, healthy volunteers designed to provide sufficient information about the drug's safety, pharmacokinetics, and pharmacological effects and culminates with pivotal controlled trials in large populations of the desired treatment group(s). However, failure to establish an early correlation of biological and clinical activity which can either be predicted by comparison of the engineered product to a natural product or observed *in vitro* has resulted in numerous early development failures.

At this stage of development, the product candidate may either be identical* to a natural product or a new biological entity. In either instance, the ideal situation is for the engineered product to closely mimic the activity of a natural substance or exert a predicted action within a biological system. This may be easier to achieve with a product that closely approximates the structure and function of a natural biological product, but it may also be achieved with a new entity. Lack of measurable biological activity at this stage of development signals the end of that product candidate.

Detection of a desired and measurable activity should be followed by the development of a biological assay procedure which is predictive of a

*The term "identical" in this context is not intended to require exact structural and conformational identity between a natural and an engineered biotechnology product. Modifications encountered during production frequently result in alterations, which may or may not have a detrimental effect on product activity. These alterations may be difficult to detect. For these reasons alternate terminology (such as similarity or comparability) has been utilized in recent FDA guidance documents.

desired clinical effect. For example, the antiviral activity of interferon is measured by its cytopathic effect on host cells. Such biological potency assays are intended to quantify the activity of a product and ensure its efficacy during development as well as for batch-to-batch release. There are three major types of potency assays utilized for products of biotechnology: animal model assays, cell-line derived assays, and *in vitro* biochemical assays. The selection of an appropriate assay is based, in part, on the ability to demonstrate a correlation with a desired clinical activity.

During this early stage of development, this correlation can be based on animal model studies which are predictive of activity in humans. Validation of the selected animal model should include documentation of its potential applicability to a clinical effect in humans.

2.2.1.4. Biochemical and Biophysical Characterization

At this stage of development, consistent production of an engineered biotechnology product which has a valid potency assay correlated with anticipated clinical activity should have been demonstrated in validated systems (genetic stability, cell expression, and assay systems). These efforts should be correlated with the production of product on a laboratory scale, which may be utilized for preclinical studies if production follows appropriate good laboratory practice (GLP) regulations. Failure to validate development conditions or follow the documentation requirements of the GLP regulations may result in use of a product in preclinical studies

Table II
Biochemical and Biophysical Characterization Methods

Attribute	Impact on manufacturing or testing	Assay method
Biological activity		
Immunological identity		
Size		
Charge		
Hydrophobicity		
Composition		
Binds to glass		
Forms aggregates		
Hydrolyzes		
Oxidizes		
Needs appropriate 3D structure		

which cannot be considered comparable with subsequently manufactured batches.

At this stage, early characterization studies designed to determine molecular attributes, assay methodology, and impurities should be conducted prior to or concurrently with preclinical studies, but before early studies in humans (Phase I) are initiated. Table II is intended to serve as a template for determining the assays appropriate for biochemical and biophysical characterization. Selection of appropriate biophysical and biochemical characterization parameters begins with a listing of the attributes of the engineered molecule, which can be derived from knowledge of the structure, activity assays (developed in a prior development step), or previous experience with comparable products.

The impact of each identified molecular attribute on selection or conduct of potential manufacturing or testing operations should be tabulated. If the impact is negligible (i.e., has no adverse effect on the safety, purity, or potency of the molecule), further characterization of that attribute is unnecessary. Attributes impacting or potentially impacting manufacturing or testing should be described and appropriate assay methodology should be developed to characterize the attribute. Molecular size serves as an example of this type of process. Molecular size can contribute to potential crosslinking, cleavage, aggregation, or dissociation during production. Processing conditions should be selected to minimize the potential for such events and assays developed to determine the effect of selected processing conditions.

2.2.1.5. Manufacturing Process Optimization

The process described above for defining biochemical and biophysical characterization facilitates the next stage of biotechnology product development, manufacturing process optimization.

Manufacturing process optimization should consider the product attributes, information derived from characterization studies, and experience with previous laboratory scale production. This information should be combined in a table (Table III) which describes the process or potential alternative processes, manufacturing process interactions (MPIs) anticipated either from the process itself or product attributes, the potential effect of the MPIs, assay methodology for detection of a potential manufacturing interaction, and indication of whether the detection of such an interaction is a critical or noncritical in-process control.

A development plan which considers the potential impact of process and product attributes together has two advantages. First, it facilitates

Table III
Manufacturing Process Optimization

Process/alternate processes	Manufacturing process interactions (MPI)	Potential effect of MPI	Assay methodology	Critical or noncritical in-process control
Expression systems				
Microbial fermentation				
Bacterial				
Yeast				
Mammalian				
Continuous cell lines				
Insect cell lines				
Hybridomas				
Murine				
Humanoid				
Chimeric				
Harvesting				
Inclusion product				
Secreted product				
Isolation				
Inclusion product				
Secreted product				
Purification				
Affinity chromatography				
Monoclonal/??				
Ion-exchange chromatography				
Size-exclusion chromatography				
Postpurification processing				
Pooling				
Bulking				
Adjuventing				
Cross-Linking				
Chemical Modification				
Compounding				
Final processing				
Filling				
Lyophillization				
Container/closure integrity				
Compatibility studies				

consideration of alternative manufacturing protocols designed to avoid manufacturing conditions which adversely affect the product, including the potential to easily validate production and testing conditions. Second, it assists in defining critical in-process testing parameters designed to set specifications which must be achieved prior to continuing production at any point. Failure to define appropriate in-process parameters and determine which are critical to continuation of manufacturing results in all in-process testing being considered critical by investigators from regulatory authorities.

2.2.1.6. Formulation Design and Optimization

As indicated previously, it is imperative to have a defined formulation for a product which has been characterized to the extent that its activity has been defined and correlated with expected clinical activity in humans prior to initiation of preclinical studies intended to define the pharmacology and toxicology. Failure to appropriately describe a product at this point may invalidate these supportive animal studies.

Product and manufacturing process characterization should be defined to the extent that the process results in a consistent, reproducible, and defined product prior to first administration in humans (usually Phase I). Complete product characterization and formulation optimization is not required until the initiation of pivotal clinical trials.

The biotechnology product obtained following characterization of the source production, product, and process is similar to a product derived from traditional drug manufacturing. Exceptions to this premise are generally due to issues associated with stability or unpredictable decreases in biological activity. Products at this stage of development may otherwise undergo formulation optimization utilizing principles applied to traditional drug manufacturing. These studies begin with preformulation screening of excipients, additives, and stabilizers based on molecular attributes defined during early characterization studies. Literature review is the first step in conducting preformulation studies. However, information regarding product failures resulting from formulations is sparse due to the proprietary nature of such intelligence. Solution pH is one of the most important factors in protein stability. Many biotechnology-derived proteins or their natural analogues have optimum pH ranges reported in the literature. From this information, properly designed preformulation studies will yield the optimum pH and buffering conditions.

Most methods and agents utilized to stabilize proteins result from trial and error studies based on the molecular attributes of the product. Classes of molecules including albumins, sugars, surfactants, polyols, and amino acids have been successfully used to stabilize proteins.

The table of molecular attributes previously constructed to assist in the selection of appropriate characterization tools, can therefore be utilized in the selection process for potential excipients and stabilizers. The next step in preformulation studies involves conducting excipient screening studies using a matrix design to evaluate the effects of the agents individually and in combination on final product integrity. Three classes of inactive ingredients* are usually evaluated in these studies:

1. Antimicrobial preservatives
2. Tonicity-adjusting agents
3. Protein stabilizers

Following preformulation studies, several candidate formulations are selected as prototypes for optimization studies. The formulation of biotechnology-derived products is as much art as science. As such, a rule derived from trial and error experimentation of such products is that a formulation is considered to be appropriate for consideration if either an analytical assay which is predictive of potency or a bioassay of the final product is within 15% of a drug substance control preparation without excipients.

Optimization studies should result in a formulation and packaging that confers the highest level of product integrity (defined as stability, potency, and consistency). Following selection of the optimum formulation, the process that yielded the optimum data package should be documented and translated to manufacturing scale.

2.2.1.7. Final Product Specifications

Final product specification setting appears at the end of the biotech PDP, but should be based on data developed and documented from the earliest identification and characterization studies. These data, including methodology selected for characterization and critical in-process testing parameters, should be considered along with a listing of available test methodology to define specifications for release of product following manufacture and at the end of the stability dating period.

A table defining potential analytical methods for setting final product specifications is shown in Table IV. Such a table should define the potential analytical method, the final product attribute detected by such a method (if any), the value of information provided by utilizing the method, and an indication of whether such information is critical to the release or stability of the product. This developmental process accomplishes two objectives. First, it permits selection only of methods which are appropriate for describing the final product and eliminates unnecessary tests and controls. Second, it permits discrimination between critical and noncritical parameters, which

*In legal terms, all components of a final drug product that do not directly contribute to the defined activity of a product are considered to be inactive ingredients.

Table IV
Analytical Methods for Setting Final Product Specifications

Analytical method[a]	Detects	Information	Critical or noncritical
Potency			
SDS–PAGE			
RP-HPLC			
IEF, IEC			
N-Terminal sequencing			
C-Terminal characterization			
Peptide map			
ELISA, RIA			
CD (far-UV)			
UV absorbance			
Fluorescence			
CD (near-UV)			
SEC			
Ultracentrifugation			
Light scattering			

[a] SDS–PAGE, sodium dodecyl sulfate–polyacrylamide gel electrophoresis; RP-HPLC, reversed-phase high-performance liquid chromatography; IEF, isoelectric focusing; IEC, ion-exchange chromatography; ELISA, enzyme-linked immunosorbent assay; RIA, radioimmunoassay; CD, circular dichroism; UV, ultraviolet; SEC, size-exclusion chromatography.

may be utilized to assess the integrity of the final product before and after changes in process or product. If postapproval changes can be accomplished without the need to repeat and revalidate all methods, considerable time and expense can be spared.

2.2.2. UTILIZATION OF THE BIOTECHNOLOGY PDP

The biotechnology PDP may be compared to the formulation development report which has been requested (not required) by FDA field investigators during preapproval inspections for traditional drugs as a mechanism to describe and justify product development and formulations utilized at different preclinical and clinical stages. Development reports for drugs and biological products are based upon the same principles of drug development (see Fig. 2). Biological and biotechnology product development reports should begin with considerations for genetic stability and integrity and end with a process for establishing rational final product specifications.

The PDP concept was refined in parallel with research on characterization of complex biotechnology products and components of this algorithm have been incorporated into numerous FDA guidance documents

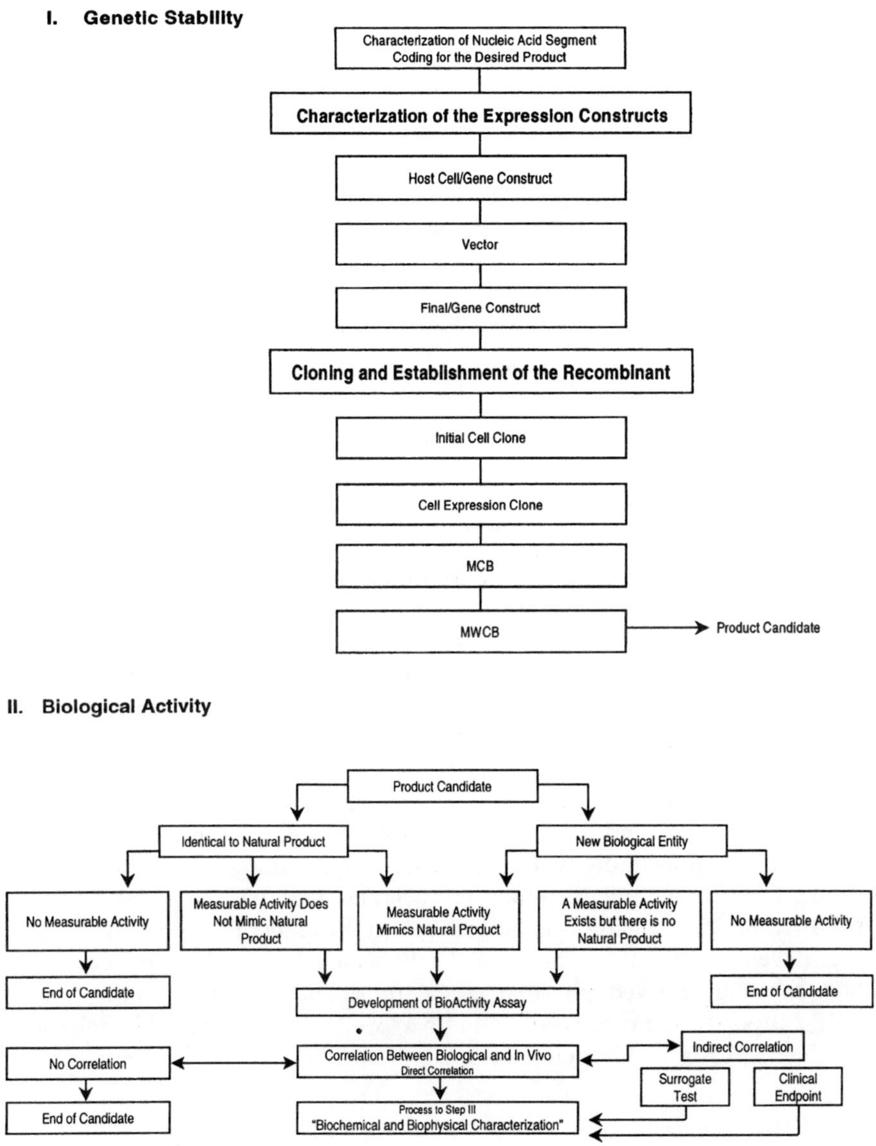

Figure 2. Biotechnology product development profile/flow sheet for recombinant products.

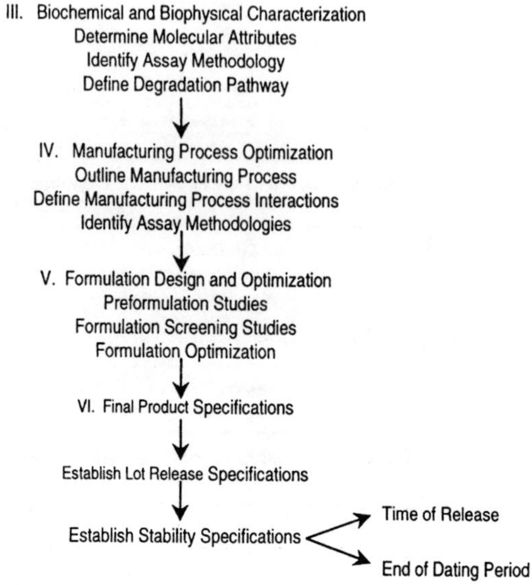

Figure 2. (*continued*)

and discussed at national and international pharmaceutical meetings since its inception in the mid 1980s.

These early studies designed to develop the PDP also supported the concept of defining a class of biotechnology products which would be considered "well-characterized." Adequate product characterization as well as the design of an effective tool for documentation of all phases of product development could only be realized if there existed a thorough understanding of the nature of the product and manufacturing process, including product attributes and production variables that underpin the manufacture of complex biotechnology products. The following sections describe the product attributes and production variables that were considered in the production and research on biotechnology products and ultimately led to the development of a rational PDP which could be applied to all biotechnology-derived pharmaceutical products. The development of the PDP supported the process of defining a "well-characterized" biotechnology product of drug or biological origin.

2.2.3. HYPOTHESIS

The hypothesis was that a new regulatory paradigm could be developed which would permit well-characterized biotechnology products to be manufactured and reviewed based on the same principles as applied to traditional drug product development.

The initial strategy, therefore, was to address problems such as those previously identified which are associated with the review of new technology products. Due to the restrictive regulations in place when the project started and the difficulty inherent in attempting to revise or eliminate laws and regulations, the initial rationale was to develop and publish guidance documents utilizing sound scientific principles as a vehicle to broaden the interpretation of existing regulations. Studies integrating newly evolving scientific methodology with existing principles of product development assisted in defining a new class of "well-characterized" biotechnology products, an outcome which was the basis for creation of the new regulatory paradigm.

Two elements were needed to achieve this hypothesis. The first was access to and interpretation of all of the information related to the successful and failed development of biotechnology products. The second element in this endeavor was development of a broad understanding of the acceptable current bioanalytical methodology utilized to evaluate such products at all stages of development (e.g., intermediate and final products).

The fact that a single test cannot describe the activity of the complex molecules which have been developed utilizing biotechnology processes means that variables associated with molecular attributes, production, and testing need to be examined in detail.

The following discussion outlines the key principles which emerged from biotechnology product development research, starting with a basic concern for the method of production and continuing with discussions about regulatory and scientific aspects of genetic engineering, biological diversity and heterogeneity, solubility, formulation optimization and the setting of final product specifications. The production of a recombinant DNA-derived protein will be referenced, because it is the process utilized in the development of interferon α, one of the first therapeutic biotechnology products to be approved by the FDA.

3. PRODUCTION VARIABLES

Most biotechnology products are derived from biological systems which are complex and inconsistent. The production sequence may include one or more of the following steps: isolation and identification, propagation of an organism or virus, growth on an expanded scale, cell disruption and/or product recovery, purification, concentration, formulation, filtration, and aseptic packaging.

Major problems can be encountered following the cell growth phase with regard to separation of the major biological component(s) from nutrients or contaminating proteins. If continuous cell culture techniques are utilized, the additional problem of the continuous elimination of contaminated matter requires initial validation, constant monitoring, and ongoing cyclic validation studies. When the complex growth, purification, and formulation process is completed, the resultant final mixture is often complex, and the clinically active biological component is frequently a very small proportion of the final product, which may be difficult or impossible to isolate and analyze in the final container.

The above translates into a product which, in its intermediate and/or final form, may be difficult or impossible to assay and quantify on a consistent basis. In addition, direct correlation of a biological assay with the product's biological and clinical activity may be nonachievable. Therefore, if the manufacturing process has not been adequately validated, both quantifiable and nonquantifiable product characteristics may change with each lot or batch produced. In fact, this is a common observation for products of biotechnology submitted to the FDA's CBER.

The approval pathway for a product with this level of variability may be difficult to impossible. Adequate process validation should precede and/or be conducted in parallel with studies appropriate to characterize the product. Products which have been manufactured using appropriate process validation and are capable of being thoroughly characterized are most likely to be considered well-characterized.

Process validation must include all steps in manufacturing, because products of biotechnology may also be subject to inactivation or loss of activity as a result of chemical or physical factors encountered during long and complex manufacturing processes. Factors such as exposure to heat and shear can lead to an irreversible decrease in activity due to denaturation or conformational changes which affect activity.

In addition, the residence time for the product in nutrient-rich media and the numerous transfers of additives and sampling events during manufacture increase the probability of multisource contamination. The effects of contamination may not be detectable until several downstream steps have been completed and may challenge the capability of the purification steps beyond their validated capacities.

Table V outlines the steps involved in the manufacture of a typical rDNA product. This differs from the steps involved in chemical synthesis of a drug product in that, once the impurity profile of a drug substance

Table V
Steps in the Manufacture of a Typical Recombinant DNA-Derived Product

Synthesis of the DNA fragment encoding the product
Nucleotide sequence and restriction enzyme map
Construction of expression vectors
Description of the host cell
Transformation of host cell by expression vector(s)
Selection and cloning of cells expressing the product
Fermentation
Cell separation/harvesting
Purification
Concentration or dilution
Formulation
Aseptic processing
Final product testing
Cloning and genetic engineering

had been established and sensitive and specific assay procedures developed, lot-to-lot consistency is considered to be relatively easily quantified and maintained, thereby eliminating the need for extensive process and postapproval regulatory documentation (Beatrice, 1993).

Another important distinction between rDNA-derived products and pure chemical drugs is that the types of information regulatory agencies need to evaluate an rDNA product are not all related to characterization of the final product. As previously described, it is important to establish that the product can be consistently produced and that it will not be contaminated by unwanted materials. This is also true for conventional drug products; however, the emphasis begins much earlier, with the cloning and genetic engineering steps for products of biotechnology. Thus a description of the manufacturing process and facility as well as the materials and methods used to construct a recombinant gene and the final production strain are considered vital to an assessment of the consistency and safety of the final product.

4. MANUFACTURING PROCESS ISSUES

Once the genetic engineering steps have been optimized, the manufacturing process should be developed, which incorporates equipment and procedures that will not result in negative product interactions.

4.1. Validation of the Fermentation Process

Because large-scale fermentation processes begin with an inoculum of stock culture, the maintenance of the master seed must be considered when addressing validation of fermentation. A master seed is a single culture grown from a selected host cell colony that contains the recombinant DNA expression vector, such as a plasmid. This master seed is subdivided into ampoules and stored at the temperature of liquid nitrogen to maintain viability and purity. Each time a master seed is prepared it must be fully tested to demonstrate that it contains only the appropriate genetic and biochemical characteristics. These tests generally include identification of genotypic and phenotypic markers, analysis of cellular proteins, determination of culture sterility, specific growth rate, and concentration of product, restriction enzyme mapping of the plasmid DNA, and verification of the sequence of DNA segments coding for the product and regulating its expression. The stability of the master seed in storage must also be validated by appropriate testing. As an alternative during the production stage, a second-generation stock culture (or working seed bank) can be grown from a single colony selected by plating out the contents of one ampoule of the original master seed bank. Occasionally a small preparation of a third-generation stock culture (a working bank) is grown from the working seed bank, which minimizes repetitive cloning. Like the master seed bank, the working seed bank and the working bank must be properly characterized and their stability during storage must be properly monitored.

By the time FDA inspectors arrive at the manufacturing facility, the cloning process has been long completed and the seed banks have been established (Chiu, 1988). The inspectors and reviewers must rely on documentation of the processes to ensure that they were validated and conducted in compliance with established standard operating procedures and GMP principles. In addition, since many products are acquired from an innovator following discovery and some amount of animal or early human testing, meaningful documentation must be available to describe all aspects of the genetic product design.

Equipment utilized for fermentation is also critical to the process and should be designed to maintain the integrity and sterility of the culture medium and permit easy cleaning and removal of residual product followed by sterilization. The equipment design should also allow the contents to be readily mixed and aerated and evenly heated and cooled. The equipment should resist corrosion and should permit sampling and nutrient addition without contamination.

Raw materials used to prepare the culture medium must be carefully selected to provide a high rate of growth and be of sufficient purity and quality to permit consistent product yields. They should not contain any toxic components that could be carried through the purification process to the final product. An adequate supply of selected amino acids must be provided in the nutrient mix to prevent errors in the biosynthesis process.

Air is another essential ingredient during fermentation. It should be filtered before delivery into the fermentor, introduced in measured quantities appropriate to support cell expansion, and validated to be of sufficient purity.

As previously stated, the production process begins with the preparation of a fermentor inoculum, which is a culture grown from stock culture, seed, or working cell banks. Prior to inoculation into the fermentor, the culture may pass through several growth and containment vessels, including flasks and seed tanks. To comply with containment guidelines issued by regulatory authorities and protect culture integrity, transfers from one vessel to another must be performed in a manner that minimizes the release of aerosols or the contamination of exposed surfaces. Cell stability is measured by the lack of mutation of the recombinant DNA-containing microorganisms and plasmid stability for multiple generations of cell growth beyond cell expansion and fermentation processes.

Proper temperature, air flow rate, pressure, pH, agitation speed, and amount of foaming must be selected to maintain a high rate of growth and productivity. Other parameters that must be monitored during fermentation include feed rate, rate of oxygen uptake, evolution rate of carbon dioxide, cell density, and product concentration. A profile of each of these parameters should be established after a successful fermentation run is completed and any significant deviation that occurs from established profiles during subsequent production runs should be investigated. These parameters form the basis of the initial specifications, which must be maintained to yield consistent product of adequate purity and potency. A fermentation run with a low product yield should not be accepted unless the degree of purification required in such instances has been demonstrated not to exceed the capacity of a validated purification process.

The previously discussed limitations to final product testing necessitate that the culture must be sampled at least once—preferably immediately prior to harvesting—and tested to demonstrate the absence of contamination by non-host microorganisms. This can be accomplished by utilizing an immunoassay that distinguishes the residual protein antigens originating from the recombinant DNA-containing host microorganism from those of non-host microorganisms. To conduct the immunoassay, a placebo clone is prepared that is identical to the production clone except that the plasmid does not contain the DNA sequence coding for the product. After fermenting the placebo clone, antiserum is raised by extracting cellular proteins and using them to immunize animals. The antiserum collected is the reagent used in the immunoassay to determine the residual protein antigens that are present in the fermented product. This antiserum will not be able to recognize antigens produced by other microorganisms that contaminate the culture medium during a production fermentation run.

4.2. Cell Separation/Harvesting

Once fermentation is complete, the cells must be separated from the culture medium. Cell products may be harvested after an inactivation process (e.g., heat treatment) or directly from live *contained* cells. The appropriate containment measures as prescribed by regulatory authorities must be in place when working with live, recombinant organisms to protect the environment, workers, and product. Regulatory reform efforts initiated in 1996 will permit certain cell lines and host organisms with a documented history of safe use in large scale fermentation to be exempt from strict containment requirements (National Environmental Policy Act: Proposed Revision of Policies and Procedures, April 1996). However, if there exists a potential for aerosolization (e.g., from centrifugation or other processes) and/or if the fermentation or separation areas will be used for more than one organism, rational containment measures should be instituted.

4.3. Purification

Once separation and harvesting have occurred and the extracellular or intracellular products have been released, the next critical phase in processing is purification. A large-scale purification procedure should be designed to maintain a high degree of product purity as well as support a high product yield. Scaling up the purification process to accommodate

Table VI
Purification: Quality Assurance/Quality Control Concerns for Columns

Single use vs. multiple use
Maintenance and storage
Bioburden monitoring
Endotoxin monitoring

larger volumes of material needed for later stage clinical studies involves optimizing the process with regard to packing materials, column lengths, flow velocities, and buffers.

Quality control of the columns is critical and includes consideration of the elements listed in Table VI.

Single-use columns are expensive and impractical in many large-scale purification operations. The alternative is to dedicate columns to specific products to minimize cross-contamination. Multiple use of columns requires establishment of a database demonstrating product yields and purity profiles over time. This database can be used to determine the shelf life of the column, giving the length of time after which it must be replaced. Proper maintenance and storage of columns is also essential to maintain integrity and efficiency. Ordinarily columns should be conditioned after use, identified, protected from environmental bioburden, and stored in a cold room.

In order to minimize bioburden form all sources, buffers should be aseptically produced using water for injection USP as a vehicle and reagents of the highest quality and purity. Endotoxin (utilizing the *Limulus* amebocyte lysate, LAL, test) and sterility testing of buffers provides greater assurance of integrity. Several instances have been noted of contaminated buffers reducing the efficiency of columns which were validated to operate under optimal conditions.

5. PRODUCT QUALITY ATTRIBUTES

From a regulatory viewpoint, three product attributes are considered vital to development of a bioanalytical profile, an essential early component of the product development profile in support of the development of a well characterized product. Identification of the product attributes and process interactions provides a unique fingerprint for each product. A properly constructed bioanalytical profile considers such factors while addressing the regulatory requirements necessary to assess the purity, potency, and quality of each biotechnology product. These three elements are fundamental to the regulatory review of applications for marketing of biotechnology products and must be assessed by appropriate, validated methods. The regulatory interpretations of these scientific concepts were introduced by this author in the Quality Section of the International Conference on Harmonization at the Orlando, Florida, International Meeting in October 1993 and accepted by all participating government and regulatory authorities as basic tenets of the quality assessment for products of biotechnology.

5.1. Purity

The term "purity" is defined in the U.S. regulations for biological/biotechnology products at CFR 21, Section 600.3(r), as follows:

> Purity means relative freedom from extraneous matter in the finished product, whether or not harmful to the recipient or deleterious to the product. Purity includes but is not limited to relative freedom from residual moisture or other volatile substances and pyrogenic substances.

In spite of the fact that the regulatory definition of purity is applicable to the finished product, the appearance of degradation products, fragmentation, or aggregation at any stage of manufacture can compromise final product purity. Since chemical and physical changes cannot always be predicted or prevented, consideration should be given to potential hazards and to the need for identification of such impurities. Acceptable limits for degradation products or fragmentation should be proposed and justified.

The use of relevant physico-, bio- and immunochemical analytical methodologies (e.g., molecular sizing, charge determination, hydrophobicity, and changes which may occur during storage) will facilitate comprehensive product characterization. As previously indicated, methods that will contribute to adequate characterization include electrophoresis (sodium dodecyl sulfate–polyacrylamide gel electrophoresis, immunoelectrophoresis, Western blot, isoelectric focusing), high-resolution chromatography (gel-filtration, ion-exchange, affinity chromatography) and end-terminus peptide mapping.

5.2. Potency

The term "potency" is defined in 21 CFR, Section 600.3(s), as follows:

> The word potency is interpreted to mean the specific ability or capacity of the product, as indicated by appropriate laboratory tests or by adequately controlled clinical data obtained through the administration of the product in the manner intended, to effect a given result.

At the point in development when the intended use of a product is linked to a definable and measurable biological activity, testing for potency determination should be part of the product development documentation. The correlation between potency, biological activity, and clinical activity must be established prior to introduction of any product into human subjects. Potency studies should be performed using appropriate reference preparations calibrated, whenever possible, in units of biological activity (in-house, national, or international as applicable). The confidence limits of the assays and, where applicable, the coefficient of variation of the potency testing system should be indicated. In many biological products potency is dependent upon the conjugation of the active component(s) to a second moiety or binding to an adjuvant, for example, vaccines produced by adsorption of the antigenic substance to aluminum substrates. Leakage of the active compound(s) from the carrier used in conjugates or from adjuvants should be examined in real-time/real-temperature studies. The maintenance of potency (i.e., assessment of the stability of such products) is critical and may be associated with difficulties, because, in some cases, *in vitro* tests for biological activity and physicochemical characterization are impractical or provide inaccurate results. Appropriate strategies (e.g., release of the active compound from the second moiety, *in vivo* assays) should be considered to overcome these difficulties through the use of an appropriate surrogate test.

5.3. Quality

No regulatory definition exists for quantifying product quality. Quality is related to characteristics of the product in final containers, which are useful in describing attributes of both small and large molecules. The product characteristics in Table VII, though not unique to biological products, are indicators of final product quality.

Concentration of excipients or inactive components may change because of chemical reaction or degradation and therefore may have to be monitored. The nature of these inactive components determines to what extent their levels should be assessed.

The container/closure system has the potential to affect the product such that changes which occur in one configuration will go undetected if one

Table VII
Product Quality Attributes

Visual description of the product in the final container (color and opacity for a solution and the color and texture and dissolution time for a powder)
Particulates in solutions/suspensions or after the reconstitution of powders
pH
Loss on drying or moisture level of powders
Initial and end-of-shelf-life sterility
Preservative content (if present)

assumes that all container/closures are equivalent. Container/closure physical characteristics which should be considered include glass type, resin type for plastic containers, container thickness, cleaning methods, and pretreatment (depyrogenation, sterilization, siliconization, etc). Closure configurations such as sealed ampoules, stoppers, and caps should be considered regardless of whether or not seal types are identical. Vial liners and method of sealing must also be considered.

The above considerations related to definition of product attributes, analytical methodology, and manufacturing process interactions are based on experience in the review of biotechnology-derived products submitted to FDA since 1980.

6. EVOLUTION OF THE NEW REGULATORY PARADIGM

The FDA and biopharmaceutical industry have made enormous progress in the design, development, and evaluation of products of biotechnology from the early 1980s, when the first approval was granted, to the mid 1990s. Biotechnology, as we know it today, has evolved over the past 20 years from the initial gene cloning of interferon and insulin to the development of many new and important pharmaceuticals. These include products such as vaccines, therapeutic agents, replacement therapies, and diagnostic test kits.

Production and testing techniques have also evolved and matured, especially in the field of protein biotechnology. This is especially true of recombinant DNA and monoclonal antibody technology, which generates products capable of being considered "well-characterized" in a manner similar to conventional drug products.

The origin of the era of more flexible regulation for products derived from biotechnology can be traced back to March 1982, when the Office of Biologics of the National Center for Drugs and Biologics, FDA (now CBER, FDA) published the first Points to Consider document, entitled *Points to Consider in the Manufacture of in Vitro Monoclonal Antibody Products*. This type of document was considered to be an interpretation of specific criteria related to products and manufacturing processes that would be acceptable to the agency, based on good science. It did not have the force of law, as is the situation for a regulation or guideline. Manufacturers were asked to respond with their ideas, recommendations, and comments concerning this PTC document. In those responses, many manufacturers emphasized that hybridoma technology was a fast-moving field and that premature development of restrictive guidelines or requirements could adversely affect product development and ultimate product availability.

In response to these comments, and after much consideration of how existing regulations could be flexibly interpreted, the FDA published a Notice (Federal Register, 1983) acknowledging that monoclonal antibody products intended either for *in vitro* use of for *in vitro* testing of a licensed biological product were considered to be biological products and as such required licensing under Section 351 of the *Public Health Service Act* (PHS Act) [U.S. *Code*, Title 42, Section 262(a)]. As previously stated, the existing interpretation of requirements for licensing biological products was very restrictive in terms of requiring demonstration of control over all manufacturing steps to qualify for a license. However, through this interpretative Notice, flexibility was extended for the first time to permit manufacturers seeking a license for a final monoclonal antibody dosage form to obtain partially processed monoclonal antibody products from another legal entity, a concept that became known as "shared manufacturing." The FDA maintained the requirement that all legal entities participating in the manufacture of the product at any stage submit and receive approval for licenses for all intermediate and final products and the establishments in which they were made. At that time, FDA interpreted the PHS Act to require manufacturers of biological products to obtain separate establishment licenses for all legal entities participating in product manufacture and separate product licenses for all intermediate and final products. In fact, these separate license applications for all products and establishments were required to be submitted, reviewed, and approved simultaneously.

The Establishment License Application (ELA) was unique to the application process for biological products. The ELA form had specific questions and requested detailed information related to the design, construction, validation, and operation of facilities in which biological products were made as well as documentation of strict control over all manufacturing steps. Control over the manufacturing process at that time was interpreted to mean that the applicant also be the owner of all facilities and the employer of all personnel engaged in significant manufacturing. Significant manufacturing was understood to include all steps in the manufacturing process, from propagation of the source material to preparation of the final dosage form, an interpretation that would become part of the FDA's cooperative manufacturing policy, which was extended to all biological products in 1992. Limited contract operations were permitted, including filling, lyophilizing, labeling and packaging, after a final stabilized dosage form was prepared.

The Product License Application (PLA) consisted of information which was required to be submitted on "forms prescribed for such purposes" [21 CFR, Part 601.2(a)] by an applicant that demonstrated full supervision and control over all manufacturing steps. Clinical data in support of the PLA was required to be submitted in a manner identical for drugs or devices, following the regulations at 21 CFR, Part 312 using the investigational new drug (IND) mechanism for drugs or 21 CFR, Part 812, the investigational device exemption (IDE) for medical devices and diagnostic products. All review, inspection, approval, and compliance activities for biological products continued to be the responsibility of the FDA component responsible for biological products (CBER). This constituted another departure from the regulation of similar drug products, in that the FDA field was responsible for facility inspection and GMP compliance for drug products.

Given the wide range of biological product categories, from human blood for transfusion or preparation of blood derivatives to vaccines and therapeutic agents, to tissues and somatic cell and gene therapy products, there were 19 separate product application forms and one establishment application form. These separate PLA forms were not designed to be consistent with each other or with the ELA. Therefore, a considerable amount of redundancy existed between the PLA and ELA and between PLAs for different products.

An applicant for approval of a traditional drug product under the NDA process did not need to document facility details in an ELA or Master File.* Details related to compliance with standards for drug facilities and manufacturing processes were reviewed by the FDA Field Operations under the direction of the Office of Regional Operations, and inspections for evaluation of compliance with (GMP) regulations (21 CFR, Part 211) were conducted at some point during NDA review.

Biotechnology assisted in the transition of biological products from products for prevention of disease (vaccines) and replacement therapies (e.g., transfusable blood, albumin, plasma protein fraction, etc.) to a multitude of therapeutic products (e.g., interferons, interleukins, tissue plasminogen activator, etc). This transition brought many traditional drug manufacturing companies under the jurisdiction of biologics regulation. These manufacturers discovered that they needed two groups of regulatory experts with two distinct fields of expertise, drugs and biologics, to bring their full range of products to market, even though the manufacturing processes and product characterization studies were almost identical. For example, recombinant human growth hormone, a drug product regulated under the NDA system due to historical precedent, was reviewed based on the principles established in the FDA Points to Consider documents without the burden of (1) submitting an additional application (the ELA), (2) maintaining strict control over all manufacturing steps in facilities under the applicant's supervision and control, and (3)

*A Master File is a submission of information to the FDA by a person (the drug Master File holder) without the need to disclose the information to any person outside of the FDA (21 CFR, Part 314.420).

obtaining separate but simultaneously issued licenses for the establishment and product.

This "uneven playing field" became more pronounced with the advancement of the science of product development and characterization methodology, leading to a change in the political climate and increasing pressure to institute regulatory reform.

During the same time period, industry, academia, and government were making tremendous strides in developing enhanced process control and validation methodology, refined analytical techniques, and highly specialized and sophisticated manufacturing equipment. These advances, coupled with experience gained from years of research and regulation, accelerated the FDA's efforts to increase regulatory flexibility, initially by publishing a series of interpretative guidance and policy documents and eventually leading to proposed changes in the regulations and legislative reform efforts. Guidance and Points to Consider documents are not regulations and do not have the force of law. They are an attempt to interpret broad regulations and statutory (law) language based on the most current thinking of industry, government, and academia. However, FDA expects any activities or procedures which are described in such documents to either follow the guidance or be performed in a manner which is equal to or better than that described.

From a policy viewpoint, language in the FDA's Points to Consider, guidance and policy documents reflected the regulatory interpretation that existed at the time such documents were written. There was a gradual change in the regulatory paradigm based on evolving science. For example, a 1983 National Center for Drugs and Biologics document (NCDB, 1983) maintained the following in the "Consistency of Manufacture" Section:

> Biological products derived from living cells are subject to variabilities inherent in complex biological systems. Product Consistency should, therefore, be maintained by close attention to consistency of manufacturing procedures as well as analysis of the final product.

A 1985 Points to Consider (U.S. FDA, 1985) stressed product characterization and the necessity to conduct clinical trials for safety and efficacy assessment, but did not discuss whether these requirements were applicable to new biological entities and/or after changes were made during product development.

The 1989 *Points to Consider in the Manufacture and Clinical Evaluation of In Vitro Tests to Detect Antibodies to Human Immunodeficiency Virus, Type 1* discouraged process changes by indicating that:

> A fundamental change in the product, e.g., a change to a different recombinant antigen construct or a different solid phase, necessitates a separate IND application.

and

> Any change in the product during the clinical trial could be problematic because clinical data obtained with variants of the product described in the PLA might not be acceptable to support licensure of that PLA.

The 1990 *Cytokine and Growth Factor Pre-Pivotal Trial Information Package* guidance document stressed that the commercial product be "manufactured and formulated in a manner which is essentially identical to the product tested in the pivotal trial."

The 1991 *Points to Consider in Human Somatic Cell Therapy and Gene Therapy* appeared to support the position that such new technology products are similar to traditional biologicals, as follows:

> Biological products are often complex mixtures that cannot be completely defined. Quality control of the manufacturing process as well as the final product is necessary. Poor control of production processes can lead to introduction of adventitious agents or other contaminants, or to inadvertent changes in the properties or stability of the biological product that may not be detectable in final product testing.

Up to this point, the published policy and guidance supported the traditional regulatory paradigm for new technology products, even though greater flexibility was given to monoclonal antibody products in 1983. The apparent regression of policy in the face of evolving science continued until 1992, when a policy document that began the true paradigm shift was published.

This FDA document (Federal Register, 1992) had the effect of extending the same flexibility that was given to monoclonal antibody products in 1983, the ability to utilize shared manufacturing arrangements, to all biological products. This was based in part on product development research, which helped to demonstrate the maturity of biotechnology product development. Products of biotechnology can now be characterized to a state approximating drug products, and flexible manufacturing arrangements can be entered into without adversely affecting the safety, purity, or potency of such products. Publication of this policy required reinterpretation of regulations governing how products were licensed, redefining the extent of control necessary to be exercised by the manufacturer of the final dosage form and changing the labeling requirements for intermediate and final biological products.

A shift in policy related to product comparability was also becoming evident due to advance in the power of analytical methodology, as indicated in a 1994 Points to Consider (U.S. FDA, 1994); which stated that products under investigational studies should be manufactured "using the procedures, scale of production, and facilities that are intended to be used commercially" (the traditional biologics approach). However, the document went on to say the following:

> Scale-up programs and contemplated changes in product manufacture should be addressed before Phase 3 trials. If major manufacturing changes (e.g., change of cell bank, change from serum-containing to serum-free medium, or change of site or portion of site of manufacture) are made after the phase III trial is completed, clinical studies may be necessary depending upon the information yielded by pre-clinical biochemical, pharmacokinetic and functional comparisons.

The next step in the formation of the new regulatory paradigm was to clarify the requirements needed to demonstrate that products made by old and new manufacturing procedures are comparable after product and/or process changes are made. Early FDA policy and guidance documents used the term "equivalent" for biotech products. However, parameters used to evaluate bioequivalence, as defined for pure chemical drug products, may not be relevant or meaningful for biological drugs. Therefore, the term "comparability" was proposed by FDA as an alternative. The FDA and the ICH Biotechnology Safety Expert Working Group had been engaged in discussions on equivalence programs for biotechnology products in clinical development with the intention of introducing a globally harmonized guidance document. In addition, FDA outlined current product equivalency issues in a 1995 paper of the same title authored by the CBER Director of the Division of Cytokine Biology (Gerrard, 1995). The guiding principle for this policy centers around the extent to which a product can be defined by biochemical characterization. If the product can be adequately characterized, meets consistent specifications from lot to lot, and has a biological activity that correlates with a specific and sensitive potency assay, comparability can be demonstrated without head-to head clinical studies. This was described in a guidance document on product comparability published in April 1996 (U.S. FDA, 1996).

The most dramatic policy shift was initiated as a result of a meeting of the Drug Information Association in January 1995, where this author presented FDA's current comparability policy as described above. In response to this topic, industry representatives expressed the belief that a comparability policy which could favorably compare two well-characterized products without head-to-head clinical studies was desirable, however, they continued to believe that FDA required biological products intended for pivotal clinical trials and/or in support of an application to be made in commercial manufacturing facilities, which was not required for traditional drug products under the NDA system.

This was not the situation for well-characterized biotechnology products, which, at the time, were mainly therapeutic recombinant DNA-derived and monoclonal antibody products. Correction of this misconception became the cornerstone of the first reinventing government initiative and resulted in the publication of an appropriate in 1995 FDA guidance document (Federal Register, 1995a).

Government reinvention initiatives began on September 10, 1993, when the Vice President of the United States published a document (Gore, 1993) which stressed performance-based management for all government agencies and outlined the following four basic principles, which became a part of Reinventing Government I (Rego I):

First: Cut obsolete regulations
Second: Reward Results, NOT red tape
Third: Get out of Washington and create grassroots Partnerships
Forth: Negotiate, don't dictate

The biologics regulatory paradigm shift gained momentum on September 30, 1993, with Executive Order number 12866, which asked federal agencies to retrospectively review existing regulations for the purpose of eliminating or revising burdensome requirements. The CBER of the FDA had already formed two task groups to begin this process, and in June 1994 public comment was solicited to help define and revise or eliminate the outdated biologics regulations (Federal Register, 1994).

REGO II initiatives were initiated by the Clinton Administration in a public meeting on March 16, 1995. They consisted of several major regulatory reform initiatives, including the following, which came to be known as the REGO IIa initiatives:

1. Eliminating many requirements for FDA preapproval of manufacturing changes for drugs and biologics
2. A new policy to permit the use of small-scale and pilot manufacturing facilities during development of biological products in support of eventual product approval
3. Revision of labeling requirements for biological products to partially harmonize drug and biologics labeling regulations
4. Revision of requirements to submit an environmental assessment analysis for all application submissions

The policies on use of pilot facilities and reduction of postapproval reporting requirements were clearly a result of scientific advances in the characterization of drug and biological products and experience in the regulation of such products.

The effort to eliminate many requirements for FDA preapproval of manufacturing changes was a first effort for biological and biotechnology products due to the existence of a regulation which required preapproval for all important proposed changes in manufacturing methods and labeling (21

CFR, Part 601.12). The use of ambiguous terms such as "important" in the regulation resulted in a strict interpretation that any change which had a potential to alter the safety, purity, or potency of a product be submitted as a supplement to the approved application and await review and approval. Until this regulation could be changed, the FDA chose to provide guidance documents which interpreted the level of importance for changes in manufacturing methods. This permitted certain changes to be implemented without preapproval. Realizing that this effort was inconsistent with the drug regulation at 21 CFR, Part 314.70, which permitted three categories of change, the next set of reforms focused on revising or removing restrictive regulations.

REGO IIb was the next cooperative initiative between FDA and the Clinton Administration, which followed in November 1995 with publication of a document entitled *Reinventing the Regulation of Drugs Made from Biotechnology*. This announcement signaled the irreversible forward momentum of the biologics new regulatory paradigm made possible by evaluation of scientific advance and regulatory experience gained after almost two decades of experience with biotechnology products. REGO IIb contained the following significant reform proposals:

1. Elimination of the requirement for an establishment license application for therapeutic biotechnology drugs (and biological products) that are considered to be specified (previously well-characterized)
2. Elimination of the requirement that FDA, CBER release every lot of all specified biotechnology products after application approval
3. Development of a harmonized application format, initially for all specified drug and biological products, and eventually (following development of specific guidance documents) for all human drug and biological products
4. Elimination of the requirement to preapprove promotional labeling for biological products to be consistent with the promotional labeling and advertising requirements for drug products
5. A commitment by FDA to review and respond to data submitted in response to an agency-instituted hold on a clinical study within 30 days of receipt of the submission
6. Revision of the restrictive requirement to name a single individual, the responsible head, to exercise control of a biologics manufacturing establishment in all matters relating to compliance with the regulations and communication with the FDA

These initiatives not only represented a radical departure from the traditional biologics regulatory paradigm, they also served to harmonize the regulatory requirements for drugs and biologics that had similar characteristics, that is, that were capable of being "well-characterized."

REGO IIb contained the most significant reform effort since the regulation of biotechnology in pharmaceutical practice was initiated, the effort to define a specified (well-characterized) biological product. This effort began with a publication in December 1995 which defined the attributes of a well-characterized biological product and announced an international workshop to discuss the definition and implementation of a new regulatory paradigm (Federal Register, 1995b).

The characteristics of a well-characterized product were considered to constitute an interim definition for a well-characterized therapeutic recombinant DNA-derived and monoclonal antibody biotechnology product. The definition was divided into several parts as follows:

Interim definition of a well-characterized therapeutic recombinant DNA derived and monoclonal antibody biotechnology product

A chemical entity(ies) whose identity, purity, impurities, potency, and quantity can be determined and controlled

Identity:

Recombinant DNA Biotechnology Products

The primary structure is known (i.e., Amino Acid Sequence)
The second structure is known (i.e., Disulfide Linkage)
Post-translational modifications are known (e.g., Gyclosylation)

Monoclonal Antibodies

The identity can be determined by rigorous physicochemical and immunochemical characterization without fully knowing its chemical structure

Purities and Impurities

The purity is quantifiable
The impurities are quantifiable, and identified if possible

Potency and Quantity

The biological activity is measurable
The quantity is measurable

The international workshop which followed publication of the above interim definition was attended by recognized national and international experts in biotechnology product development. Interest in this topic was global, because FDA had committed to reduce the regulatory burden on all products which met the definition of well-characterized, regardless of where they were manufactured. This burden reduction consisted of the regulatory flexibilities described under the REGO IIb initiatives above, which were intended to harmonize the regulation of biotechnology products regardless of whether they were regulated as drugs under the NDA system (e.g., human growth hormone and insulin) or biological products under the PLA/ELA system (e.g., interferon, TPA).

The original intent of the FDA was to require manufacturers of such products to submit information to FDA which described the specific characterization studies performed and await approval prior to utilizing the single application or the use of the other regulatory flexibilities. This proved to be an undesirable situation, which would have resulted in products of the same class being regulated differently, depending upon the quality of an additional regulatory submission. In addition to this, FDA had not approved any product, regardless of whether it was a traditional biological product or recombinant DNA-derived or monoclonal antibody, unless it was characterized to the fullest extent possible.

Therefore, topics such as the practicality of a universal definition, use of a product development profile as a rational tool for optimizing characterization efforts, the choice of bioanalytical methodology, and the scope of such regulation became the major focus of discussion. The result was universal agreement on the following points:

1. The definition was too restrictive and subject to inconsistent interpretation.
2. Regulatory flexibility should be extended to a broader range of products than the recombinant DNA and monoclonal antibody products described in the initial proposal.
3. Products which fell into certain distinct categories should be specified as adequately characterized and afforded all of the regulatory flexibilities described as a part of the December 8, 1995, *Federal Register* proposal (Federal Register, 1995b).

The result of these efforts was the publication of another final rule on May 14, 1996 (Federal Register, 1996), which extended all of the regulatory flexibilities previously described to the following four categories of specified products:

1. Therapeutic DNA plasmid products
2. Therapeutic synthetic peptide products of less than 40 amino acids
3. Monoclonal antibody products for *in vitro* use
4. Therapeutic recombinant DNA-derived products

The other significant effort that became part of the REGO IIb initiative was the commitment by FDA to harmonize the requirements for postapproval changes for drugs and biological products and revise the restrictive regulation at 21 CRF, Part 601.12, which required preapproval of all important changes to manufacturing methods and labeling.

6.1. Implementation of the New Regulatory Paradigm

Development of the new regulatory paradigm was made possible by thoughtful integration of regulatory and laboratory science based on experience gained in over two decades of biotechnology-derived product regulation. The steps involved in this paradigm shift included redesigning, based on rational regulatory principles, laws and regulations which had been promulgated in the 1940s and applied to new technology products and accepted without change in 1984 in a study commissioned by the White House Office of Science and Technology Policy (Federal Register, 1984). This would not have been possible without research, conducted at the FDA and supported by results from the biopharmaceutical industry, into the characterization of these new technology products.

These research efforts yielded a biotechnology product development profile (Fig. 2), which was the first rational tool for documentation of the product attributes necessary for supporting production of an optimized biotechnology product at all stages of investigation and for justifying process changes that may occur at any phase of product development. This PDP has been discussed at several national and international meetings and relevant elements of it incorporated into several guidance documents for the biotechnology industry, including the following:

1. The supplement to the *Points to Consider in the Production and Testing of New Drugs and Biologicals Produced by Recombinant DNA Technology: Nucleic Acid Characterization and Genetic Stability* (1992)
2. The updated *Points to Consider in the Characterization of Cell Lines Used to Produce Biologicals* (1993)
3. The *Points to Consider in the Manufacture and Testing of Monoclonal Antibody Products for Human Use* (1994)
4. The *Points to Consider for the Evaluation and Testing of Combination Vaccines: Production, Testing and Clinical Study* (1995)

It was also utilized at the International Workshop on Characterization of Biotechnology Pharmaceutical Products (U.S. FDA, 1995) as one model for design of an acceptable testing algorithm in support of the definition of "well-characterized" products.

Another major aspect of the paradigm shift was initiated in December 1995 with publication of the proposal to eliminate the Establishment License Application for biotechnology products which are specified (well-characterized) (Federal Register, 1995c). In place of the ELA, the CBER, FDA proposed evaluating the adequacy of manufacturing facilities and process validation through inspections for compliance with GMPs and by reducing redundancy of information required to be submitted in applications. Reduction of redundancy would be achieved by incorporating important questions about manufacturing into a new Chemistry, Manufacturing, and Controls Section (CMC) of a harmonized product application form.

Elimination of the requirement that CBER, FDA release each lot of all specified biotechnology products was implemented in December 1995 after the identification of most of the approved therapeutic recombinant DNA products as specified. FDA now believes that once a specified biotechnology product has been approved for marketing and its manufacturing process validated, the manufacturer should be responsible for quality control and assurance that each lot or batch meets established specifications for product release without additional oversight by the agency.

The harmonized application form was made available to all manufacturers of well-characterized biotechnology products through the International Workshop on Characterization in December 1995.

6.2. Future Steps Necessary for Completing the New Regulatory Paradigm

The new regulatory paradigm was designed to bring the requirements for drugs and biological products into close harmony. In addition to defining the procedures that the agency will adopt to implement this new paradigm, several major steps remain to complete the transition, including the following:

1. Finalizing the CMC Guidance documents for nonspecified biological products
2. Determining if the scope of products currently defined as specified should be extended to products other than the list included in the May 14, 1996 *Federal Register* Notice
3. Determining if the same or a similar paradigm should be applied to more traditional biological products which are characterized to the best extent scientifically achievable prior to approval
4. Optimization of the CMC Section of the new harmonized application format for nonspecified biological products and publication for review and comment

5. Refining the harmonized application format to be applicable to the submission of all drug and biological products, regardless of their ability to be characterized
6. Harmonization of application review and inspection procedures for drugs and biological products

Completion of the above elements will permit applicants to utilize the same regulatory pathway with identical requirements for submission of applications to the agency for all drugs, biologicals, and biotechnology products. In addition, it will bring the requirements for traditional, nonbiotechnology-derived drugs and biologicals into closer harmony based on utilization of a single application format, design guidance documents with similar formats, and establishment of identical procedures for meeting compliance with GMP and establishment standards.

7. CONCLUSION

A new regulatory paradigm has been initiated for products derived from biotechnology processes which are specified by the FDA. This paradigm shift was made possible by the following:

1. Thoughtful interpretation of existing regulations, and publication of such interpretations in flexible guidance documents
2. Fundamental research at CBER, FDA keeping up with advancements in biotechnology
3. The Clinton Administration's reinventing government initiatives, which facilitated the retrospective review and revision of outdated regulations
4. Transition of biotech products from preventative and replacement therapies to therapeutic agents, resulting in traditional pharmaceutical industry involvement in biotechnology policy
5. Industry and trade association proposals for legislative reform to the *Federal Food Drug and Cosmetic Act* and *The Public Health Service Act*

In conclusion, several important aspects of the new regulatory paradigm for well-characterized biological products have been implemented and remain permanently in place due to changes which have been effected in the regulations and published guidance documents. The steps necessary to complete the paradigm shift have been publicly discussed and committed to by the Commissioner of FDA and the Director and Deputy Director of the

Center for Biologics Evaluation and Research. All that remains are the details and administrative procedures related to implementation of the new paradigm, which, although a major initiative and resource drain on the agency, will be accomplished to meet the commitments documented in the regulatory and legislative proposals of late 1995 through 1997.

REFERENCES

Beatrice, M., 1993, Biotechnology facility design and process validation, in: *Biotechnology*, 2nd ed., Vol. 3, *Bioprocessing* (G. Stephanopoulos, ed.), VCH, Mannheim, Germany.

Bonk, D. T., 1987, Address at the annual meeting of the Food Drug and Cosmetic Law Section of the New York State Bar Association (January 29, 1987).

Chiu, Y. H., 1988, Validation of the fermentation process for the production of recombinant DNA drugs, *Pharm. Technol.* **12:**132–138.

Chiu, Y. H., Gueriguian, J. L., and Sobel, S., 1987, Chemistry and manufacture of the biotechnologically derived drugs, in: *Biotechnology Derived Medical Agents* (J. L. Gueriguian, V. Fattorusso, and D. Poggiolini, eds.), Raven Press, New York.

EC Committee for Proprietary Medicinal Products, 1994, *Regul. Affairs J.* **5:**574.

Federal Register, 1983, Licensing of a biological monoclonal antibody product prepared from hybridoma technology, *Fed. Reg.* **48:**50795–50796.

Federal Register, 1984, Proposal for a coordinated framework for regulation of biotechnology, *Fed. Reg.* **49:**50856–50858.

Federal Register, 1985, Coordinated framework for regulation of biotechnology: Establishment of the biotechnology science coordinating committee, *Fed. Reg.* **50:**47174–47176.

Federal Register, 1986, Coordinated framework for regulation of biotechnology, *Fed. Reg.* **51:**23302–23309.

Federal Register, 1992, FDA's policy statement concerning cooperative manufacturing arrangements for licensed biologics, *Fed. Reg.* **57:**55544–55546.

Federal Register, 1994, Review of general biologics and licensing regulations *Fed. Reg.* **59:**28821–28823.

Federal Register, 1995a, FDA guidance document concerning use of pilot manufacturing facilities for the development and manufacture of biological products, *Fed. Reg.* **60:**35750–35753.

Federal Register, 1995b, Interim definition and elimination of lot-by-lot release for well-characterized therapeutic recombinant DNA-derived and monoclonal antibody biotechnology products, *Fed. Reg.* **60:**63048–63049.

Federal Register, 1995c, Interim definition and elimination of lot-by-lot release for well-characterized therapeutic recombinant DNA-derived and monoclonal antibody biotechnology products, *Fed. Reg.* **60:**63048;–63049.

Federal Register, 1996, Elimination of the establishment license application for specified biotechnology and specified synthetic biological products, *Fed. Reg.* **61**:24227–24233.

Garnick, R. L., Solli, N. J. and Papa, P. A., 1988, The role of quality control in biotechnology: An analytical perspective, *Anal. Chem.* **60**:2546.

Gerrard, T. L., 1995, Current product equivalency issues. *Food Drug Law J.* **50**:451–453.

Gore, A., Jr., 1993, *From Red Tape to Results, Creating a Government that Works Better and Costs Less: Reeingineering through Information Technology,* Accompanying report of the national performance review, U.S. Government Printing Office, Washington, D.C.

Marino, M. H., 1989, Expression systems for heterologous protein production, *BioPharm* **2**:18.

Marshak, D. R., and Liu, D. T., 1988, *Banbury Report No. 29: Therapeutic Peptides and Proteins: Assessing New Technologies*, Cold Spring Harbor Laboratory, Cold Spring Harbor, NY.

McPhie, P. 1995. *Proceedings of the Workshop on Characterization of Biotechnology Pharmaceutical Products*, CBER Fax Information Services.

NCDB, 1983, *Points to Consider in the Production and Testing of Interferon Intended for Investigational Use in Humans*, Office of Biologics, National Center for Drugs and Biologics, Bethesda, MD.

U.S. Food and Drug Administration, 1985, *Points to Consider in the Production and Testing of New Drugs and Biologicals Produced by Recombinant DNA Technology*, U.S. FDA, Rockville, MD.

U.S. Food and Drug Administration, 1991, *Draft Guideline for Submitting Supporting Chemistry Documentation*, Center for Drug Evaluation and Research, U.S. FDA, Rockville, MD.

U.S. Food and Drug Administration, 1992, *Supplement to the Points to Consider in the Production and Testing of New Drugs and Biologicals Produced by Recombinant DNA Technology*, U.S. FDA, Rockville, MD.

U.S. Food and Drug Administration, 1994, *Points to Consider in the Manufacture and Testing of Monoclonal Antibody Products for Human Use*, CBER, U.S. FDA, Bethesda, MA.

U.S. Food and Drug Administration, 1995, CBER International Workshop on Characterization of Biotechnology Pharmaceutical Products, December 11–13, 1995, Washington, D.C.

U.S. Food and Drug Administration, 1996, *Guidance Concerning Demonstration of Human Biological Products Including Therapeutic Biotechnology-Derived Products*, CBER (FDA) fax information system.

U.S. Food and Drug Administration, 2001, Title 21, *Code of Federal Regulations*, U.S. Government Printing Office, Washington, DC.

Werner, R. G., Langlouis-Gau, H., Walz, F., Allgaier, H., and Hoffman, H., 1988, Validation of biotechnological production processes, *Arzneim. Forsch. Drug Res.* **38**:855–862.

Index

Pharmaceutical Biotechnology

Chronological Listing of Volumes

Volume 1 PROTEIN PHARMACOKINETICS AND METABOLISM
Edited by Bobbe L. Ferraiolo, Marjorie A. Mohler, and Carol A. Gloff

Volume 2 STABILITY OF PROTEIN PHARMACEUTICALS, Part A: Chemical and Physical Pathways of Protein Degradation
Edited by Tim J. Ahern and Mark C. Manning

Volume 3 STABILITY OF PROTEIN PHARMACEUTICALS, Part B: *In Vivo* Pathways of Degradation and Strategies for Protein Stabilization
Edited by Tim J. Ahern and Mark C. Manning

Volume 4 BIOLOGICAL BARRIERS TO PROTEIN DELIVERY
Edited by Kenneth L. Audus and Thomas J. Raub

Volume 5 STABILITY AND CHARACTERIZATION OF PROTEIN AND PEPTIDE DRUGS: Case Histories
Edited by Y. John Wang and Rodney Pearlman

Volume 6 VACCINE DESIGN: The Subunit and Adjuvant Approach
Edited by Michael F. Powell and Mark J. Newman

Volume 7 PHYSICAL METHODS TO CHARACTERIZE PHARMACEUTICAL PROTEINS
Edited by James N. Herron, Wim Jiskoot, and Daan J. A. Crommelin

Volume 8 MODELS FOR ASSESSING DRUG ABSORPTION AND METABOLISM
Edited by Ronald T. Borchardt, Philip L. Smith, and Glynn Wilson

Volume 9 FORMULATION, CHARACTERIZATION, AND STABILITY OF PROTEIN DRUGS: Case Histories
Edited by Rodney Pearlman and Y. John Wang

Volume 10 PROTEIN DELIVERY: Physical Systems
Edited by Lynda M. Sanders and R. Wayne Hendren

Volume 11	INTEGRATION OF PHARMACEUTICAL DISCOVERY AND DEVELOPMENT: Case Histories Edited by Ronald T. Borchardt, Roger M. Freidinger, Tomi K. Sawyer, and Philip L. Smith
Volume 12	MEMBRANE TRANSPORTERS AS DRUG TARGETS Edited by Gordon L. Amidon and Wolfgang Sadée
Volume 13	RATIONAL DESIGN OF STABLE PROTEIN FORMULATIONS: Theory and Practice Edited by John F. Carpenter and Mark C. Manning
Volume 14	DEVELOPMENT AND MANUFACTURE OF PROTEIN PHARMACEUTICALS Edited by Steven L. Nail and Michael J. Akers